W0235459

ELECTROCHEMICAL PROCESSES FOR CLEAN TECHNOLOGY

Electrochemical Processes for Clean Technology

Keith Scott
Department of Chemical and Process Engineering
University of Newcastle
Newcastle upon Tyne NE1 7RU
UK

ISBN 0-85404-506-6

A catalogue record of this book is available from the British Library.

© The Royal Society of Chemistry 1995

All rights reserved.

Apart from any fair dealing for the purposes of research or private study, or criticism or review as permitted under the terms of the UK Copyright, Designs and Patents Act, 1988, this publication may not be reproduced, stored or transmitted, in any form or by any means, without the prior permission in writing of The Royal Society of Chemistry, or in the case of reprographic reproduction only in accordance with the terms of the licences issued by the Copyright Licensing Agency in the UK, or in accordance with the terms of the licences issued by the appropriate Reproduction Rights Organization outside the UK. Enquiries concerning reproduction outside the terms stated here should be sent to The Royal Society of Chemistry at the address printed on this page.

Published by The Royal Society of Chemistry,
Thomas Graham House, Science Park, Milton Road, Cambridge CB4 4WF, UK

Typeset by Turn-Around Typesetting, Maulden, Bedfordshire, UK
Printed by Hartnolls Ltd., Bodmin, Cornwall, UK

To my wife, Jean, and four children, Tamsin, Alexander, Sophie and Amy.

Preface

Electrochemistry is a science which is widely utilised and applied in most industries. Applications embrace sensors, monitors, control, analysis, corrosion (and its protection), surface technologies, energy generation, materials and chemical manufacture, recycling, and effluent treatment. These applications can loosely be classified into two general areas: indirect and direct processes. Indirect can be defined as processes which occur without the specific imposition of a potential field whereas direct processes imply application of a potential field. This is clearly a broad division, as in practice overlaps in areas will occur, for example in energy production (*e.g.* batteries). Nevertheless, this distinction allows the author to define the focus of the book. This book is concerned with the direct applications of electrochemistry in industry as methods for the manufacture of materials and chemicals, as means of separation of specific species for product recovery and recycling, and as methods for effluent and waste treatment. It is fair to say that the main emphasis of the book is in the latter two areas of recycling and effluent treatment – three chapters are devoted to these topics. This is quite intentional as these are relatively new and rapidly expanding areas for electrochemical technology. This expansion is due to several factors: increased awareness of the importance of better environmental protection; tighter controls and more stringent requirements on the discharge of potential pollutants and toxic materials; appreciation of the value (monetary and otherwise) gained in the recovery and recycling of valuable chemicals and reagents; and advances made in the science, engineering and technology of the discipline. Electrochemical technology has developed to a point where materials for cells, particularly electrodes and membranes, are readily available to meet most applications. Electrochemical cells and reactors are no longer confined to single or speciality applications but can be purchased from several manufacturers as off-the-shelf modules in a range of sizes from small laboratory units to plant scale reactors. The significance of this fact is that these cells have found thousands of commercial applications worldwide.

Electrochemistry has always been used as a method for chemical manufacture in traditional areas of metal winning, chlorine and caustic manufacture, and other inorganic compounds. More recently the benefits of electrochemistry as a clean technology at the site of manufacture for the synthesis of a range of organic chemicals and intermediates in the fine chemical, pharmaceutical and agrochemical industries has seen over sixty commercial processes emerge. Additionally several chemical reagents are now manufactured, and also recycled, by electrochemical processes. An interesting trend which seems to be developing in industry is the move away from bulk storage of chemical reagents towards safe, on-site, on-demand production. Electrochemistry is ideally suited to this move offering the appropriate supply at the flick of a switch. Small fully functional plants can now be purchased for the supply of species such as chlorine, hydrogen and arsine *etc.*

The overall objective of this book is to describe the wide range of applications of electrochemistry in what is rapidly becoming known as clean technology, *i.e.* those processes which operate without (or with minimal) generation of waste products, spent reagents *etc.* The direct application of the electron as a reagent is seen as the intrinsically clean processing method. With modern electrochemical technology and engineering now available, the overall process can be suitably designed as Clean Technology. Finally a question one might rightly ask is why write a book on 'Clean Technology the Electrochemical Way'? Well the original intent (in 1991, the time of proposal) was to bring together the pertinent technological aspects of this rapidly developing field, focusing on newer processes and developments whilst adhering to important state-of-the-art materials. To underpin the subject, for the benefit of readers unacquainted with electrochemistry or with little knowledge of engineering, I felt it important to build the book from basic fundamental science, relevant to rate processes in electrochemical cells, and to integrate these concepts with basic engineering and technology. Overall the aim is to provide a text which informs the reader about electrochemical (clean) technology but also enables the reader to appreciate the principles of science and engineering behind the technology. The book consequently has been written for a broad spectrum of scientists and engineers at both undergraduate and graduate level as well as for industrialists, and academics, who may have little experience or knowledge of electrochemical technology and processes or their design.

Keith Scott

University of Newcastle upon Tyne

June 1995

Contents

Chapter 8   Electrochemical Synthesis

CHAPTER 1

Introduction

The applications of electrochemistry are quite diverse and span over a wide range of the process industries. The technology is used as a means of producing chemicals, materials and other species, for the storage and production of electrical energy and also as a means of monitoring and analysing the process or system operation.[1] Table 1 indicates the areas of application of electrochemistry of relevance to the process sector, classified under six broad headings: corrosion, electrical energy storage and generation, sensors and monitoring, synthesis, effluent treatment and recycling, metals and materials processing.

Corrosion is an area of utmost importance to industry as great efforts and cost are required to minimise its effects. The vast majority of corrosion processes involve some form of electrochemical reaction, or process, of a metal component which leads to a gradual loss of the function or property. Although the subject of corrosion is not within the scope of this book its influence is important in the context of the selection of electrode materials. The corrosion of both cathode and anode may occur both in 'current-on' conditions and in standby, or open circuit, conditions. Current-on conditions can induce cathodic reduction of coatings and anodic dissolution of so called 'inert anodes', whilst at standby conditions the electrodes are at different potentials to the current-on situation. This can result in a loss of protection of the material as it takes up an active state in the electrolyte or cell environment and can be aggravated by the presence of corrosive chemical products and reagents, dissolved oxygen or other impurities. It should also be recognised that electrochemical processes, by their very nature, involve the flow of current which can induce leakage or parasitic currents and thus cause and accelerate corrosion of other parts of the electrolytic plant.

Energy storage and generation through the application of electrochemical processes constitutes the domains of the battery and the fuel cell. The generation of electrical energy is caused by two redox reactions (with a negative free energy), which occur spontaneously within the battery. There is a wide range of batteries for low, medium and high power applications and three of the most common are lead/acid, nickel/cadmium and zinc/carbon.

Fuel cells are devices for generating electrical energy by the continuous

1

Table 1   *Application areas of electrochemistry*

Electrosynthesis	*Metal and materials processing*
inorganic chemicals	electrochemical machining
organic chemicals	electroforming
metals and alloys	etching of metals
semiconductors	etching of semiconductors
conductive polymers	electrophoretic painting
composites	electroplating and anodising
Electrical power storage and conversion	*Effluent treatment and recycling*
fuel cells	metal ion removal
storage batteries	removal of inorganic and organic
redox batteries	contaminants
solar cells	water purification
	recycling of redox and other reagents
	gas cleaning
	salt splitting
	electroflotation
Sensors and monitoring	*Corrosion*
ion-selective electrodes	protection
polarography	prevention
electrochemical detection	monitoring
biomedical applications	control
in-vivo sensors and monitors	
in-vitro diagnostics	
slow-release drug delivery devices	

supply of this fuel to one electrode. There are a variety of devices operating at both low and high temperatures. High-temperature cells operate at temperatures up to 1000 °C, as in the case of the molten carbonate and the solid oxide cells. Low-temperature cells include the phosphoric acid and the solid polymer electrolyte cells which operate at temperatures of 200 °C and below. The electrodes in fuel cells are different to those in batteries as they must generally be permeable to the gas feed. In low-temperature cells catalytic gas diffusion electrodes are therefore used, which are typically a composite structure of electrocatalyst, carbon, hydrophobic binder and coating. Applications of fuel cells are in large scale power generation, vehicle traction and small scale remote site energy generation. The electrode technology has been adapted to attempt to improve the energy efficiency of several electrochemical processes that consume electrical energy. This has essentially involved the introduction of oxygen cathodes and hydrogen anodes which are discussed further in this book.

A rapidly developing application of electrochemistry is in sensors and in monitoring. Major areas are in polarography and anodic stripping voltammetry for trace metal ion analysis, ion selective electrodes, electrochemical biosensors and as detectors in high-performance liquid chromatography (HPLC) *etc.*

Electrochemistry can be used to advantage in the metal and material processing industries producing components, which are virtually impossible to produce by mechanical means to the exacting standards often required. The methods include machining, grinding, deburring and etching. In addition electrochemistry is used in the finishing of many components through the deposition of coatings (*e.g.* metallic and polymer) and through anodising to produce surface oxide films. Electrophoretic painting is used widely in the motor industry for bodywork protection. Other applications are in electropolishing and in electrochemical cleaning, pickling and stripping.

This chapter introduces the subject area of electrochemistry in relation to the applications in synthesis and effluent treatment and recycling. This is done by providing an overview of the application areas and, in a wider context, particularly in the area of waste treatment, discussing competing technologies.

1.1 ELECTROCHEMICAL SYNTHESIS

The area of electrochemical synthesis covers the production of many chemicals, both organic and inorganic. The two dominant industries in terms of tonnage are, by far, chlor-alkali and aluminium electrowinning. These two processes use radically different cell technology, aluminium production being based on molten salt electrolysis at temperatures around 1000 °C, whilst the chlor-alkali industry, for the production of chlorine and caustic soda, offers three cell technologies based on the electrolysis of aqueous brine solutions at relatively low temperatures. Other inorganic electrochemical processes include the production of fluorine, peracids, sodium chlorate and bromate, manganese dioxide, copper(I) oxide and water electrolysis. Several metals other than aluminium are electrowon from molten salts and include sodium, lithium and magnesium. Electrowinning from aqueous electrolyte is carried out to a large extent, particularly in the case of copper, zinc and nickel. The purification of several metals is carried out by a process referred to as electrorefining in which the impure metal is dissolved anodically in an electrolyte bath and the pure metal simultaneously electroplated onto a cathode in a pure form.

The electrosynthesis of organic chemicals is largely located in the fine chemical industries where, over the last few decades, the advantages of this technique have seen somewhere in the region of a hundred or so processes developed in industry. Although this is a relatively small number in terms of the many thousands of fine chemicals produced world wide, the technology is still relatively new and unfamiliar to many fine chemical manufacturers. It is anticipated that in the future many other syntheses will be carried out electrochemically when the environmental advantages, and the relatively safe operating conditions, are realised

along with the fact that the electron is an inexpensive reagent. The estimated cost of the electron as a reagent at approximately £8000 per tonne
equivalent, compares favourably with for example zinc dust, at £29000,
sodium borohydride, at £59000, and sodium dichromate at £164000, per
tonne.[2]

In bulk organic chemical manufacturing, electrochemistry is up against
strong competition where gas phase catalysis has been researched and
engineered to a sophisticated level. The relatively large cost associated
with the direct electrical energy input into the reaction step is often cited
as being the major economic factor against frequent adoption of electrochemical processes in this area. The most important exception to this is
the production of adiponitrile from acrylonitrile, which is an intermediate in the production of nylon. This process competes well on the open
market with a gas phase catalytic route.

A developing area for the application of electrochemical technology is
that associated with effluent treatment, water purification and the recycling of material from process streams. There are several established
processes such as those for metal recovery by electrodeposition, ion separation by electrodialysis, water treatment using hypochlorite and the treatment of liquors bearing chromium species. Many other processes are
actively being researched at a pilot and laboratory scale and include the
destruction of toxic organic compounds, cyanide species and the desulfurisation of flue gases. Thus it is apparent that electrochemistry is used for
both liquid and gaseous effluents and this can be achieved by the use of
both direct and indirect electrochemical reactions. For example sulfur
dioxide can be oxidised either directly at an anode to sulfuric acid or
indirectly through the electrolytic generation of bromine. Furthermore
many species such as hydrogen peroxide and ozone, can be advanta-
geously produced on-site electrochemcially to carry out appropriate treatment without the need for bulk storage of hazardous reagents.

An important part of electrochemical technology is that associated with
the use of ion-exchange membranes. Membranes are both a vital component of many electrolytic cells and also a means of carrying out specific
separations of ionic and non-ionic species and through the specificity of
transport of certain species can cause the formation of chemical products.
Membranes find uses in effluent treatment and recycling and electrosynthesis. Membranes can, economically, carry out the separation of salts and
other species from aqueous streams. The membrane process, which selectively removes ions, is called electrodialysis. Market areas are in the desalination of brackish water, in desalting foods, in effluent treatment and
notably in Japan for the manufacture of salt. Competition for electrodialysis comes not only from separations such as evaporation but also from
hyperfiltration. The economic advantage gained with electrodialysis is
through its specificity and efficiency of separation achieved at low temperatures.

1.2   EFFLUENT TREATMENT AND RECYCLING [3–6]

The process industries are under environmental and economic pressure to make more effective use of the material resources used in the manufacture of commercial products. This impacts in several areas of process and product planning; the selection of the most appropriate starting materials and end product and the overall design of the process steps to implement this transformation. Owing to the inherent inefficiencies of physical and chemical processes there will be species and streams generated that are not a desirable part of the envisaged manufacturing process. If these materials are seen to have some immediate economic value, methods will be implemented to recover and re-use them. If the economics of re-use are not directly apparent then procedures are generally adopted which dispose of the material at short term minimum cost. In many cases these materials are potential pollutant and/or hazardous materials and their disposal should be looked at in a much larger context. The safe management of these materials, especially the more toxic and hazardous, is a major problem and issue. This is a complex problem fraught with many issues; economic, social, political and technological. Nevertheless there are methods and procedures currently available, and in use, which can lead to improved approaches to waste management based on strategies of re-use, recycling and recovery. The recovery of hazardous materials from process streams and effluents followed by recycling is a good method for reducing the quantity of hazardous or toxic waste. The concept of recovery of a material is different to that of recycling. The recovered material must either be disposed of in a suitable way, or alternatively utilised, in for example the incineration of organics to produce process heat.

Electrochemical methods can be applied to the recycling of many species present in solid, liquid or gaseous phases. Such processes are clearly in competition with many other methods not based on electrochemistry and it is therefore useful, and informative, to indicate the methods for recovery and recycling in a more wider context. Existing technologies for the recovery of raw material can be classified into three general categories (see Table 2).

(i)   *Physical separations.* These methods include gravity settling, filtration, flotation, flocculation and centrifugation. They are specifically used for the separation of fine and suspended solids and dispersed liquids from liquid.

(ii) *Component separations.* These technologies distinguish between constituents by virtue of differences in some physical property, electrical charge, boiling point, miscibility *etc.* Methods include evaporation, distillation, solvent extraction, absorption, ion exchange and reverse osmosis.

(iii) *Chemical transformations.* These methods require a chemical reaction to remove specific constituents and examples include, precipitation, electrolysis, electrodialysis, and oxidation and reduction reactions.

Table 2   *Description of technologies for recovery of materials* (adapted from ref. 4)

Technology and description	Types of waste streams	Separation efficiency [a]	Typical industrial applications
Physical separation Gravity settling: Tanks, ponds provide hold-up time allowing solids to settle	Slurries with separate phase solids, such as hydroxide	Limited to solids (large particles) that settle quickly (less than 2 h)	Industrial wastewater treatment, first step
Filtration: Liquid passes through and solids are retained on porous media	Aqueous solutions with finely divided solids; gelatinous sludge	Good for relatively large particles	Various tannery, water
Flotation: Air bubbled through liquid to collect finely divided solids that rise to the surface with the bubbles	Aqueous solutions with finely divided solids	Good for finely divided solids	Refinery (oil/ water mixtures); paper waste; mineral industry
Flocculation: Agent added to aggregate solids together which are settled easily	Aqueous solutions with finely divided solids	Good for finely divided solids	Refinery; paper waste; mine industry
Centrifugation: Centrifugal force causes separation by different densities	Liquid/liquid or liquid/solid separation, *i.e.* oil/ water resins; pigments from lacquers	Fairly high (90%)	Paints
Component separation Distillation: Boiling off materials at different temperatures (based on different boiling points)	Organic liquids	Very high separations achievable (99+% concentrations) of several components	Solvent separations: chemical and petroleum industry
Evaporation: Solvent recovery by boiling off the solvent	Organic/inorganic aqueous streams, slurries, sludges, *i.e.* caustic soda	Very high separations of single, evaporated component achievable	Rinse waters from metal plating waste

cont.

Table 2 continued

Technology and description	Types of waste streams	Separation efficiency [a]	Typical industrial applications
Stripping: (air or steam)	Dissolved organics	Good	Solvent separations
Ion exchange: Waste stream passed through resin bed, ionic materials selectively removed. Ionic exchange materials must be regenerated	Heavy metals aqueous solutions cyanide removed	Fairly high	Metal-plating solutions
Ultrafiltration: Separation of molecules by size using membrane	Macromolecules. Heavy metal aqueous solutions	Fairly high	Metal-coating applications
Reverse osmosis: Separation of dissolved materials from liquid through a membrane	Heavy metals, organics; inorganic aqueous solutions	Good for concentrations <300 ppm	Secondary treatment process such as metal-plating, pharmaceuticals
Carbon/resin absorption: Dissolved materials selectivity adsorbed in carbon or resins. Adsorbents must be regenerated	Organics/inorganics from aqueous solutions with low concentration *i.e.* phenols	Good, overall effectiveness dependent on regeneration method	Phenolics
Solvent extraction: Solvent used to selectively dissolve solid or extract liquid from waste	Organic liquids, phenols, acids	Fairly high loss of solvent may contribute to hazardous waste problem	Recovery of dyes
Chemical transformation Precipitation: Chemical reaction causes formation of solids which settle	Lime slurries	Good	Metal-plating waste water treatment
Electrodialysis: Separation based on differential rates of diffusion through membranes. Electrical current applied to enhance ionic movement	Separation/concentration of ions from aqueous streams; application of chromium recovery	Fairly high	Separation of acids and metallic solutions

cont.

Table 2 continued

Technology and description	Types of waste streams	Separation efficiency [a]	Typical industrial applications
Electrolysis: Separation of positively/negatively charged materials by application of electric current	Heavy metals: ions from aqueous solutions; copper recovery	Good	Metal plating
Reduction: Oxidative state of chemical changed through chemical reaction	Metals, mercury in dilute streams	Good	Chrome-plating solutions and tanning operations
Chemical dechlorination: Reagents selectively attack carbon–chlorine bonds	PCB[b]-contaminated oils	High	Transformer oils
Thermal oxidation: Thermal conversion of components	Chlorinated organic liquids; silver	Fairly high	Recovery of sulfur, HCl
Chemical oxidation: Chlorination, ozonation	Dissolved organics, inorganics	High	Metal plating

[a] Good implies 50–80% efficiency, fairly high implies 90%.   [b] PCB = polychlorinated biphenyl.

An area not included in this list is biological processes. The conventional biological processes have been used in industrial waste treatment for many years and utilise either aerobic or anaerobic bacteria. They are effectively used for treatment of dissolved organics and sludges, and involve destruction of the contaminant by microbial catabolism. There are continued developments in new microbial strains for improved degradation of recalcitrant compounds, with improved tolerance to more severe operating conditions and increased rates of degradation.

1.2.1   Treatment of Dilute Wastes and Solutions

Electrochemistry is of particular use in the treatment and recovery of dissolved species from dilute solutions. In the treatment of both inorganic

Table 3 *Recovery/recycling technologies under development*

Technology	Potential application
Ion exchange	Chromium recovery; metal-plating waste
Adsorption various substrates	Organic liquids with or without metal contamination; pesticides
Electrolysis	Metallic/ionic solution organics
Extraction Liquid membranes	Extraction of metals with acids organics
Ultrafiltration	Organic
Reverse osmosis	Salt solutions
Evaporation	Fluorides from aluminium smelting operation
Stripping (air or steam)	Organics
Reduction	Mercury, chromium, inorganics
Chemical dehalogenation	Halogenated organic
Oxidation: chemical, catalytic	Organics, cyanide

and organic species, there is direct competition from several techniques, which are summarised in Table 3. Many of these methods can be used as recovery and recycling processes, particularly ion-exchange, electrolysis, extraction and evaporation. Others however, are used mainly for waste or effluent treatment and particularly for inorganic species, which is also an important area of activity for electrochemical methods, either for dilute or more concentrated solutions, which are classified as hazardous. Inorganic hazardous waste can be classified generally as wastes containing corrosives, metals and cyanides *i.e.*:

- Liquids having a pH < 2;
- Liquids, including free liquids associated with any solid or sludge, containing free cyanides at concentrations > 1000 mg dm^{-3};
- Liquids, and free liquids associated with any solid or sludge, containing metals such as: arsenic, mercury, cadmium, nickel, chromium, selenium and lead.

Technologies that are used extensively for the treatment of dilute aqueous inorganic wastes are generally applicable to the treatment of the more concentrated wastes and include neutralisation, precipitation, oxidation and reduction.

1.2.1.1 Neutralisation. Neutralisation of an excess of acid waste can be accomplished by addition of hydrated lime, caustic soda, or alkaline waste materials such as carbide lime. Neutralisation reactions are always exothermic, which is an important factor in the neutralisation of strong corrosive wastes. At acid concentrations greater than about 5%, careful control of the reaction is necessary to avoid excessive temperature rise that could damage tanks, pumps, piping or other equipment. Another consideration is the potential emission of acid fumes, nitrogen oxides or volatile constituents in the waste, prior to adding acid waste.

1.2.1.2 Heavy Metals Precipitation. The removal of heavy metals from solution is usually accomplished by precipitation of the metals as insoluble compounds, *e.g.* hydroxide or sulfide precipitation. The effectiveness of the hydroxide process is a function of the solubility of the metal hydroxides, which in turn is a function of pH and water quality. Usually a pH in the range 8.5–11 is found to give the best overall effluent quality when treating a waste containing several metals. The hydroxide precipitation process is capable of achieving soluble metal concentrations of less than 1.0 mg dm^{-3}, and often less than 0.1 mg dm^{-3}, approaching the theoretical solubilities, when treating dilute aqueous. The removal of the precipitated metal hydroxides is accomplished by a settling and typically dewatering device such as a filter press. Some metal-containing wastes contain constituents that inhibit the precipitation of metals, *e.g.* chelating agents used in metal plating bath. This can seriously affect the efficiency and applicability of the hydroxide precipitation process.

Sulfide precipitation, is effective even in the presence of certain complexing and chelating agents and theoretically, lower effluent metal concentrations are achievable. However, in sulfide precipitation there is potential for substantial evolution of H_2S at the point of addition of a soluble sulfide salt to an aqueous waste due to rapid hydrolysis. Thus, the use of iron(II) sulfide, which is only sparingly soluble, has been advocated. The solubility of iron(II) sulfide is too low to produce significant H_2S but sufficiently high to precipitate other heavy metals.

Generally if the waste contains a high concentration of organics, the precipitated sludge could contain a substantial amount of organics, which by itself may render the sludge hazardous.

1.2.1.3 Complexed Metal Wastes. Many chemical compounds are used as complexing agents, which inhibit or prevent the use of conventional precipitation methods. The more common complexing agents encountered are ammonia, cyanide and ethylenediaminetetraacetic acid (EDTA), and treatment techniques which can be used include:

(i)   pretreatment, *e.g.* of ammonia by air or steam stripping, cyanide by alkaline chlorination, can be employed to remove these materials prior to hydroxide precipitation.
(ii) EDTA can be precipitated as the free acid at low pH.

1.2.1.4 Hexavalent Chromium Reduction. The common method for treatment of hexavalent chromium is to convert it to the trivalent state, and then precipitate the trivalent chromium as chromic hydroxide. Reducing agents commonly employed are sulfur dioxide, sodium metabisulfite or hydrosulfite and iron(II) sulfate. The pH of the waste is adjusted to less than 3.0 to speed the rate of reaction.

1.2.1.5 Cyanide Treatment. The most widely used method for treating aqueous cyanide wastes is alkaline chlorination. Although theoretically 6.83 parts of chlorine are required per part of cyanide oxidised, in practice because of the presence of other oxidisable materials more chlorine is usually required. The alkaline chlorination process can be accomplished with chlorine or with hypochlorites. Effluent concentrations of 1 ppm or less can be achieved. Concentrated cyanide wastes may contain high concentrations of organics. Where this occurs, there is a high probability that chlorinated organics will be formed by alkaline chlorination of the waste.

REFERENCES

1   D. Pletcher and F. C. Walsh, 'Industrial Electrochemistry', Blackie Technical, 2nd edn., 1993.

2   'ICI Applied Electrotechnology bulletin',ICI Chemicals and Polymers Ltd, Runcorn Cheshire, UK, WA7 4QG, 1988.

3   P. S. Cartwright, '44th Purdue Industrial Waste Conference Proceedings', 1990, 343.

4   'Technologies and Management Strategies for Hazardous Waste Control', Congress of the United States, Office of Technology Assessment, ed. J. Hirschorn, US Government printing office, Washington, DC 20402.

5   'Performance and Costs of Alternatives To Land Disposal', ed. T. Oppelt, B. I. Blaney and W. F. Kemner, APCA, 1987.

6   'Treatment Technologies For Hazardous Wastes', ed. H. M. England and L. F. Mafica, APCA, Pittsburgh, PA, 1987.

CHAPTER 2

The Electrochemical Cell and Reactor

Electrochemistry and its applications play an important role in the industrial and commercial world. The types of cell reactions are varied, ranging from simple single phase redox reactions, to reactions which involve a phase transition (metal deposition, gas evolution, precipitation *etc.*) to multistep reactions, which combine chemical and electrochemical processes. These reactions occur in the wide range of applications of synthesis and effluent treatment and recycling which are the focus for this book. This chapter introduces the subject area of electrochemistry; the electrochemical cell, its function and the materials which make up its structure.

2.1 THE ELECTROCHEMICAL CELL

An electrochemical cell can be formed when any two electronically conducting materials are put into an electrolyte solution. The conducting materials, which are typically metal or carbon (semiconductors and polymers are also used), are referred to as the electrodes. The electrolyte solution is an ionically conducting liquid containing a proportion of dissociating salt or acid *etc.* A simple example is an electrolyte of an aqueous solution of copper(II) ions in sulfuric acid with a copper electrode and an iron electrode placed in solution. If the electrodes are connected by an external circuit then two processes will occur at the electrodes:
At the iron electrode:

$$Fe \rightarrow Fe^{2+} + 2 \ e^- \qquad (1)$$

The iron will dissolve into the solution and this electrode is called the *anode*.
At the copper electrode:

$$Cu^{2+} + 2 \ e^- \rightarrow Cu \qquad (2)$$

The copper is deposited onto the surface and this electrode is called the *cathode*. These two processes give rise to the flow of current, *I*, in the external circuit. By this convention it is evident that the anode withdraws electrons from the cell and the anode reaction is therefore an oxidation and conversely the cathode supplies electrons to the cell and the reaction is therefore a reduction. The above example of iron dissolution and copper electrodeposition is in fact the industrially used system referred to as cementation, in which metals such as copper are extracted from liquors, containing relatively low concentrations of copper (metal) ions, by the dissolution of iron.

If in the above example cell a d.c. power supply is placed in the external circuit and current is made to flow externally from the iron electrode to the copper electrode then the copper electrode becomes the anode and the reverse of reaction (2) is made to occur. At the iron electrode, which is now the cathode, reaction (2) occurs, in which copper is deposited onto the cathode. This system is frequently referred to as an electrolytic cell in which electrode reactions occur by the application of an external cell voltage. This type of cathode reaction is common in the recovery of metal ions from aqueous solutions in effluent treatment and electrowinning.

It is apparent in the above cell that copper ions are moving from the anode to the cathode and that there is no accumulation of charge, as the amount of oxidation is equal to the amount of reduction. The movement of the ions is therefore responsible for the transfer of charge in solution from one electrode to the other. In practice the charge will be carried by several ions, both cations (positively charged) and anions (negatively charged). The rate of charge transfer in the cell is therefore the total rate of the electrode reactions, which is conveniently measured by the current. The relationship between the charge passed and the amount of reaction is found through Faraday's Law.

2.2 FARADAY'S LAW AND CURRENT EFFICIENCY

In an electrochemical cell the amount of charge passed, q, in a time interval, t, is given by

$$q = \int I \, dt \tag{3}$$

In reaction (2) the amount of the metal deposited (in mols) is calculated from Faraday's law of electrolysis as:

$$m = q / n \, F = I \, t / n \, F \tag{4}$$

where F is the Faraday constant and is equal to $96\,485$ C mol^{-1}.

The processes at individual electrodes frequently involve more than one reaction and this brings in the concept of current efficiency, which is a major criterion for measuring the performance of electrochemical systems. For example in the process for metal recovery discussed in chapter 6 a major side reaction at the cathode is the formation of hydrogen gas,

$$2 \ H_2O \ +2 \ e^- \rightarrow H_2 \ + 2 \ OH^- \tag{5}$$

which represents a loss of Faradaic, or current efficiency, CE.

The current efficiency is the yield of a process based on the charge passed and can be defined as:

$$CE = \frac{\text{charge consumed in forming product}}{\text{total charge consumed}} \tag{6}$$

Experimentally the current efficiency is obtained from the measure of the amount of product formed, or reactant consumed, m_{act}, and the theoretical amount, m, i.e.,

$$CE = m_{act}/m \tag{7}$$

From Faraday's Law this is:

$$CE = m_{act} \ n \ F/q \tag{8}$$

The total current at an electrode is the sum of the currents, I_j, of the individual reactions

$$I = \Sigma I_j \tag{9}$$

and the current efficiency at any instant can thus be expressed in terms of current as:

$$CE = I_j/I \tag{10}$$

2.3   ELECTRODE POTENTIAL AND CURRENT DENSITY

The example in section 2.1, of the dissolution of iron coupled with the formation of metallic copper, is a Galvanic process in which the free energy of the overall reaction

$$Cu^{2+} + Fe \rightarrow Cu \ + Fe^{2+} \tag{11}$$

has a negative value. From the definition of the Gibbs free energy

$$\Delta G = -n \ F \ E_o \tag{12}$$

the equilibrium cell potential, E_o (or the open cell voltage), has a positive value. The equilibrium cell potential is made up from two components, the anode and cathode equilibrium (half-cell) potentials, such that

$$E_o = (E_e)_c - (E_e)_a \tag{13}$$

When a current flows, the electrode is said to be polarised and the electrode potential departs from its equilibrium value and takes up a value E. The degree of polarisation is measured in terms of the over-potential, η, which is defined as $\eta = E - E_e$. The electrode potential is defined as the difference in the potential of the electrode surface, E_m, and the potential of the solution adjacent to the electrode surface, E_s, *i.e.*

$$E = E_m - E_s \tag{14}$$

By convention cathode potentials have negative values and anode potentials have positive values. In practice it should be noted that there is in fact a distribution of potential in solution at the electrode interface, which decays rapidly a short distance into the solution[1].

It is usual to normalise the rate of an electrode reaction with regard to the area of an electrode, A, and thus the term current density, j, is introduced:

$$j = I / A \tag{15}$$

The rate of an electrode process can be defined as:

$$r_j = I / A\, n\, F = j / n\, F \tag{16}$$

and has the units of mol m^{-2} s^{-1}.

The current density of simple reactions increases with the over-potential and is typically represented as a 'polarisation' curve in which the logarithm of current density is plotted against the over-potential (see Figure 1). It is seen that current density and over-potential are exponentially related and the rate of reaction is controlled by kinetics and is said to be under activation control. Departure from this behaviour occurs at low potentials when equilibrium is approached and at high over-potentials where the influence of mass transport of reactants becomes an important factor and other reactions will possibly occur.

2.4   THE ELECTROCHEMICAL REACTOR

The electrochemical reactor is the engineered version of the electrochemical cell. The industrial application of electrochemistry depends largely on the engineering design and implementation of an effective reactor unit.

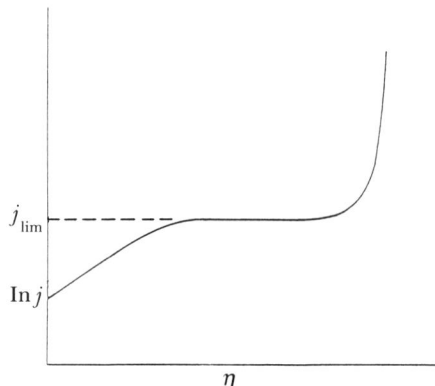

Figure 1   *A typical polarisation curve at an electrode*

The electrochemical reactor unit can be made up of many cells which are suitably connected together both electrically and hydraulically. Factors which are important in the electrochemical reactor design are:

1   productivity;
2   energy requirement and cell voltage;
3   temperature control;
4   hydrodynamics and mass transport;
5   reactor operation factors, and
6   electrode, membrane and other materials.

An important feature in electrochemical reactor design is the fact that two sets of reactions occur at the two different electrode sites in the reactor. This latter factor may, or may not, work to the advantage of the process. Generally the counter electrode reaction may

(i)   simply act to effect the flow of electrons;
(ii) be a source of reagent for reaction at the working electrode, *e.g.*, the supply of protons generated in the anodic oxidation of oxygen;
(iii) be responsible for the generation of a second product from the reactor, *e.g.*, the generation of both chlorine and caustic soda in chlor-alkali cells;
(iv) be a cause of reactant or product degradation and thus a loss in the yield and selectivity of the process.

The performance of an electrochemical reactor is in particular judged on its production capacity and its energy consumption.

2.4.1 Production Capacity

The production capacity of an electrochemical reactor is related directly to the applied current density and is conveniently expressed as:

$$\text{Capacity} = j\,\text{CE}/n\,\text{F} \quad \text{mol m}^{-2}\text{s}^{-1} \tag{17}$$

A given reactor will have a certain electrode area per unit, contained electrolyte volume ('specific area'), a, and thus the production capacity can be expressed on a volumetric basis. This in fact is the space time yield (STY) in units of mol m^{-3} s^{-1}.

$$\text{STY} = a\,j\,\text{CE}/n\text{F} \tag{18}$$

Thus the production capacity is proportional to the effective current density for the process and the electrode area per unit volume.

There are a number of interrelated factors which limit the value of specific area and hence the production capacity. These include material of fabrication, and distribution of current density over the electrode. The current density is itself limited by several factors such as reaction selectivity and efficiency. Thus it is not generally possible to arbitrarily increase the current density in order to increase the space time yield. In addition an increase in current density will also incur an increase in the energy consumption of the reactor due to the higher voltage required. Generally values of current density used vary approximately from 10–10 000 A m^{-2}, based on the superficial area of the electrode, and are frequently limited by the concentration of the reacting species.

2.4.2 Energy Requirements and Cell Voltage

The overall energy consumption of an electrolytic plant is made up of the energy used in feed preparation, reactor operation and product purification and reagent recycling. The reactor operation can constitute a large, but not always a major, part of this energy demand. However the adoption of the electrochemical route, in preference to a chemical or catalytic route, can often result in considerable reductions in the energy demand in other parts of the plant and consequently in an overall reduction in energy requirement per unit mass of product and thus in the total cost per unit mass of product.

The electrical energy consumption of an electrochemical reactor is determined by the thermodynamics of the reactions, the electrochemical kinetics and the cell design. The energy consumption is calculated from the following expression:

$$\text{Energy consumption} = n\,\text{F}\,E_{\text{cell}}/3\,600\,\text{M (CE)} \quad \text{kW h kg}^{-1} \tag{19}$$

where E_{cell} is the cell voltage.

The specific energy consumption, *i.e.* the energy per unit mass of product, is minimised by operating with a current efficiency as close to 1.0 as possible and with the lowest practical cell voltage. The components, which make up the cell voltage illustrated in Figure 2, are:

(i) equilibrium potentials of cathode and anode reactions, E_e, which make up the reversible cell voltage determined from the Nernst equation:

$$E_e = E^o - (RT/nF)\sum v_j \ln(a_j) \tag{20}$$

where, a_i, represents the activity of species i, v_j the stoichiometric coefficient and E^o is the standard electrode potential at the temperature of interest;

(ii) over-potentials at the cathode and the anode, $\eta_a + \eta_c$;

(iii) ohmic voltage losses in the electrolyte, the cell separator, electrodes and in the connections from the power supply to the electrodes, IR_{cell}.

The cell voltage is therefore given by:

$$E_{cell} = E_e + \eta_a + \eta_c + IR_{cell} \tag{21}$$

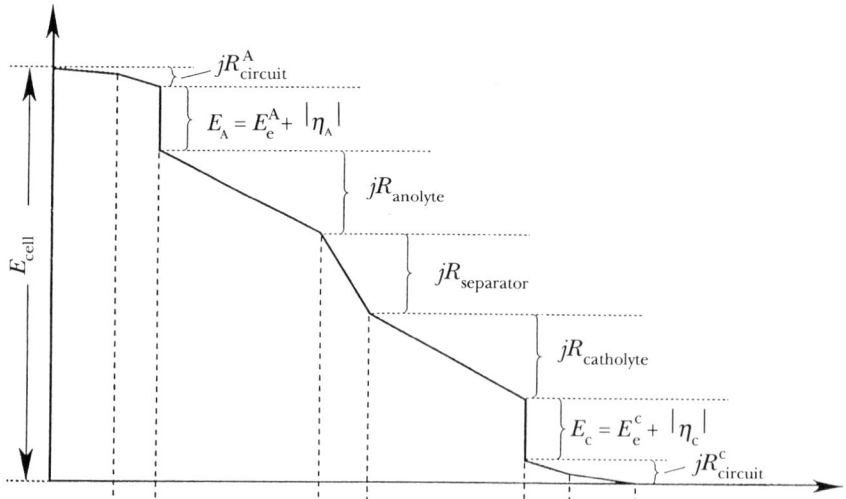

Figure 2   *Voltage components of an electrolytic cell*

The equilibrium potential determines the minimum voltage for the cell. If only one electrode is directly involved in product formation then the use of an appropriate counter electrode reaction which is thermodynamically more favourable and, in particular, by a reaction which is 'spontaneous' is desirable. A typical example is the replacement of the hydrogen reduction reaction with the reduction of oxygen in gas diffusion electrodes. The theoretical reduction in the cell voltage can amount to 1.2 V in this case. Oxygen reduction gas diffusion cathodes are available from several commercial manufacturers.

The over-potentials at electrodes arise from various polarisation phenomena and increase in magnitude as the rate of reaction, or the current density, increases. The over-potentials should, wherever possible, be minimised by the use of suitable, stable electrocatalytic materials.

The ohmic voltage losses in the cell are caused by electrical resistances in all conductors in the cell circuit. The material of the current feeders and connectors should be of highly conductive material and designed to the thickness required to carry anticipated maximum cell currents. The electrode material should also, if possible, be highly conducting and be of the required thickness. The use of conducting metal substrates may be considered for electrode surfaces which are relatively poor conductors. A low electrical resistance of the electrodes is also desirable to achieve a uniform potential distribution over the face of the electrode. The practical size of the electrodes is often limited by these two considerations.

The electrode design and material should also facilitate efficient gas release from its surface to ensure that the *IR* drop in the electrolyte and the 'bubble polarisation' is low.

The resistance of the electrolyte and associated cell separator can constitute a major component of the *IR* loss in a cell. The electrolyte conductivity should be high, using appropriate conducting electrolytes, although this must not prove detrimental to the stability of the electrodes and the separator and to the performance of the reactions. The electrolyte resistance can be minimised by using small interelectrode gaps, although it is important that the electrode configuration allow good passage of electrolytic gas from the cell. The presence of gas bubbles with near zero electrolyte conductivity can drastically increase the effective electrolyte resistance in the cell, especially if they have a large volume fraction. The effective electrolyte conductivity, in the presence of gas bubbles, can be estimated from equations of the form shown in eqn. (22).

$$\kappa_e = \kappa \ f(\ e_g\ ) \tag{22}$$

where $f(e_g)$ is a function of the bubble void fraction, e_g, for example $(1-e_g)^{3/2}$.

When cell separators are used they should be as thin as possible to minimise the electrical resistance whilst retaining structural stability. The electrical resistance will depend on the type and material used and the composition of the electrolyte solutions.

An important factor which affects the energy consumption is the number of electrons required in the relevant reaction. Multiple electron transfer reactions will clearly be heavy energy users and will generally be less competitive when compared against other routes.

2.4.3   Temperature Control

The supply of electrical energy to an electrolytic reactor usually gives rise to the generation of an excess of heat, due primarily to sensible heating of the electrolyte. The design of the reactor must consider the required heat transfer characteristics in order to limit, control or capitalise on the resulting temperature changes. The type of cell and the conditions and mode of operation play a significant role. For example in the electrowinning of aluminium temperatures in excess of 900 °C are used and the heat generated by electrolysis must balance the heat losses from the surface of the cells to, amongst other factors, maintain a stable melt.

The attainment of a good reaction selectivity may be limited to a small temperature range and thus appropriate heat removal may be required. Temperature control is often effected *via* external heat exchangers (see Figure 3) rather than built into the reactors and a flowing electrolyte is thus convenient. Other cell designs are equipped with direct electrolyte cooling to give closer control of the reaction temperature, for example in the production of fluorine gas.

Operating temperatures above ambient, caused by Joule heating, can be beneficial as they offer increased values of electrolyte conductivity, mass transport and kinetic rates. Operation of the reactor without, or with minimal, external heat exchange reduces cooling water costs and capital expenditure.

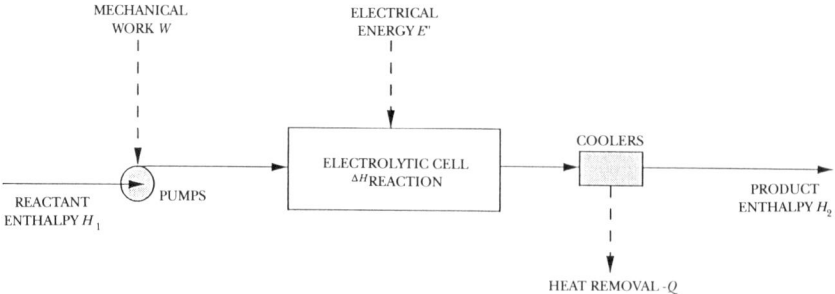

Figure 3   *Heat-transfer arrangement in electrolytic cells*

2.4.4   Hydrodynamics and Mass Transport

In the simplest case for an electrochemical process to occur a succession
of three rate steps, illustrated in Figure 4, must take place:

1   mass transport of reactants to the surface;
2   electrochemical transformations;
3   mass transport of products from the surface.

In general both the kinetics of electron transfer and the mass transport
of reacting species determine the rate. With increasing polarisation of the
electrode, as current density is increased, the electrode can become
starved of reactant and the rate of mass transport governs, and limits, the
overall reaction rate. The mass transport of ionic species is determined by
the hydrodynamics of electrolyte flow in the vicinity of the electrode, *i.e.*
the convective movement of species, and diffusion and migration, which
are discussed further in chapter 3. In general convective mass transfer is
determined by the selection of the cell or reactor and can be adjusted by
selection of suitable operating parameters, such as electrolyte velocity or
electrode rotation.

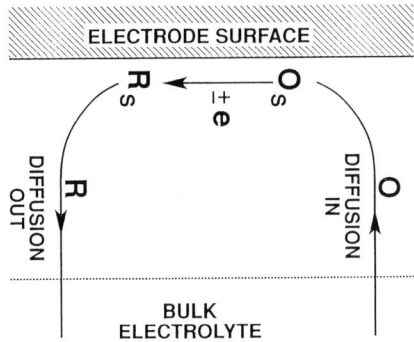

Figure 4   *Rate processes at an electrode surface*

2.4.5   Reactor Operating Factors

Electrochemical reactors can be operated in either batch or continuous
processes, with usually the scale of operation the deciding factor,
although the type of reaction can also be a contributing factor. For example in the electrowinning of metals, from aqueous solutions, batch operation is required to enable the solid metal deposit to be removed from the

cathode after a certain amount is deposited. The need to supply electrical power to the cells gives rise to two other factors;

1   The method and configuration of the electrical connection to the electrode.
2   The type of control of the cell power. Whether or not to attempt a potentiostatic control, *i.e.* operating at a fixed value of electrode potential, which can result in an improved selectivity in comparison to operation with a fixed cell current or cell voltage.

The number of cells used will depend on the scale of operation, the selected current density of operation and the size of the individual electrode. The layout and configuration of the multiple cell system is affected by factors such as the nature of the electrode materials and the type of separators used, the power rating of each cell and thus that of the appropriate combinations of cells. These factors are discussed in further detail in chapter 4 in relation to cell and reactor design.

2.5   CELL COMPONENTS AND MATERIALS

Two important material components of electrochemical cells are the electrodes and the cell separator. The developments in electrochemical cells and processes over the last decade or so has seen a range of materials become standards for many applications. This is reflected in the availability of commercial off-the-shelf cells for electrosynthesis and effluent and waste treatment applications.

2.5.1   Cell Separators

In electrochemical processes it is frequently necessary to separate the processes occurring at the anode and cathode for three main reasons:

(*i*)   to give high yield and selectivity;
(*ii*)   for product separation;
(*iii*) safety in operation.

If starting materials or products are susceptible to reaction at the counter electrode then separation of anode and cathode is required to achieve high yield. When starting reagents or products of one electrode process can react chemically with species generated at the counter electrode then again separators will be required. This for example is the case in the chlor-alkali industry (see chapter 8) where contact of brine solution, saturated with chlorine, with the hydroxide ion generated at the cathode would result in the formation of hypochlorous acid and other species.

Cell separators are often required for reasons of safety to separate the products obtained at the two electrodes in a cell, *e.g.* the electrolytic production of hydrogen and oxygen by water electrolysis (see chapter 8).

The materials used as separators are varied and may simply be microporous separators or may posses specific ion-transport characteristics. The separators used in cells are generally classified into two types, permeable and semi-permeable:

1   Permeable separators permit the bulk flow of liquid through their structure and are thus non-selective regarding transport of ions or neutral molecules. In electrochemical processes these are frequently referred to as diaphragms.

2   Semi-permeable separators permit the selective passage of certain species by virtue of molecular size or charge. In electrochemical processes these are termed 'membranes' and separation is based on the charge carried by the molecule.

To be effective as a separator a material must exhibit a range of desirable properties as listed in Table 1. Although over more recent years there has been a steady move away from the use of porous diaphragms towards ion-exchange materials the former are still used in several industries. The porous diaphragm represents a compromise between the demands for separation of anolyte and catholyte and effective electrical conductivity between anode and cathode *via* the ions in solution. A good degree of separation is achieved by using a uniformly fine porous structure which permits diffusion of material, but not mass flow. The higher the porosity (size and/or number of pores) the greater the electrical conductivity of the diaphragm, but the poorer the separation of the anolyte from the catholyte. Also, transport of materials across a porous diaphragm will be

Table 1   *Desirable properties of electrochemical separators*

Correct permeability and selectivity	Good current efficiency (CE)
Inert to cell environment: good temperature and chemical stability	Long life
Uniform properties across its surface and homogeneous flow	Good CE, good quality products
Low voltage drop	Lower energy performance
Finite thickness	Overcomes diffusion
Some physical strength: mechanical stability	Long life — easy to install in cells
Resistance to gas blinding	Low voltage — good energy performance
Low cost	Economic operation

greater the thinner the material and the higher the concentration gradient across it. Since a diaphragm is positioned in a voltage gradient the material it is made of should be an electrical non-conductor to prevent it from acting as an electrode.

There is potentially a wide range of materials that can be used as cell separators,[2] which are divided into three types, organic, inorganic and composites.

2.5.1.1 Diaphragm Materials. Asbestos fibres are widely used in the chlor-alkali industry and in water electrolysis because they possess the required chemical and physical stability in alkaline media and can be engineered to give the required permeability. In the chlor-alkali industry the dia-phragm is constructed onto the electrode itself. Chrysotile asbestos [approximate formula $Mg_3(Si_2O_5)(OH)_4$] fibres are slurried with caustic soda solution and the slurry sucked on to a mild steel mesh cathode which collects the asbestos fibre. An alternative diaphragm material is asbestos paper made from specifically chosen chrysotile asbestos fibres with the addition of binding agents. As a general diaphragm asbestos is not an ideal material; it is not resistant to very acid conditions, it is not physically robust and it is an environmentally unacceptable material outside the cell. Thus in other applications polymer or ceramics materials are used.

In the case of ceramics processing the material into a porous structure, suitable for a diaphragm, can be difficult with the resultant material being quite brittle.

Installation in parallel plate electrolysers presents mechanical difficulties, which outweigh advantages of mechanical strength, and temperature and chemical stability. In the case of alkaline water electrolysis some flexibility in the structure is introduced by sintering the porous ceramic onto a supporting metal net. Ceramic materials are available as membranes for microfiltration and ultrafiltration, typically in tubular form, and thus only a few cells can capitalise on this technology. One example of its use however is a design for a fluidised bed electrode, discussed later in chapter 6.

The fabrication of polymers in usable forms for diaphragms does not create any unusual problems. In principle polymeric membranes can be selected from a wide range of materials although operating conditions in electrolysis usually involve extremes of pH and/or organic solvents. Thus in practice the materials are limited to polymers of ethylene, propylene, vinyl chloride and tetrafluoroethylene *etc.* In manufacturing, an open porous structure can either be created at the time of fabrication of the sheet or by the incorporation of a removable filler that is leached out at a later stage.

Several polymer materials have been tested in alkaline water electrolysis and in chlor-alkali cells. In these applications the materials used must be hydrophilic so that they are completely wetted and thus not blocked by gas bubbles. In the case of fluoropolymers this can be achieved by adding suitable wetting agents, *e.g.* ZrO_2, onto the diaphragm structure.

2.5.1.2 Semi-permeable Membranes – Ion-exchange Membranes. The disadvantages and limitations of permeable separators in electrosynthesis has focused interest on one range of semi-permeable separators *i.e.* ion-exchange membranes. Ion-exchange materials have the characteristic property of being able to distinguish between cations and anions. Thus the interest in ion-exchange membranes has arisen because of two, in-built specific functions. First they can be used to stop, selectively, either anions or cations from transferring from one cell compartment to the other, as shown in Figure 5. Secondly they allow electrolysis to be carried out under close control of compartment pH. For example the use of an anion-exchange membrane (AEM) will prevent the transfer of H^+ ions, generated at an anode during oxygen evolution, into the catholyte chamber and thus allow a pH differential to be set up in the cell.

The main properties required of ion-exchange membranes for them to be successful in technical processes are:

1   Low electrical resistance, *i.e.*, the permeability for the counter-ions under the driving force of an electrical potential gradient should be as high as possible, to minimise the membrane *IR* losses.
2   High permselectivity, *i.e.*, it should be highly permeable for counter-ions, but should be highly impermeable to co-ions, and to non-ionised molecules and solvents.
3   Good mechanical stability, *i.e.*, it should be mechanically strong and be dimensionally stable.
4   Good chemical stability, i.e., it should be stable over a wide pH-range and in the presence of oxidising agents.
5   Good operating characteristics *i.e.* it should be capable of operating over a wide range of current densities and under varying conditions of temperature, current density, pH *etc.*

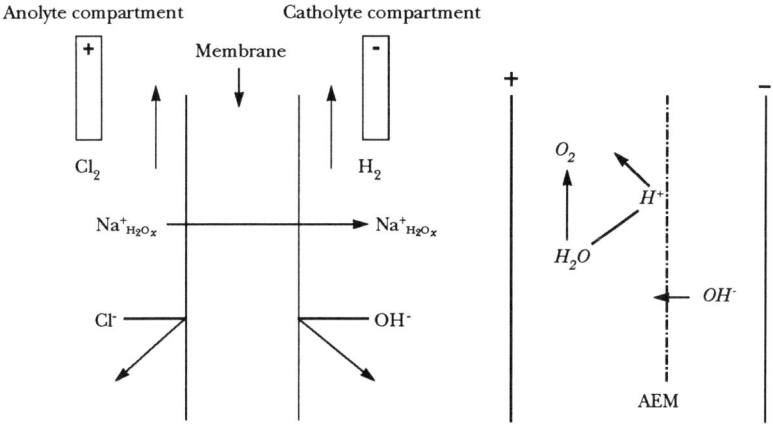

Figure 5   *Function of ion-exchange membranes as separators*

Cation-exchange membranes, that is, those selectively permeable to cations, are made from strong or weak acids in a polymeric sheet, and anion-exchange membranes are similarly made of strong or weak bases. The structure of these sheets is the same as the corresponding ion-exchange material, the ion type not transferred (anion in the case of cation-exchange material) forming a matrix by covalent attachment to a polymer. The properties of ion-exchange membranes are determined by two parameters, *i.e.*, the basic polymer matrix and the type and concentration of the fixed ionic group. The basic polymer matrix determines to a large extent the mechanical, chemical and thermal stability of a membrane. An ion-exchange membrane should be insoluble but be capable of swelling to a certain degree. The electric properties and the selectivities of ion-exchange membranes are determined mainly by the type and the concentration of the fixed ionic charges in the polymer matrix. There are a series of different cationic and anionic groups that can be introduced into a basic polymer matrix (Table 2). Sulfonic acid and carboxylic acid groups are most commonly used for the preparation of cation-exchange membranes. The sulfonic acid groups are strong acids, completely dissociated over nearly the entire pH range and carboxylic acid groups are weak acids and undissociated at values of pH of 6.

A typical structure of a cation-exchange membrane is illustrated schematically in Figure 6. The membrane is composed of a polymer matrix, which contains fixed negatively charged groups. For reasons of electroneutrality these negatively charged groups are counterbalanced by mobile, positively charged, cations which are usually referred to as counter-ions. Mobile anions, usually called co-ions, are more or less excluded by electrostatic forces since they are carrying the same charge as the fixed negatively charged groups. This process is referred to as Donnan exclusion. Owing to the exclusion of the co-ion a cation-exchange membrane, which carries negatively charged fixed groups, permits the passage

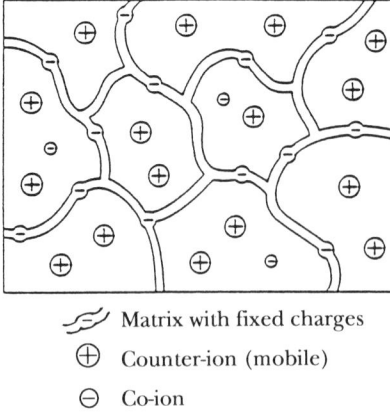

⟋⟍⟍ Matrix with fixed charges

⊕   Counter-ion (mobile)

⊖   Co-ion

Figure 6   *Structure of a cation-exchange membrane*

Table 2   *Ion exchange groups used in membranes*

Functional group	Structure	Properties
Cation-exchange membranes		
Sulfonic acid	$-\overset{\displaystyle O}{\underset{\displaystyle O}{S}}-O^{-}$	Strongly acid
Carboxylic acid	$-\overset{\displaystyle O}{\underset{\displaystyle O}{C}}^{-}$	Weakly acid
Phosphonic acid	$-\overset{\displaystyle O}{\underset{\displaystyle O}{P}}-O^{2-}$	Strongly acid
Phosphinic acid	$-\overset{\displaystyle O}{\underset{\displaystyle H}{P}}-O^{-}$	Weakly acid
Arsonic acid	$-\overset{\displaystyle O}{\underset{\displaystyle O}{As}}-O^{2-}$	Weakly acid
Anion-exchange membranes		
Quaternary amine	$-\overset{\displaystyle R}{\underset{\displaystyle R}{N}}-R^{+}$	Strongly basic
Tertiary amine	$-\overset{\displaystyle H}{\underset{\displaystyle R}{N}}-R^{+}$	Weakly basic
Secondary amine	$-\overset{\displaystyle H}{\underset{\displaystyle H}{N}}-R^{+}$	Weakly basic
Primary amine	$-\overset{\displaystyle H}{\underset{\displaystyle H}{N}}-H^{+}$	Weakly basic
Quaternary phosphonium	$-\overset{\displaystyle R}{\underset{\displaystyle R}{P}}-R^{+}$	Weakly basic
Tertiary sulfonium	$-\overset{\displaystyle R}{\underset{\displaystyle R}{S}}^{+}$	Weakly basic

of cations only. Anion-exchange membranes carry positive charges fixed on the polymer matrix and therefore exclude all cations allowing the passage of anions only.

Anion-exchange membranes mostly use the quaternary ammonium group in the polymer matrix leading to a strongly basic membrane. Tertiary, secondary and primary amines and to a lesser extent phosphonium and sulfonium groups, are used as less strongly basic functional groups. In some cases different ionic groups may be used in the same polymer matrix leading to bi- or poly-functional membranes.

An important factor in the design of cells with membranes is the transport of solvent, which accompanies the transferring ions. In aqueous systems the transport of water can be significant, *e.g.* between 3–5 water molecules accompany one sodium ion in chlorine cells. If hydrogen ions are transferred then typically two molecules of water are transferred per ion. This factor has considerable implications in the engineering of the process where one stream is concentrated due to water loss and the other stream becomes diluted.

2.5.1.3 Preparation of Ion-exchange Membranes. There are several methods to produce a fixed charge carrying polymer matrix with the desired properties.[3-5] The manufacture of commonly used ion-exchange membranes can be conveniently divided into two categories, perfluorinated and non-perfluorinated. The latter are based on cross-linked polystyrene and vinylbenzene and are usually less expensive and typically find applications in electrodialysis. The perfluorinated group of membranes have been developed to cope with the operating conditions encountered in chlor-alkali cells. The films are made from copolymers derived from the copolymerisation of tetrafluoroethylene and perfluorovinyl ethers and have the following general formula:

$$\left[(CF_2 CF_2)_x - CF_2CF \right]_y$$
$$|$$
$$(OCF_2 CF)_m \quad O(CF_2)_n{}^{-p}$$
$$|$$
$$CF_3$$

where $x = 6 - 13$
$m = 0 \text{ or } 1$
$n = 1 - 5$
$p = \text{precursor of ion-exchange group}$

The copolymer is semicrystalline and the structure of the crystalline phase has been shown to resemble that of polytetrafluoroethylene. In the above form the material is melt processable, a necessary feature to allow the fabrication of thin films and multilayer membrane structures. The material was first developed by DuPont under the Trade name of Nafion. Other manufacturers include Dow Chemicals, USA and Chlorine Engineers in Japan. The manufacture of ion-exchange membranes is dominated by the USA and Japan as can be seen in Table 3.

2.5.1.4 Anion-exchange Membranes. The application of anion-exchange membranes in electrochemical processes has been limited by the poorer chemical stability of this material, especially in alkaline conditions, in comparison to cation-exchange membranes. However, perfluorinated anion-exchange membranes with the chemical stability of the cationic perfluorinated membranes have recently become available. They are available from Tokuyama Soda Co. Japan, under the trade name of

Table 3 *Commercial manufacturers of ion-exchange membranes*

Type (trade name)	Producer
Cation–P^a (XF–OCF$_2$CO$_2$H/SO$_3$H) NP^a (Aciplex)	Asahi Chemical Ltd
Cation–P [XF–O(CF$_2$)$_n$CO$_2$H] NP (Flemion, Selemion)	Asahi Chemical Ltd
Cation–P (XF–OCF$_2$CF$_2$SO$_3$H) [Nafion (1000 series)]	DuPont
Cation–P (XF–OCF$_2$CF$_2$SO$_2$NH$_2$) (Nafion sulfonamide)	DuPont
Cation–P (XF–OCF$_2$CF$_2$CO$_2$H/SO$_3$H) (Nafion 901)	DuPont
Cation–P [XF–O(CF$_2$)$_n$CO$_2$H/SO$_3$H] NP (Neosepta-F, Neosepta), B	Tokuyama Soda
Anion–P (R$_1$NR$_3^+$X$^-$) (Tosflex)	Tosoh
P,B^a	WSI Tech
P cation, anion	Solvay–Morgane
P, B, cation, anion	Aquatech
P, B, cation, anion	Stantech
NP, cation, anion,	Ionics Inc.
NP, cation, anion	Membranes International
NP, cation, anion (RAI)	RAI Research Corp.
NP, cation, anion (Ionac)	Sybron Chemicals Inc.

^{a} P: Perfluorinated, NP: non-perfluorinated, B: bipolar

Neosepta ACM and from Morgane, France. These membranes have a similar structure to Nafion but have the sulfonic acid group replaced by the $-NR_3^+$ group. The membrane minimises the back migration of protons across the membrane towards the cathode.

2.5.1.5 Selection of Ion-exchange Membranes. The selection of an ion-exchange membrane for an electrochemical process can initially be made by consultation with commercial membrane suppliers, from knowledge of the intended conditions of operation and the electrolytes to be used. Following this, the application of the ion-exchange membrane(s) in electrosynthesis generally relies on the appropriate screening of the material(s) at laboratory scale and then pilot scale tests. Clearly many companies have gone through this process as will be evident by the number of applications of membrane electrolysers and electrodialysis cells discussed in this book. Two companies involved in electroorganic synthesis carried out this task in the late 1970s for particular industrial syntheses and reported the results.[6] BASF (Germany), investigated cation-exchange membranes for the electrosynthesis of dihydrophthalic acid. In this case the catholyte contained dioxane which on migration to the anode caused operating problems with the lead dioxide anodes. Of the membranes tested, Asahi Chem — Aciplex, DuPont — Nafion, Ionac — MC 3470, Ionics — 61 AZG and Tokuyama — CLE-E, only the last two were found to be satisfactory, giving anode off gases containing 1% of CO_2. Although the other materials were not necessarily unstable they allowed too high a transport of the organic solvent phase in the electrolyte.

2.6   ELECTRODE MATERIALS

A crucial aspect of an electrochemical process is the selection of the appropriate materials for the electrodes. Working electrode and counter electrode materials cannot always be selected independently as there will be important interactions of the cell chemistry to consider. There is currently a large body of knowledge about the suitability of electrode materials in a wide range of electrolysis conditions; electrolytes, pH, organic solvents and temperatures (see ref. 7). This can be put under two quite broad headings, chemical suitability and engineering suitability, although not surprisingly considerable interaction occurs between them. Under these two headings several criteria, shown in Table 4, are applied and in the development of the process re-applied until the most suitable combination of anode and cathode is achieved.

   The search for appropriate electrode materials starts in the laboratory and although guided by the criteria it generally is empirical or semi-empirical. At this stage selected materials are tested in terms of their chemical performance to eliminate unsuitable materials and to choose the materials for testing at the pilot plant stage. The selected materials at

Table 4 *Criteria for electrode material selection*

1	Suitable electrocatalytic and electrochemical properties
2	Chemical and electrochemical stability
3	Physical and thermal stability
4	Suitable physical form and fabrication
5	Good electrical conductivity
6	Low over-voltage
7	Environmentally suitable (non-polluting/non-contaminating)
8	Low cost

this stage will be engineered into suitable cell designs. Extensive testing of the reactor performance will then be undertaken to establish the influence of time and process interactions on the stability of the electrodes, the yield and current efficiency performance and the consequences of product recovery and purification.

2.6.1 Chemical Suitability

The standard electrode potentials will indicate the likely range of operating potentials for the cell. The possibility of reactions involving solvent or supporting electrolytes can therefore be considered. The thermodynamic stability of the electrode may also have some bearing on the selection. However, the kinetics of the processes can be of over-riding concern when dealing with processes that involve anodic dissolution or hydrogen evolution.

The chemical stability of the material can frequently be related to the chemical reactivity with intermediates, products and reactants. The formation of metal oxides on the surface or the use of alloying may either impart appropriate chemical stability, for example in the case of anode materials such as lead/lead dioxide, or may render them electrochemically inactive. The composition of the electrolyte, its pH and the temperature of operation are important factors to address. Corrosion of the surface may result from the intercalation of ions, for example in the case of graphite or by metal abstraction involving a discharged species and the electrode material.

For the electrode material to be suitable it should exhibit the required stability both during off-load (non-Faradaic) and on-load conditions. In certain cases it is recommended that the electrode when 'off-load' be held at an appropriate over-potential to avoid degradation. The material yield and current efficiency of the process is a prime consideration. This factor is influenced by the mechanism of the reactions, whether it involves simple electron transfer between reactant and surface or whether the reactions involve adsorbed species, intermediates and electrocatalytic surfaces. For example the influence of tetraalkyl salts on the surface adsorption processes in the electrohydrodimerisation of acrylonitrile is known to be responsible in giving high yields of the dimer, adiponitrile.

2.6.2   Engineering Suitability

The engineering aspects of the selection of electrode materials, which have a significant bearing on the electrolytic reactor design and thus the capital cost of the process, are:

*   Physical stability
*   Suitable fabrication or physical form
*   Cost
*   Physical properties
*   Environmental suitability

An electrode material must, like all engineering material, exhibit suitable mechanical strength in the form it is employed in operation. It should be resistant to erosion by the electrolyte materials under all operating conditions. It should enable suitable fabrication in the required form: as thin sheets, meshes, complex shapes, suitable coatings *etc.* The material must allow appropriate mechanical sealing in the reactor and connection to the external power supply.

The cost of an electrode material is linked to its operating life. Certain bulk metals, typically used as cathodes, may only require an initial capital investment and be effective throughout the operating life of the reactor with only a small amount of maintenance. Other materials may be less stable and require regular maintenance and replacement and thus be more expensive in the process operation. This is particularly true in the case of anodes where precious metals and their oxides are frequently used to give the required chemical stability and electrochemical activity. These materials are commonly applied as coatings to less expensive and chemically stable substrates such as titanium in applications such as chlor-alkali cells and as oxygen evolving anodes in acid electrolytes. It is desirable that the operating life be several years even when the material is applied as thin coatings. Several companies produce coated electrodes for laboratory and plant operation.

The important physical properties which affect selection of an electrode material are the density, resistivity, linear expansion and thermal expansion. The material should be as light as possible to aid in cell installation, operation, handling and maintenance. This is generally of secondary importance especially when either thin electrodes or coated electrodes are used.

The electrical resistivity should ideally be low when fabricated as thin electrodes with a current path of 1.0 m or more. The materials must exhibit only a small voltage drop to keep power costs low and to maintain a 'uniform' potential distribution. Most metals fall into this category, although there are limitations with the use of materials such as titanium, carbon and alloys such as Hastelloy. In such cases the electrodes are connected in a bipolar configuration (see chapter 3) or are relatively short or may be backed or supported by a more conducting material. Bipolar connection of electrodes may require a physical attachment of different materials and thus the possible expansion of these materials under thermal cycling should be considered. This applies also to coated electrodes comprising of mixtures of metal oxides, metals and suitable doping materials that impart a relatively high degree of conductivity. For example doping of antinomy into tin oxide coated electrodes imparts a significant amount of electrical conductivity into what otherwise is a non-conducting material. This material has potential as an anode in the treatment of waste streams containing dissolved organic compounds.

A factor, which is becoming of greater concern in the process industries, is the use of materials that are potential pollutants or hazardous. This especially applies to electrode materials that are likely to undergo significant amounts of corrosion. The control of toxic heavy metals during such processes may add a large factor to the overall process costs.

2.6.3 Electrodes Materials in Synthesis and Effluent Treatment

Many of the developments in electrode materials and the understanding of electrode processes themselves have come as a result of many years of effort into the chlor-alkali, water electrolysis and metal winning industries. This effort has resulted in a range of materials for chlorine evolution and hydrogen, and oxygen evolution, which can be applied effectively on a commercial scale. A valuable consequence of this is the range of materials that can be used as counter electrodes in many processes (see Table 5). Generally in the selection of electrode materials for any given application there is often more than one choice and the final selection is made usually on the basis of a series of screening tests.

2.6.3.1 Materials for the Counter Electrode Supporting Reaction. Hydrogen evolution is a frequently encountered cathodic process used as a supporting electrode reaction as it involves negligible reagent costs, the product separation is relatively easy and it can also maintain the correct pH balance in operation. In alkaline solution the preferred materials are nickel and steels. In acid solutions lead and lead alloys, carbon, and stainless steel are often used. If a large over-potential is a major concern then cathodes used in water electrolysis cells can be employed. Porous, sintered and Raney nickel coatings are sometimes used. In certain cases cathodes

Table 5 *Electrode materials used in synthesis and effluent treatment processes*

Reaction	Materials
Cathodes	
H_2 evolution	Steel, Ni, Ni-coatings, precious metal coating
O_2 reduction	Dispersed Pt on (Teflon loaded) high area carbon (other electrocatalyst)
Other reactions	High H_2 over-potential metals, *e.g.* Hg, Pb, Cd. Alloys *e.g.* Pb/Sb; other metals, *e.g.* Ni, Cu, Ag, steels, stainless steels, Hastelloy[R] (Ni–Mo–Fe or Ni–Mo–Cr), graphite, other carbons; conducting ceramics, *e.g.* TiO_x (Ti_4O_7, Ti_5O_9), Raney Ni, Pt/Pt, Pd/C for electrohydrogenation
Anodes	
O_2 evolution	IrO_2-coated Ti PbO_2 on Ti or C Pb or Pb alloying H_2SO_4 Steel in a neutral and basic medium Ni and spinels in basic medium
Cl_2 evolution	RuO_2-based coatings on Ti (DSA^R) Other oxides based on Co_3O_4 and PdO_2, and graphite
Other reactions	Pt, Pt/Ti, Ir/Ti, Pt–Ir/Ti and on other substrates PbO_2 on Ti, Nb or C, SnO_2 on Ti Fe or Pb in acid sulfate media Carbons Ni and spinels in basic media Conducting ceramics, *e.g.* TiO_x (Ti_4O_7, Ti_5O_9), ZrO_3, TaO_x Mg, Zn and other sacrificial anodes Dispersed Pt, Ru *etc.* on C

coated with precious metals employing a range of substrates including Ti and Ni are used. The problems associated with hydride formation should however be considered with such materials.

Oxygen evolution is, like hydrogen evolution, a convenient counter electrode reaction. In basic media nickel is the preferred material whilst steel and nickel are used in less basic electrolytes. In basic media a reduction in over-potential can be achieved by coating the electrode surface with transition metal oxides, spinels and perovskites. In acid electrolyte lead or lead dioxide coated (Ti, C) electrodes are frequently encountered because of there relatively low cost. The use of lead alloys has been applied to reduce the corrosion and the over-potential of lead. An alternative to lead dioxide, with offers greater corrosion resistance, is IrO_2-coated titanium. In some applications platinised titanium has been used. The selection of the appropriate anode will often depend upon the presence of contaminants such as organics, which may accelerate corrosion, or anions, such as chloride, where electrocatalysts such as IrO_2 will favour chlorine evolution.

In applications where the electrolyte can be a particularly aggressive medium the use of ceramic substrates may have several possibilities. There is currently great interest in Ebonex[R] a non-stoichiometric titanium oxide (approximately Ti_4O_7) which, although itself relatively inactive electrochemically, is an excellent substrate for several coatings.

In attempts to improve the energy consumption of electrolytic cells significant interest is being shown in the cathodic reduction of oxygen and the anodic oxidation of hydrogen. In both cases these thermodynamically favourable reactions offer significant savings in cell voltage. In both cases the products of the reactions are the same as the corresponding gas evolving reactions *i.e.* H^+ and OH^- ions. The electrodes are based on gas diffusion electrode technology using Teflon bonded high surface area carbon electrodes loaded with a suitable electrocatalyst. In the case of oxygen reduction this is commonly platinum, although other much cheaper materials are actively being researched. Several applications of this technology have been proposed, *e.g.* in chlor-alkali cells, ozone generation *etc.* The oxidation of hydrogen has been considered in the context of metal winning processes, recycling and organic electrosynthesis.

2.6.3.2 Electrodes for Electrosynthesis. In the selection of a cathode for use in an aqueous electrolyse an often over-riding consideration is the use of a material with a high hydrogen over-potential to prevent hydrogen evolution occurring. Thus materials such as Pb, Cd, Zn, Sn and C are considered, with the latter being particularly attractive owing to its non-toxic nature. For conditions where lower over-potentials are acceptable a wide range of materials can be considered where the over-riding factors are a high material yield coupled to a stable operation. The selection of suitable materials is determined from fundamental laboratory investigations.

The prime considerations of an anode material in electrosynthesis are the attainment of a high yield with a stable operation under generally thermodynamically unfavourable operating conditions. In addition a material with a high oxygen over-voltage is required. In the chlor-alkali and related industries this is achieved with coated titanium electrodes using platinum group metal oxide catalysts. Commercial electrodes typically use RuO_2 combined with other oxides *e.g.* TiO_2 and IrO_2. In the majority of other oxidation reactions the initial choice is likely to come from platinum coated titanium (or Ta or Nb), PbO_2 and carbon. In non-acidic media materials such as nickel and monel have also been used.

REFERENCES

1   D. R. Hibbert, 'Introduction to Electrochemistry', Macmillan Press, London, 1993.

2   M. I. Ismail, N. F. White, H. P. Dhar and S. D. Gupta, *Polym. Plast. Technol. Eng.*, 1980, **15,** 61.

3   'Permselective Membranes', ed. C. E. Rogers, Marcel Dekker, New York, 1983.

4   'Perfluorinated Ionomer Membranes', ed. A. Eisenberg and H.L. Yeager, ACS, Washington DC, 1982.

5   R. Rautenbach and R. Albrecht, 'Membrane Processes', J. Wiley and Sons, England, 1989.

6   N. C. Weinberg and B. V. Tilak, 'Technique of Electro-organic Synthesis, Part III Scale-up and Engineering Aspects', J. Wiley and Sons, England, 1982.

7   A. M. Cooper, D. Pletcher and F. C. Walsh, *Chem. Rev.,* 1990, **90,** 837.

CHAPTER 3

Electrode Processes and Membrane Transport

Electrochemical processes are driven by the application of a potential field, the magnitude of which will generally determine the rate of the relevent process; charge transfer and ionic flux. This chapter describes and quantifies the relevent electrochemical rate processes, which are important in the design and the eventual implementation of the operating cell or reactor. This unit may be an electrolytic device for generating specific chemicals directly at an electrode or may generate reagents for external use or reaction. The cell may also function as a means of driving specific ions, selectively, through appropriate membranes as a means of separation and concentration of ionic species or of initiating ionic reaction.

3.1 TYPES OF ELECTROCHEMICAL PROCESSES

Electrochemical reactions are surface processes, which are instigated by a suitable charge transfer at a fluid–solid interface. The physical chemistry of this charge transfer is complex, although it is convenient to present a simple picture of the interfacial processes which will allow a qualitative classification of electrochemical processes to be made. When two electrodes are placed in an ionic conducting solution and are connected externally they become charged. Thus locally at the solution–electrode interface there is a large potential difference over a molecular scale of a few nanometres. A simple model of this situation[1] consists of a double layer, comprised of a plane of closest approach [inner Helmholtz plane (IHP)] and a diffuse layer or outer layer (see Figure 1). The equilibrium established at the interface is electrostatic, something analogous to that in a capacitor. The electrostatic interactions determine the distribution of the potential and the potential difference, which constitutes the driving force for electrochemical (Faradaic) reaction when a current flows through the circuit. An increase in magnitude of the potential applied between the two electrodes, increases this potential difference at the interface. This interfacial potential difference is the driving force for the electrochemical reaction. The mathematical description of the relationship between the electrode–solution potential difference (measured as the electrode potential) and the current density is the basis of electrode kinetics.

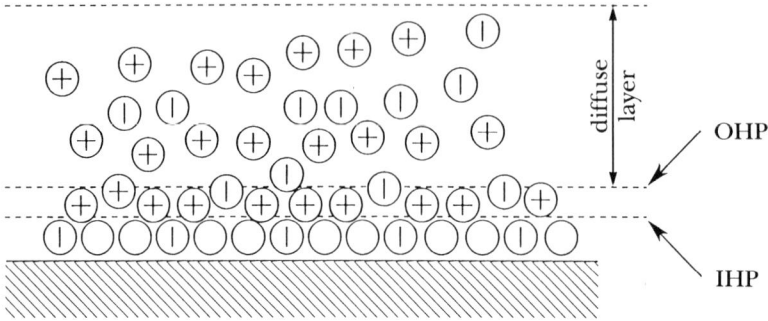

Figure 1   *Model of the electrode–solution interface and simple electrochemical processes*

The types of electrochemical reaction, which can occur at the electrode solution interface, can be divided into four general headings:

1   electrochemical reaction
2   heterogeneous electrocatalysis
3   heterogeneous redox electrocatalysis
4   homogeneous redox catalysis

3.1.1   Electrochemical Reaction

A simple electrochemical process [outer Helmholtz plane (OHP)], shown in Figure 2(a), is when a reactant undergoes a transformation to a product by the transfer of an electron from the electrode to the species in solution, without contacting or interacting with the surface in any significant way. Thus the model of the process is that of an electron hopping from the electrode to the reactant species, whilst it is within molecular dimensions of the surface. The role of the electrode is essentially that of a source or sink for electrons, and it does not influence the type of final product species. The nature of the product species is thus determined by the chemistry which occurs between the species, which has undergone charge transfer, and the electrolyte solution.

    In practice the role of the electrode in this model is through its influence on the structure of the interfacial double layer. This stucture will depend on the way solvent, ionic and neutral species in solution interact with the surface and thus on the local distribution of potential in the double layer. In addition a charged reactant species will be at a different

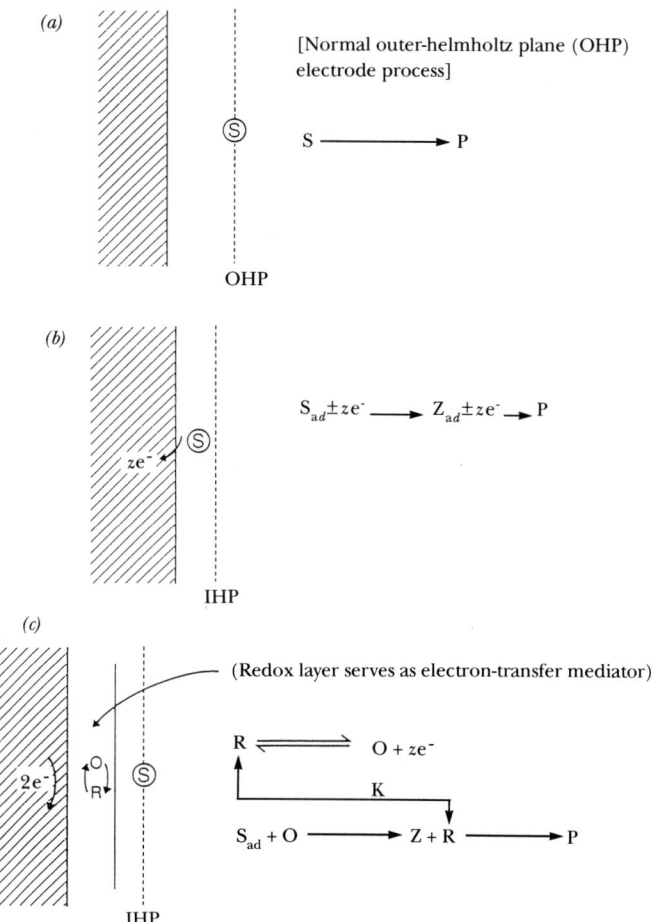

Figure 2 *Types of electrochemical processes* (a) *electrochemical reaction;* (b) *electrocatalysis;* (c) *heterogeneous electrocatalysis*

concentration at the double layer than in bulk solution because of the potential field. This is referred to as the double layer effect. Overall, therefore, these factors result in the variation of the standard rate constants for simple reactions at different electrode materials. Another factor, which can either decrease or increase the rate of electron transfer by virtue of its effect on the rate constant for the reaction for simple processes, is the differences in adsorption behaviour of neutral molecules and non-reacting ionic species at different electrode surfaces.

3.1.2   Electrocatalysis

In heterogeneous catalysis the strong adsorption of starting material(s) at the surface reduces the activation barrier for the reaction to proceed *via* an intermediate species:

$$S_{ads} \rightarrow Z_{ads} \rightarrow R \tag{1}$$

Partial charge transfer may occur as part of the process, but it is completely recycled on the formation of the product R. In electrocatalysis a similar principle holds, but there is a net flow of electronic charge. The charge transfer is an inner Helmholtz plane process between the adsorbed species and the surface [see Figure 2(*b*)].

$$S_{ads} \pm n \text{ e}^- \rightarrow Z_{ads} \pm n' \text{ e}^- \rightarrow R \tag{2}$$

The chemical steps in the process, rather than the electrochemical steps, are catalysed by the interaction with the surface. This, therefore, distinguishes between the electrochemical reaction, which is affected by adsorbed species, and electrocatalytic reactions where adsorbed electroactive intermediates are involved directly. The rate constants of electrocatalytic reactions will exhibit a wide variation with the type of electrode (electrocatalyst) used and the associated characteristic kinetic coefficients may even change with electrode material, indicating that an alternative reaction mechanism may be occurring. This has been observed in the case of the hydrogen evolution reaction which proceeds by the formation of adsorbed hydrogen atoms.

$$H^+ + \text{e}^- + M \rightarrow M\text{--}H \tag{3}$$

The formation of hydrogen then either occurs by a second charge-transfer process of the adsorbed intermediate involving hydrogen ions or by a bimolecular reaction of the adsorbed intermediate.

The formation of adsorbed hydrogen is also responsible for many reduction reactions of organic species. These reactions occur between the adsorbed hydrogen and the organic molecule (probably also adsorbed) and are only possible on electrode materials, which exhibit a significant coverage of hydrogen atoms. The chemical process is closely related to catalytic hydrogenations and the materials used are often similar, *e.g.*, Pt, Ni, Rh, Co and Fe. The process itself is not restricted to aqueous electrolytes, it can occur in other solvents/electrolytes such as methanol or acetic acid.

3.1.3   Heterogeneous Electrocatalysis

The performance of electrochemical processes can be significantly affected if the electrode surface is modified by a layer, at least a monolayer, of a redox system. This redox sysem can serve as a mediator for the reactants, S, as well as a new surface for adsorbed interactions with reagents and other molecules:

$$R = O + ne^-$$

$$S_{ads} + O \rightarrow Z + R + P \tag{4}$$

The mediator, R, exhibits typical catalytic behaviour, and is continuously regenerated as the reaction occurs [see Figure 2(b)]. This type of catalysis is termed redox catalysis and is an inner Helmholtz plane process between the adsorbed reactant and the immobilised redox mediator.

The behaviour of heterogeneous redox catalysis is analogous to the case of homogeneous redox catalysis. In the latter the reaction between the redox mediator and the reagent is not at the surface but somewhere in the bulk solution. An example is oxidation involving redox couples such as Br_2/Br^-, where the generation of free bromine takes place at an appropriate anode and the oxidation of species, *e.g.*, SO_3^{2-} is essentially a chemical process:

$$Br_2 + SO_3^{2-} + H_2O \rightarrow 2\ Br^- + 2\ H^+ + SO_4^{2-} \tag{5}$$

The chemical reaction can be either in the cell (in-cell process) or in a second reactor (ex-cell process), external to the electrochemical cell.

From a mechanistic view heterogeneous redox catalysis is an electrochemical process, deriving chemical transformation directly at the surface by the charge-transfer process. It has inherent advantages over homogeneous redox catalysis in that the reaction mixture is not contaminated by the redox reagent and thus product separation is potentially less troublesome. In practice, this requires the redox agent to have the required stability in the electrolytes.

In subsequent chapters, examples of reactions involving the above four types of electrochemical processes are discussed in more detail. This chapter continues by looking at the basic model, which is used to represent the rate processes associated with electrochemical reactions.

3.2   A SIMPLE MODEL OF ELECTRODE KINETICS

For the simple reversible electron transfer between two ions in solution
and an electrode

$$O + ne^- \overset{k_+}{\underset{k_-}{=>}} R \tag{6}$$

the rate constants of reaction for the cathodic and anodic processes are
usually written as:

$$k_- = k_o \exp\left[-\alpha\, n\, F/R\, T(E\text{-}E^o)\right] \tag{7}$$

$$k_+ = k_o \exp\left[(1\text{-}\alpha)\, nF/R\, T(E\text{-}E^o)\right] \tag{8}$$

where k_o is the standard rate constant and α is the transfer coefficient for
the reaction.

These expressions are obtained on the basis of the absolute rate theory
with the assumption that the free energy of activation of the electrode
reaction varies linearly with the electrode potential. The rates of the
cathodic and anodic processes of reaction (6) are:

$$j_- = -nFk_-\, C_O \tag{9}$$

$$j_+ = nFk_+ C_R \tag{10}$$

The overall rate of reaction, or current density, is made up of the sum
of the anodic and cathodic contributions,

$$j = j_- + j_+ \tag{11}$$

which from eqns. (10) and (11) is expressed as:

$$j = n\, F\, (k_+\, C_R - k_-\, C_O) \tag{12}$$

For the equilibrium condition of zero net rate, eqn. (12) gives

$$k_+\, C_R = k_-\, C_O \tag{13}$$

The partial current density at equilibrium is called the exchange current density, $j_o = j_- = j_+$, when the potential E is equal to the equilibrium
potential E_e. Thus from eqns. (7) and (8) the exchange current density is
given by:

$$j_o = n \, F \, k_o \, C_O \exp[-\alpha \, n \, F/R \, T \, (E_e - E^o)]$$

$$= n \, F \, k_o \, C_R \exp[(1-\alpha) \, n \, F/R \, T \, (E_e - E^o)] \tag{14}$$

Simplification of eqn. (14) gives a Nernst equation, with activity replaced by concentration:

$$E_e = E^o + (R \, T/n \, F) \ln(C_O/C_R) \tag{15}$$

Thus the kinetic model of the electrode process is consistent with the thermodynamic model for the equilibrium situation. It is often convenient in experiments to choose the equilibrium potential as a reference point and to write kinetic expressions in terms of the over-potential, $\eta = E - E_e$.

$$j = j_o \{\exp[(1-\alpha) \, nF\eta/RT] - \exp(-\alpha nF\eta/RT)\} \tag{16}$$

This equation is known as the Butler–Volmer expression for simple reversible electrode kinetics. Its characteristics are determined by the value of the transfer coefficient and the exchange current density. The value of the exchange current density depends on the standard rate constant for the reaction and the concentrations C_O and C_R.

The Butler–Volmer equation has analogous forms for multistep fast electrochemical reactions. In these cases the equation will not exhibit symmetry, *i.e.*, $\alpha = 0.5$, and the reaction order may not be equal to 1.0. For example for an overall reaction

$$mO + ne^- = w \, R \tag{17}$$

the rate of reaction at high electrode potentials is written as:

$$j = nFk_O \, (C_O)^p \exp(-\alpha^- \, nFE) \tag{18}$$

$$= nFk_{f1} \, (C_O)^p$$

where, p, is the reaction order and, α^- is used to reflect the lack of symmetry.

The order of reaction is thus obtained from:

$$\left(\frac{\partial \ln j}{\partial \ln C_O}\right)_E = p \tag{19}$$

3.2.1   Experimental Behaviour

Figure 3 shows the influence of the transfer coefficient and exchange current density on the current density–over-potential curves for a simple redox reaction. High values of exchange current density shift the curves to lower overpotentials at the same values of current density. This is the desired effect when electrocatalysts are used. An increase in the value of the transfer coefficient increases the magnitude of the anodic current density, at a fixed potential, whilst decreasing the magnitude of the cathodic current density.

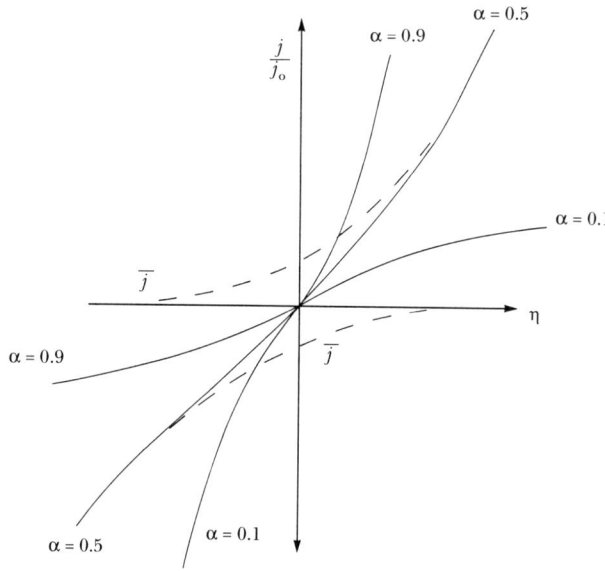

Figure 3   *Influence of exchange current density and transfer coefficient on the behaviour of the Butler–Volmer equation; values of α on figure*

The experimental determination of the kinetic parameters of the Butler–Volmer equation can conveniently be achieved from data obtained at relatively high over-potentials. Inspection of eqn. (17) shows that at values of over-potentials $>70/n$ mV, the equation reduces to the form, for a cathodic reaction:

$$j = j_0 \exp(-\alpha f \eta) \tag{20}$$

where $f = n F / R T$.
This is the Tafel approximation, which is often written as:

$$\eta = (-2.3/\alpha f) \log(j) + (2.3/\alpha f) \log(j_0) \tag{21}$$

A similar expression is obtained for the anodic process. Thus a graphical plot of over-potential against the log of the current density will give a slope (see Figure 4) of

$$d[\log(j)]/d\eta = -\alpha f/2.3 = 1/b \tag{22}$$

where b is referred to as the Tafel slope. The value of the Tafel slope can often be used as a diagnostic tool. A value of approximately 120 mV per decade gives a transfer coefficient of 0.5.

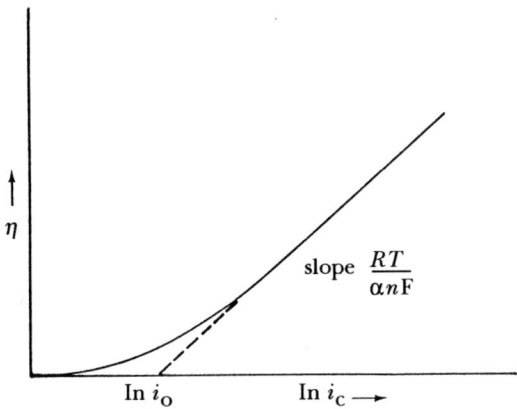

Figure 4 *Experimental behaviour of the Tafel equation*

3.3 MASS TRANSPORT AND ELECTROCHEMICAL REACTIONS

The aspect of mass transport is of considerable importance in electrochemical systems and is related directly to the transport laws of solutions and to fluid mechanics.

3.3.1 The Flux Equation of Mass Transport

The consumption of a species at a electrode must be initiated by the application of the current and will thus instigate the development of concentration profiles of species in the electrolyte solution near the electrode. In an unstirred solution the concentration profile will develop furthur into solution as time passes. However if a condition of steady convection is introduced into the system then the convective flow terms will effectively dampen down this concentration profile development and a steady state profile will be obtained as shown in Figure 5.

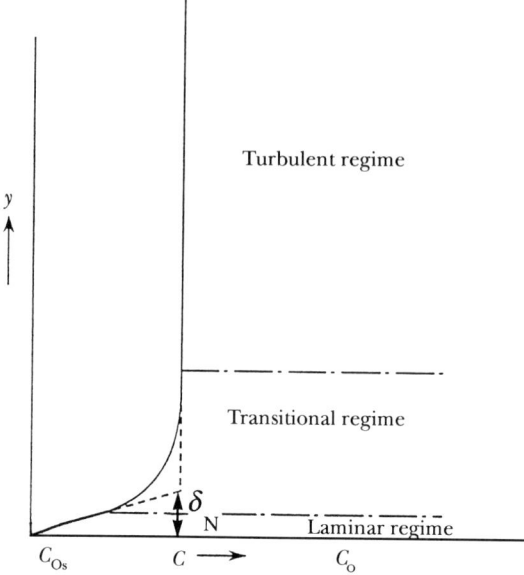

Figure 5   *Concentration profile obtained near an electrode surface*

In electrochemical systems the convection can be introduced by mechanical stirring, rotation or other movement of the electrode and by the flow of electrolyte adjacent to the electrode surface. In the absence of turbulence the flux of an ionic species in the electrolyte, as depicted in Figure 5, is attributed to the sum of three effects:

1   Migration. This is the transport of charged species under the influence of a potential gradient $(d\phi/dx)$.
2   Diffusion. This is the transport of species owing to the existence of a concentration gradient. The consumption of reactant and the formation of product at an electrode set up the appropriate concentration gradients by which reactant and product move towards and away from the electrode, respectively.
3   Convection. This is the transport of species associated with the bulk movement of the electrolyte. It is effectively the mass flow of a given species in the electrolyte.

The flux of an active species towards an electrode, at the steady state, is defined (for a one-dimensional model) by

$$N_j = -D_j \, dC_j/dx + C_j \, u - z_j \, U_j \, C_j d\phi/dx \tag{23}$$

The current density in the electrolyte is given by the sum of the fluxes of all the charged species in the electrolyte.

$$j = F\Sigma \; z_j \; N_j \tag{24}$$

An important simplification of eqn. (23) is when an excess of supporting electrolyte is present in the solution. In this case the flux of an electroactive species attributed to migration becomes negligible and the total flux is a combination of diffusion and convection:

$$N_j = C_j \, u \; - D_j \, dC_j / dx \tag{25}$$

Near an electrode surface, (x=0) where the convective flux term is eliminated, the flux is given by Fick's First Law of Diffusion:

$$N_j = -D_j \, dC_j / dx \tag{26}$$

Therefore, from eqn. (24) the current density at the electrode surface is given by:

$$j_j = -n_j F \, D_j \, dC_j / dx \tag{27}$$

In a region close to the electrode surface where convection and migration are effectively absent, called a diffusion layer, of thickness δ, eqn. (27) can be integrated to give:

$$\delta = -n_j \, F \, D_j / j_j \; (C_j - C_{js}) \tag{28}$$

where C_j and C_{js} represent bulk and surface concentrations.

The Nernst model considers the concentration to be linear near the electrode surface (see Figure 5) whereas in practice the concentration will asymptotically approach the value in the bulk. This diffusion layer offers the sole resistance to mass transfer and is, in the first instance, of a constant thickness over the electrode surface. By replacing the actual concentration profile with a linear approximation, the current density is written in terms of a mass transfer coefficient k_{lj}

$$j_j / \; n_j \, F = k_{lj} \; (C_j - C_{js}) \tag{29}$$

where $\quad k_{lj} = D_j / \delta$

The mass transfer coefficient depends upon the flow conditions, the cell geometry, the diffusion coefficient and the electrolyte solution. For example in the case of flow over a flat surface, the viscous forces arising between the surface and the fluid (frictional drag forces) cause the fluid to slow as the surface is approached and at the surface the fluid has a zero velocity (see Figure 6). Thus a 'boundary' layer is formed, which is the region near the surface, where there is a distribution of the fluid velocity. The value of the boundary layer thickness is determined by the flow con-

ditions, which are conveniently defined in terms of a dimensionless parameter, referred to as the Reynolds number, given by:

$$\mathrm{Re} = \rho\, u\, l/\mu \tag{30}$$

where, ρ is the density, μ is the viscosity, u is the velocity and, l is a characteristic length for the system geometry.

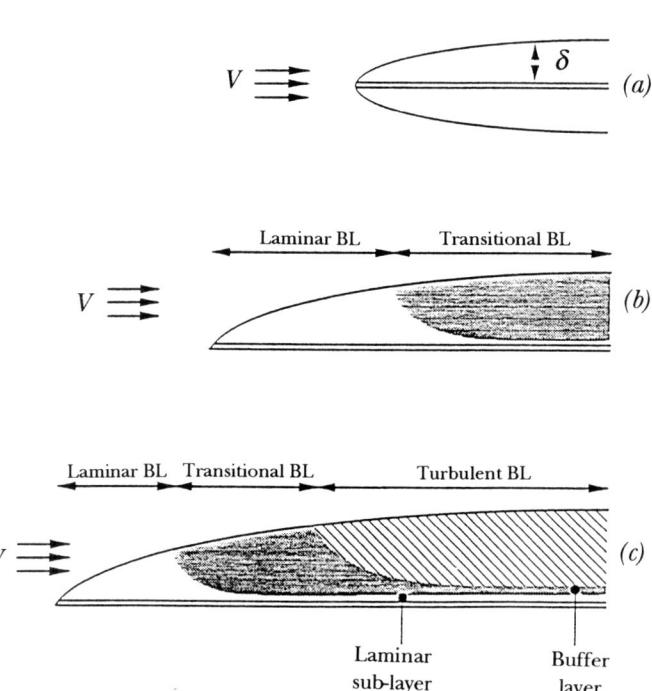

Figure 6   *Boundary layer (BL) for flow over a surface*

As the Reynolds number increases, the boundary layer becomes thinner
and causes compression of the thickness of the diffusion layer. In turbulent flow the boundary layer is comprised of several regions, and notably a
laminar sublayer at the surface, referred to as the Prandtl layer, of thickness, δ_{Pr}. The thickness of the Nernst diffusion layer is determined by the
Schmidt number for the electrolyte, $\mathrm{Sc} = \mu/\rho D$, through the relationship:

$$\delta/\delta_{\mathrm{Pr}} = \mathrm{Sc}^{1/3} \tag{31}$$

The Sc number for most electrochemical systems is large, of the order
of 10^3, and thus the diffusion layer is approximately a tenth of the thick-

ness of the laminar sublayer, of the order of 0.1 mm. The concept of a mass transfer coefficient has, through experimental and theoretical developments, enabled mass transfer to be based on empirical, or semi-empirical, expressions of the type[2]

$$Sh = function(Re, Sc) \tag{32}$$

where $Sh = k_1 \, l/D$, the Sherwood number.

An important case of diffusion at an electrode is without a supporting electrolyte.

3.3.2 The Transport Number

The fraction of the total current which corresponds to the ionic flux, N_i, in the bulk of the electrolyte is called the transport number, t_j,

$$t_j = N_i \, n \, F/j \tag{33}$$

By applying eqn. (24) at the electrode surface in the absence of convection leads to:

$$j = -D \, n \, F/(1-t_j) \, dC/dx \tag{34}$$

The transport number is determined by conditions at the electrode surface, although as an approximation conditions in the bulk are often used. This is equivalent to writing the total flux as the sum of diffusional mass transport and ionic migration. From the Nernst linear approximation this gives:

$$j = n \, F \, k_1 \, (C - C_s)/(1 - t_j) \tag{35}$$

This is a convenient approximation for the effect of ionic migration on the mass transfer of an ionic species to an electrode surface.

3.3.3 Electrode Kinetics and Mass Transport

At an electrode surface, at the steady state, the supply of reacting species to the surface is equal to the total rate of reaction of that species at the surface. Thus in the case of diffusion alone the rate of mass transport and electrochemical reaction are, for Tafel high field conditions, equated as:

$$j/nF = k_1 \, (C - C_s) = k_+ \, C_s \tag{36}$$

Eliminating the surface concentration in eqn. (36)

$$j = j_o \exp(-\alpha n F \eta) / [1 + j_o \exp(-\alpha n f \eta) / n \; F \; C \; k_1] \tag{37}$$

At high potentials, the system comes under mass transport control and the 'mass transport limited' current density is given by:

$$j_l = n \; F \; k_1 \; C \tag{38}$$

Near the equilibrium potential the Tafel approximation is not valid and the Butler–Volmer equation should be applied. The behaviour of the current–potential response around the equilibrium potential depends on the magnitude of the standard rate constant or the exchange current density. If the standard rate constant is large ($> 2 \; 10^{-4} \; m \, s^{-1}$) then the system is said to be reversible and the rates of the anodic and cathodic redox processes are in equilibrium and continuity in the current response is seen. If the standard rate constant is small ($< 5 \; 10^{-5} \; m \, s^{-1}$) then the system is said to be irreversible and the redox processes are not in equilibrium, and a point of inflection is seen in the current response.

In the analysis of electrochemical current density–potential curves equations of the form of (36) and (37) can be suitably rearranged to enable linear correlation of appropriate data, *e.g.*,

$$\ln[j \, j_l / (j_l - j)] = \ln j_k = \ln j_o - \alpha n f \eta \tag{46}$$

3.4   REACTION RATE MODELS

The simple reaction rate model for an electrode process combines the electrochemical kinetic rate with a mass transfer parameter, which on experiencing increased convection increases in value due to a thinning of the stagnant diffusion layer. This model is widely adopted and significant efforts have been made by researchers to produce correlations for the prediction of mass transport coefficients for many practical systems. The model has limitations when for example mass transport of ionic species is more appropriately described by the convective diffusion equation. In addition the total supply of a species by mass transport from the bulk is equal to the total consumption of a species at the electrode by all relevant reactions; which will also include all non-electrochemical processes. When reactions are not confined to the electrode surface then the simple model is not applicable. In this case the behaviour will depend upon the relative rates of the electrochemical, chemical and mass transport processes. Homogeneous reactions can occur at a comparable rate to diffusion of the associated species and this may occur mainly in the diffusion layer and thus simultaneously with the mass transport process. The reaction model then becomes one of electrochemical surface behaviour combined with that of the concurrent diffusion and reaction.[3]

Another category of reaction is when the chemical reactions are slow in comparison to mass transport and thus the majority of reaction takes place in the bulk electrolyte. This system is then considered as a succession of steps:

(i)   mass transport to the electrode surface;
(ii)   electrochemical transformation;
(iii) mass transport to the bulk electrolyte; and
(iv)   chemical reaction in the bulk electrolyte.

The more general picture of the rate processes is when reaction can occur in both the bulk electrolyte and in the diffusion layer. The total chemical reaction is then determined by the sum of the rate of product species diffusing from the diffusion layer and the product formed by a homogeneous bulk reaction.

3.4.1   The Role of Adsorption

Adsorption of ionic or neutral species onto electrode surfaces can have a significant impact on the progression and direction of electrochemical processes. These species may be electroreactive reagents, intermediates or solvent molecules which form some type of bond with the surface and thereby accelerate, decelerate or alter the pathway the reaction(s) take to form products. The adsorption bond may be covalent or electrostatic or the molecule may have a preferred affinity for the surface. The role of adsorption in electrochemical processes will generally result in a modification of the kinetic pathway by avoiding a slow step in the process *i.e.* electrocatalysis and a change in the interfacial environment at the electrode which may induce depletion of particular species, thereby altering reaction pathways.

These factors are influenced by the degree of surface coverage which itself depends on the type and concentration of adsorbate, the solution composition, temperature, the electrode material and the electrode potential. The surface coverage of adsorbate is described by an adsorption isotherm, which relates the degree of coverage to the concentration of adsorbate in the electrolyte and the free energy of adsorption, *e.g.* Langmuir, based on the assumption of monolayer coverage, uniformly energetic sites of adsorption and that adsorbed molecules do not interact. Rate models for reactions influenced by adsorption are generally developed on the basis of the stationary state approximation or on the assumption of a rate determining step (rds).

An important example where adsorption can be of major significance is the cathodic evolution of hydrogen[4] where a possible mechanism involves the formation of an adsorbed hydrogen molecule, M–H, followed by a bimolecular cleavage of the bond:

$$H^+ + e^- + M \rightarrow M\text{--}H \tag{40}$$

$$2\ M\text{--}H \rightarrow 2\ M + H_2 \tag{41}$$

This type of mechanism is responsible for the variation in the rates of hydrogen generation, *i.e.* the exchange current density for the reaction, on different electrode materials. A small free energy of the M–H bond will mean a low formation of the adsorbed species, increasing the free energy of the bond will increase the adsorption of H, whilst making it more difficult to cleave the bond. Thus the exchange current density for this type of mechanism exhibits a maximum with an increase in the free energy of the M–H bond. Analysis of experimental data of the hydrogen evolution reaction is in agreement with the model *i.e.* on platinum in a sulfuric acid electrolyte the Tafel slope changes from a value of 30 mV per decade to 120 mV per decade on a more negative increase in the electrode potential.

The development of rate equations for more complex reactions results in a set of algebraic equations. Although the solution of the resulting equations is possible the 'rds' limiting forms are more convenient, both for engineering use and in experimental data analysis. Overall characteristics of the rate equations, which include adsorption, can be summarised in the general form

$$\text{rate} = \frac{(\text{kinetic factor})\,(\text{driving force})}{(\text{adsorption group})} \tag{42}$$

There are a large number of rate forms for processes such as bimolecular reactions, dissociations, reaction between adsorbed and unadsorbed species, *etc.* Although the detail of the equations is different, they can generally be expressed in the following form of electrochemical rate model:

$$\frac{j}{n_j F} = \frac{k_{fj}\,\Sigma_i\,(C_i)^{q_i}}{(1 + \Sigma C_i^q K_j)^m} - \frac{k_{bj}\,\Sigma_i\,(C_i)^{p_{i-}}}{[1 + \Sigma(C_i)^p K_j]^w} \tag{43}$$

The exponents, m and w, on the adsorption group will generally have values of 0, 1 or 2. Within this group the exponents on the concentrations, p and q, may be positive or negative and the terms K_j are lumped parameters. The lumped electrochemical kinetic terms k_{fj} and k_{bj} are based on the use of a working electrode potential *vs.* a suitable reference potential [*e.g.* normal hydrogen electrode (NHE)] and their pre-exponential rate coefficients are the values of the forward and reverse rate constants at zero potential.

3.5 HETEROGENEOUS REDOX CATALYSIS

Heterogeneous redox catalysis is a special case of electrochemical reaction involving adsorption.[5] Here the redox reagent is fixed to the electrode surface and the sum of the oxidised and reduced states is a constant value:

$$C_R + C_O = C_T \qquad (44)$$

Assuming that only the redox surface layer is electroactive and that the rate of the redox process

$$R = O - ne^- \qquad (45)$$

is faster than the reaction betwen O and adsorbed S

$$S_{ads} + O \rightarrow Z + R \rightarrow P \qquad (46)$$

the potential at the electrode is given by the Nernst equation.

$$E = E^o + R\,T/n\,F\,\ln(\,C_O/\,C_R) \qquad (47)$$

Substituting eqn. (44) gives:

$$E = E^o + RT/nF\,\ln[C_O/(C_T\text{-}C_O)] \qquad (48)$$

where E^o is the standard potential of the surface redox couple.

If the mass transport of dissolved species S is fast and its concentration is in equilbrium with the adsorbed species

$$S_{ads} = K\,C_S \qquad (49)$$

the kinetic equation of the surface reaction is:

$$j = n\,F\,k\,S_{ads}\,C_O = n\,F\,k\,K\,C_S C_O \qquad (50)$$

Substituting for C_O using eqn. (48) gives:

$$j = n\,F\,k\,K\,C_S\,C_T\,/\{1 + \exp[-n\,f\,(E\text{-}E^o)]\} \qquad (51)$$

In this equation the current density increases with a positive shift in potential but does not exceed a maximum value given by:

$$j_{max} = n\,F\,k\,K\,C_S C_T \qquad (52)$$

The reaction models for activation control of the redox reaction can be derived on a similar basis to that of eqn. (51). The introduction of mass transport control of starting reagent can also be incorporated in a similar manner as described for the non-catalytic electrochemical reactions.

3.6   EXAMPLE OF THE DEVELOPMENT OF A REACTION MODEL

As an example of the development of an electrochemical reaction model the reduction of nitrobenzene to *p*-aminophenol is considered.[6] This also serves the purpose of illustrating the use of an important electroanalytical device, the rotating disc electrode (rde).[1] The rde, shown in Figure 7, consists of a planar disc of the electrode material rotating at a constant speed about an axis perpendicular to the disc and passing through its centre.

   In operation the rotation of the disc draws up the fluid from below and throws it out radially across its surface. Thus over the complete region below the surface there is a three-dimensional distribution of fluid velocity. In the limit of an infinitely small electrode, embedded in the centre of the rotating surface, the velocity normal to the electrode depends only on the distance normal to the disc. In practice this region can be extended to an electrode diameter of approximately 0.5–1.0 cm, providing the disc is suitably designed with a large surrounding area. Consequently the concentration of species near the electrode will only depend on the distance normal to the electrode. The concentration distribution component

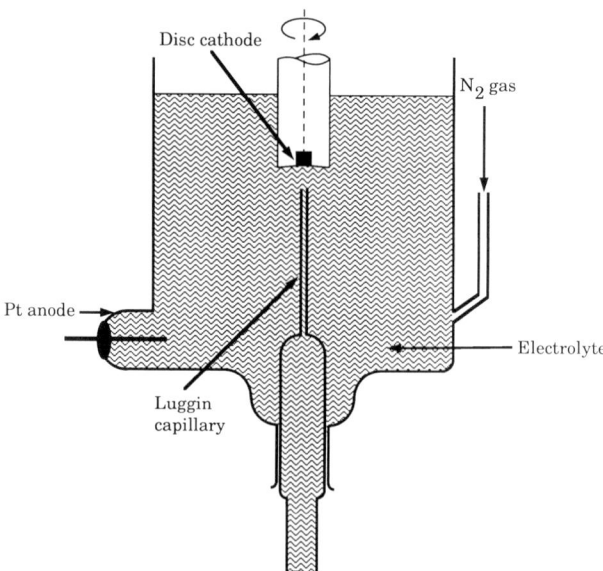

Figure 7   *Rotating disc electrode*

normal to the disc is then found from the solution of the steady state convective diffusion equation (see ref. 1). The approximate theoretical result for the thickness of the mass transfer limiting current density, j_l, is:

$$j_l = 0.62 \ n \ \text{F} \ D^{2/3} v^{-1/6} \omega^{1/2} \ C \tag{53}$$

where C is the bulk concentration, v is the viscosity and ω the angular velocity (units cm, s, mol).

With the rde the mass transfer rate can be conveniently changed by simple alteration in the speed of rotation of the disc. The range of rotating speeds of the rde are between 1 rad s^{-1} and 2000 rad s^{-1} for a disc of 1.0 cm diameter. The upper limit is imposed by the requirement of laminar flow.

3.6.1 The Reduction of Nitrobenzene

The reaction mechanism considered in the reduction of nitrobenzene on a copper electrode in a sulfuric acid electrolyte is shown in Figure 8. Two major products are formed in this scheme, aniline (AN) and *p*-aminophenol (PA) by the electrochemical reduction and chemical rearrangement of the first reduction product, phenylhydroxylamine (PH).

In the reduction of nitrobenzene three processes interact; mass transport, electrochemical kinetics and chemical reaction. The rate of chemical rearrangement is slow compared to mass transport and thus chemical reaction is confined mainly to the bulk electrolyte. The reaction model requires experimental measurements of the individual processes; electroreduction of nitrobenzene to PH, chemical reaction of PH to PA, electroreduction of PH to AN, mass transport behaviour and side reactions.

3.6.1.1 Diffusion Coefficient. The values of diffusion coeficients for nitrobenzene and PH can be determined from an appropriate plot of j_d vs. $\omega^{1/2}$, which has a slope proportional to $D^{2/3}$, from eqn. (53). This need not be carried out for the reaction in question as it is a physical property not related to kinetics, although clearly the electrolyte environment will influence it's value. Thus in the case of the reduction of PH, where the hydrogen evolution reaction masks the limiting current behaviour, its electrooxidation to nitrobenzene on platinum can be used to determine the diffusion coefficient.

3.6.1.2 Side Reactions. To determine the influence of potential side reactions a typical method is the measurement of the background current in the absence of reacting species, *i.e.* nitrobenzene. This will indicate the extent of, in this case, the hydrogen evolution. This assumes that the presence of the reagents in the main reactions do not modify the kinetic char-

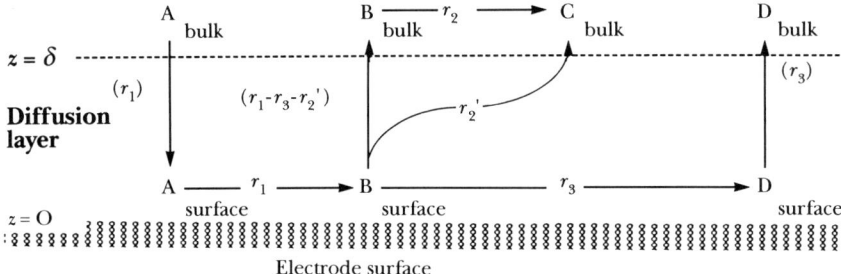

Bulk Solution

Figure 8 *Reaction mechanism and model of the electroreduction of nitrobenzene*

acteristics of the side reaction, by for example, co-adsorption with hydrogen onto the electrode surface. If there is interaction between the main reaction and the side reaction, more detailed preparative experiments are required to determine the Faradaic yields.

3.6.1.3 Electroreduction of Nitrobenzene and Phenylhydroxylamine. For the rde, eqns. (37) and (53) lead to the following expression:

$$1/j = 1/j_k + 1/(B\,C\,\omega^{1/2}) \tag{54}$$

where $B = 0.62\ n\,\mathrm{F}\ D^{2/3}v^{1/6}$

From a linear plot of $1/j$ against $1/(\omega)^{1/2}$ the intercept gives the kinetic limited current, $n\,\mathrm{F}\ k_{f1}\,C^m$ and the slope yields the value of B. The electrode kinetics can be determined as shown in section 3.3, *e.g.*, as Tafel plots, when the procedure is repeated at several electrode potentials to give the required data of E *vs.* j_k.

The rde can be used to obtain the order of an irreversible electrode reaction without the need to change the bulk concentration of electrolyte. For example when the current density is given by a Tafel equation this can be arranged as:

$$\log(j) = \log(k_f) + p \log[(j_l - j)/\omega^{1/2}] - p \log(B) \tag{55}$$

Thus a plot of $\log(j)$ against $\log[(j_l - j)/\omega^{1/2}]$ at constant potential gives the value of p without changing the concentration C.

In the electroreduction of nitrobenzene the variation of j_k with the concentration of nitrobenzene, C_{NB}, is linear and thus the reduction is a first-order process. The values of kinetic current density:

$$j_k = n F k_{f1} C_{NB}{}^p \tag{56}$$

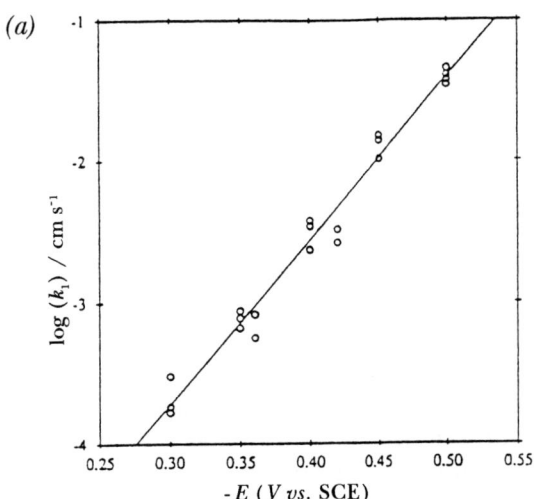

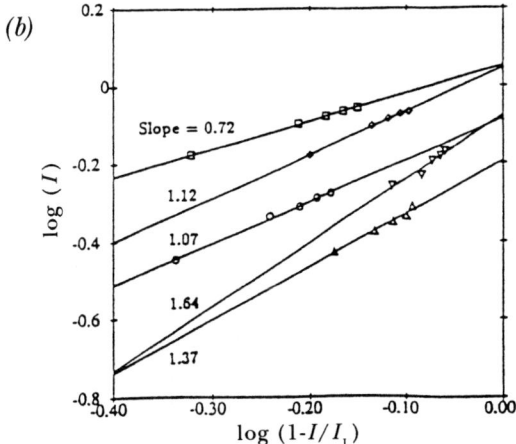

Figure 9 *Experimental correlation of data in the electroreduction of nitrobenzene (a) reduction of nitrobenzene; (b) reduction of phenylhydroxylamine*

is then obtained from the usual inverse current *vs.* $\omega^{1/2}$ plots (see Figure 9).

In the electroreduction of phenylhydroxylamine the interference of the hydrogen evolution side reaction and the parallel chemical rearrangement need to be considered. At low temperatures the latter is not a significant factor over the timescale of the experiments. The current density of hydrogen evolution was substracted from measured current densities using data determined in the absence of phenylhydroxylamine. The kinetics of the reduction of phenylhydroxylamine, were determined from the recommended plot of eqn. (55) [see Figure 9(*b*)] written in the form:

$$\log j = p \log(1 - j/j_k) + \log(j_k) \tag{57}$$

and the subsequent plot of the Tafel equation. The order of reaction with respect to phenylhydroxylamine is 1.0 as determined from the recommended plot of eqn. (57).

3.6.1.4 Chemical Reaction of Phenylhydroxylamine. The chemical reaction of phenylhydroxylamine was studied in a small batch reactor at several temperatures. From a plot of the logarithm of the concentration or pH against time the order with respect to pH was determined as 1.0. The rate constant closely follows an Arrhenius equation as demonstrated by a linear plot of $\ln(k)$ against $1/T$. The predicted influence of temperature on the production of *p*-aminophenol agrees with experiment, *i.e.*, at a higher operating temperature the chemical rearrangement to *p*-aminophenol is more favourable. Overall the reaction model of this system is shown in Figure 8.

3.7   TRANSPORT IN MEMBRANES AND DIAPHRAGMS

The prime function of a membrane or diaphragm is the separation of the chemistry of the anode and cathode regions of the cell. In electrodialysis applications the membrane function is to achieve the concentration of selected ionic species. The three important parameters, which determine the behaviour of the separator material, are the pore size, the porosity and the ion-exchange capability.

The simplest of separators are the porous separators which prevent the mixing of gaseous products (and solid particles) by the appropriate size of the pores in the structure, typically in the range 1–50 μm. They function by hydraulically limiting the movement of liquids from either side of the structure but in practical situations this is not eliminated. The separator has no inherent means of discriminating between the transport of ionic species. The current is carried unselectively by ionic motion. The resulting voltage drop in the separator can be expressed in terms of the effective specific resistance, ρ, of the electrolyte in the material.

The second property, which serves to characterise diaphragms, is the permeability (or permeability coefficient) K', which relates the pressure loss that occurs in the diaphragm with the fluid flow through its structure according to Darcy's law

$$u = K' \, \Delta P / (\mu \, d_s) \tag{58}$$

where μ is the viscosity and d_s is the diaphragm thickness.

In general with the use of diaphragms there are two modes of operation; one in which the material separates two electrolyte compartments in which the chemistry may be radically different and the second in which the flow of electrolyte goes entirely from one compartment to the other through the diaphragm. In the former case, where with a flowing electrolyte there will be hydrodynamic boundary layers set up at the diaphragm surface, the material design will attempt to negate the bulk flow of electrolyte. Ionic and material transport will ideally be due to diffusion and migration through the diaphragm. Flow disturbances and other factors that cause a differential pressure to be set up, which would tend to encourage convective flow, will be kept to a minimum by the small pore size of the material used. In certain cases a small net flow may be encouraged to prevent the transport of certain species in the opposite direction, which may upset the chemistry or adversely affect the electrode material.

Overall the separator performance will depend on its ability to control the transport of species through its structure. These species are continuously in motion, driven by diffusive and convective forces. When the structure of the diaphragm readily allows the convective transport of material under modest pressure differentials this will often be the dominant mode of ionic transport. Alternatively, in for example industrial cells in the chlor-alkali industry, the convective flow and diffusive flows are often in balance.

3.7.1 Transport Processes in Diaphragms

The three modes of transport responsible for ionic movement through diaphragms are convection, diffusion and migration. The components of the flux equation have in the case of diaphragms to allow for the structure of the material, which is conveniently done by the application of a single characteristic material coefficient. In the case of diffusional transport due to a concentration gradient, the flux can be defined in terms of an effective ionic diffusion coefficient, D_{eff}, for the material, which can be related to the ionic diffusivity using a factor referred to as the MacMullin number, N_M, which is a function of the tortuosity and porosity:

$$D_{eff} = D_j / N_M \tag{59}$$

The overall flux equation for a diaphragm is written as:

$$N_j = - (D_j/N_M) d C_j / d x - n_j C_j z_j / F + u \; C_j \tag{60}$$

One of the important industrial examples of the application of this equation is in the chlor-alkali diaphragm cells shown schematically in Figure 10. The brine electrolyte first flows into the anode compartment, where chlorine gas is generated, and flows through the diaphragm into the catholyte in which hydroxide ions are generated by the cathodic evolution of hydrogen. The principle current carrying species across the diaphragm is sodium ions. The system is designed to prevent the hydroxide ions transporting from the catholyte into the anolyte, where, otherwise, the formation of hypochlorite and chlorate would occur with a resulting loss in efficiency. Thus, in design, the net ionic flux of hydroxide ions across the separator should be zero, with a balance between the convective ion transport and the migration and diffusion transport of the ions.

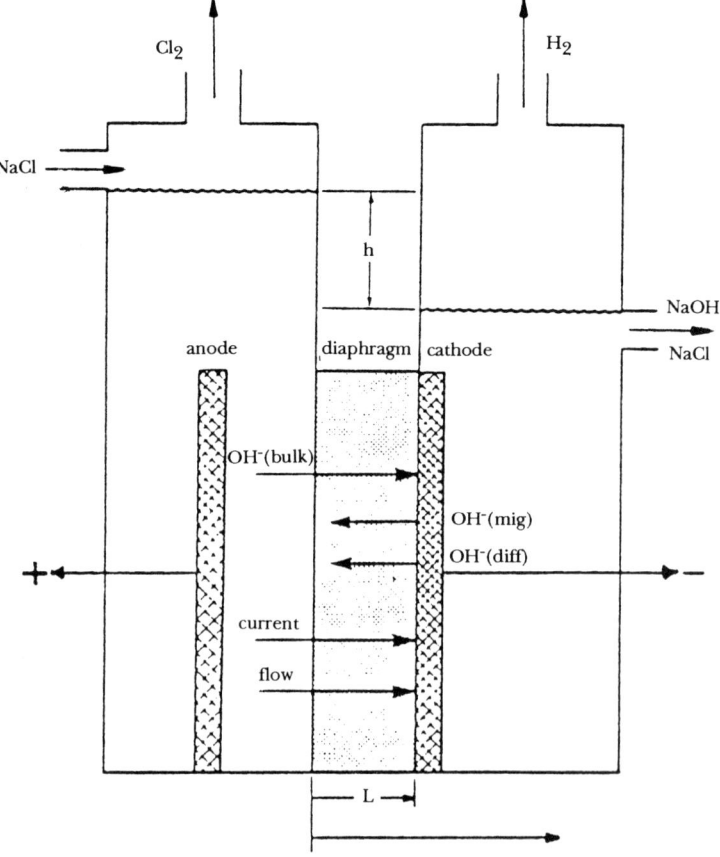

Figure 10 *Principle of operation of a chlor-alkali diaphragm cell*

3.7.2   Membranes and the Transport of Ions

A membrane is a thin wall or barrier made from a material or phase which opposes the transport of certain matter through it. The material supports an unequal resistance to the transport of different penetrants when the latter are driven across it by forces arising from imposed electrochemical or chemical gradients. The transport rate of species through the barrier is given by

$$J_j = -K\,U_j\,d\,\mu_j/dx \tag{61}$$

where μ_j is the chemical or electrochemical potential or related property, such as activity or concentration, and U_j is the mobility of the penetrant in the membrane.

This equation is applicable to the simultaneous transport of species only when there is no interaction, *i.e.*, when the transport of one species does not influence the transport of any of the other species. In practice the equation only holds for low concentrations of penetrants and in other cases deviations from this ideal are realised when the displacement of the different species become mutually dependent.

The membranes, which are of principle interest in the area of electrochemistry, are ion-exchange materials which are typically 100–200 μ thick and often of areas greater than one square metre in industrial applications.

When a membrane bearing an ionic group such as $-SO_3H$ is put in water it swells sufficiently such that the ionisable group will release the small counter ion, *e.g.* H^+, and the oppositely charged group, called the fixed ion, remains covalently bonded to the skeleton. When the membrane is placed in an aqueous electrolyte some salt will enter the membrane. The sorbed ions which have a charge similar to the fixed ions are called co-ions. The concentration of these co-ions increases with the concentration of the electrolyte. When an electrical current passes through the membrane the counter-ions can enter into it from one side and can leave it from the other side, which results in the formation of a concentration gradient. Membrane transport of ions is typically expressed in terms of an overall transport number t_m for the species in the membrane, *i.e.* at the membrane surface the flux of a particular ion is given by the following balance (for a cation):

$$\frac{j\,t_+}{F} + k_{LM}\left[C - C_m\right] = \frac{j\,t_{+m}}{F}$$

where k_{LM} is the mass transfer coefficient at the membrane surface.

At relatively low current density and with turbulence the convective effect will be small. On increasing the current density a point is reached whereby the concentration of counter-ions at the interface approaches zero and the system is then polarised.

In practice it is possible to design membranes with transport numbers greater than 0.95 for a particular ion. However, it is more difficult to obtain selectivity of for example one cation in a mixture of cations. Control of selectivity is frequently by using an excess of the particular cation in the electrolyte.

More detailed treatments of ion transport in membranes can be found in refs. 7–11.

REFERENCES

1　Southampton electrochemistry group, 'Instrumental methods in electrochemistry', Ellis Horwood Ltd, 1985.

2　F. Goodridge and K. Scott, 'Electrochemical Process Engineering', Plenum Press, NY, 1995.

3　K. Scott, 'Electrochemical Reaction Engineering', Academic Press, London, 1991.

4　J. Koryta, J. Dvorak and L. Kavan, 'Principles of Electrochemistry', J. Wiley and Sons, 1993.

5　F. Beck and W. Gabriel, *J. Electroanal. Chem. Interfacial Electrochem.*, 1985, **182**, 355.

6　T. R. Nolen and P. S Fedkiw, *J. App. Electrochem.* 1990, **20**, 370.

7　P.N. Pintauro and D. N. Bennion, *Ind. Eng. Chem. Fundam.*, 1984, **23**, 230.

8　V.K. Kadija and D.N. Bennion, in 'Electrochemical reactors . Their science and technology, Part A, Fundamentals, Electrolysers, Batteries and Fuel Cells', ed. M.I Ismail, Elsevier, 1989, ch. 9, p.210.

9　'Permselectrive Membranes', ed. C.E. Rogers, Marcel Dekker, New York, 1983.

10　'Perfluorinated Ionomer Membranes', ed. A. Eisenberg and H.L. Yeager, ACS, Washington DC, 1982.

11　K. Scott, 'Membrane Separation Technology', ed. K. Scott and R. Hughes, Blackie Academic, 1995, ch. 7.

Electrochemical Cell Design and Engineering

The application of electrochemistry in the areas of synthesis, effluent remediation and recycling has resulted in a diversity of cells and reactors. The design of these reactors has frequently had to contend with conflicting requirements in for example the balancing of efficiency against the selection of appropriate materials. The design of electrochemical reactors has learnt from general reactor engineering concepts and accommodated the need to supply one of the reagents, the electron, by the application of an external potential field. A number of these designs have been radically improved over recent years to give greater versatility and enhanced performance in terms of energy consumption and throughput, and thus have found a range of applications. Several commercial general-purpose electrochemical cells are now available, which can meet the requirements of many reactions for syntheses and other applications. The marketing of such units has taken into account the scale-up requirements of users, and has resulted in a range of sizes.

4.1 OPERATING FACTORS IN ELECTROCHEMICAL REACTOR DESIGN

Early cell and reactor designs effectively took the form of tank electrolysers operated as either batch or semi-batch devices. In practice, however, the electrochemical reactor presents a wide choice of operating factors and materials of construction (see Table 1). The choice of the electrolyte is generally determined by the requirements for reactant solubility, low electrical resistivity and a good selectivity.

4.1.1 Modes Of Operation

Electrochemical reactors, in general, can either be operated continuously as flow cells, with or without a high degree of mixing, or intermittently for the supply of product species in batches. In certain situations a semi-batch

Table 1   *Operating factors in electrochemical reactors*

Types of electrolytes	Aqueous solution, organic solvent, heterogeneous dispersion of gas and liquid, molten salt and solid polymer
Types of electrodes	Anodes: consumable, insoluble, solid polymer and gas diffusion type; Cathodes: rigid metal, insoluble, porous, liquid metal, conducting polymers, solid polymers and gas diffusion
Separators	Microporous plastic, ceramic, ion exchange, liquid membranes
Cell construction	Undivided, divided cells, tank electrolysers, flow cells, mono- or bi-polar electrodes
Modes of operation	Continuous stirred tank or plug flow reactors, batch reactors, recycle reactors and non-ideal flow patterns
Electrode configuration	Two- or three-dimensional, dynamic or static, at the separator (zero gap), on the separator (SPE) and away from the separator
Reactor products	Soluble, insoluble, gas, dispersed liquid, metal deposit and liquid metal
Reaction species	Gases, dispersions, soluble inorganic and organic, solid metals and redox species

mode of operation is adopted in which one or more reagents is supplied continuously whilst another is present as a batch *e.g.* in gas–liquid reactions. The scale of the process will often dictate the mode of operation, large production capacities generally requiring continuous operation on economic grounds. Batch operation will generally be for small scale operation and for processes where there is uncertainty in the market life of the product, thus enabling the reactor and plant to be used in alternative processes.

4.1.2   In-cell and Ex-cell Reactions

The generation of product species can arise directly from activity at the electrode in which soluble, gaseous or insoluble or immiscible components are formed. This *in-situ* generation of species in the cell can also be accomplished indirectly using mediating agents either attached to the electrode (redox electrocatalysis) or formed freely in the solution (homogeneous redox catalysis) see Figure 1. Typical examples of this (discussed

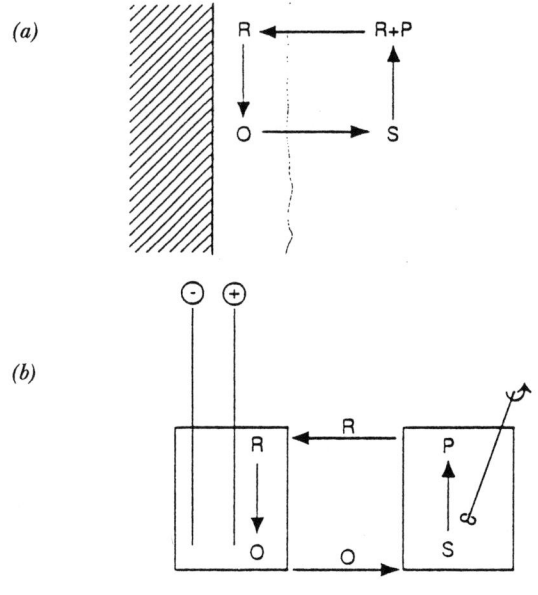

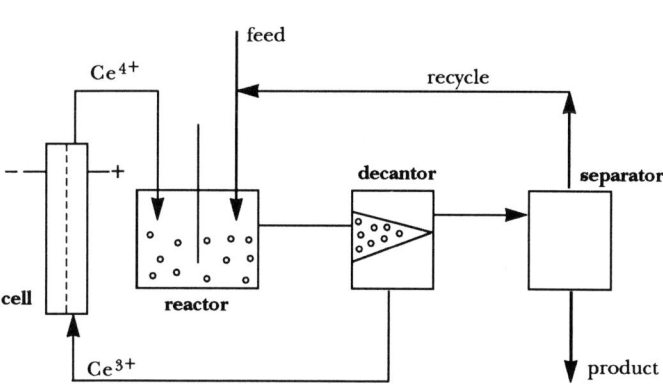

Figure 1   *Types of in-cell and ex-cell reactions:* (a) *in-cell mediated reaction;* (b) *ex-cell mediated reaction;* (c) *ex-cell reaction and phase separation*

in chapter 8) are organic oxidations using metal ion oxidants such as cerium(IV). In certain cases for example the low solubility of the organic in the aqueous phase can limit the overall rate of the reaction. Thus in this case (see Figure 1) the reagent is generated electrochemically in the cell and pumped to an external two-phase reactor. In certain cases there may be an additional requirement to remove dissolved organic material from the aqueous electrolyte prior to recycle to the cell to prevent electrode contamination.

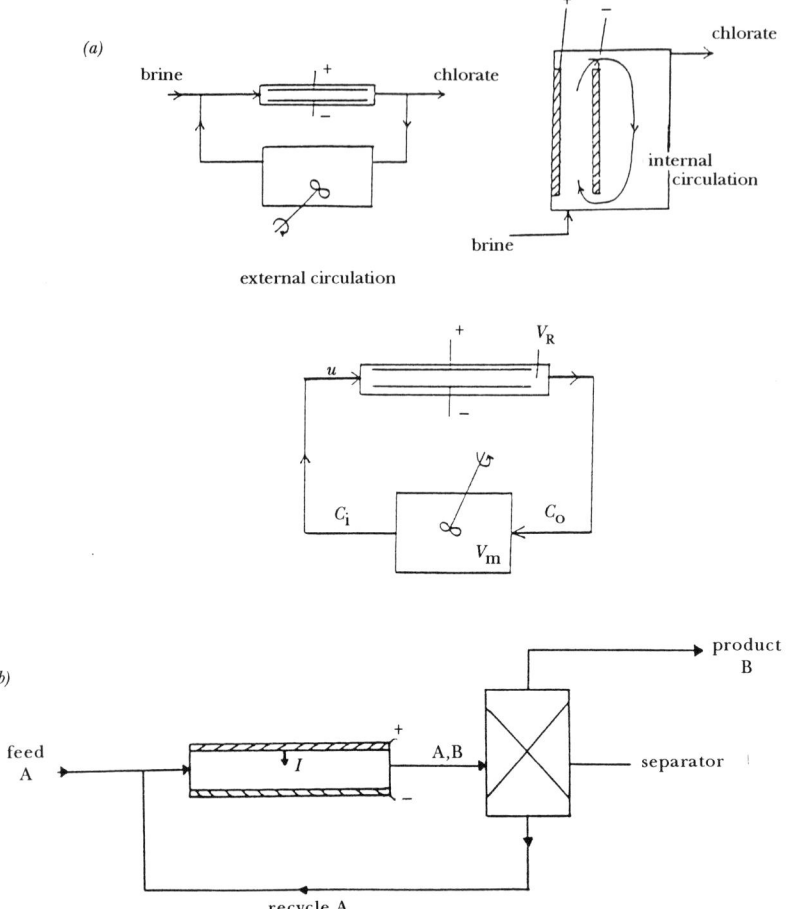

Figure 2 *Types of recycle reactors:* (a) *batch recycle or continuous recycle;* (b) *with product separation*

4.1.3   Recycle Operation

The use of recycle (see Figure 2), *i.e.* the return of part or all of the reactor product solution to its inlet, can be necessary for several reasons:

(*i*)   to achieve suitable hydrodynamic and mass transport conditions and to achieve acceptable yields and conversions;

(*ii*)   to recycle unconverted reactant *via* a phase separation in *e.g.* the formation of an immiscible organic or by an effective distillation [see Figure 2(*b*)], when the process economics may be suitably attractive to justify a low conversion of the reactant;

(iii) in batch operation as a result of the small interelectrode gap generally required to minimise energy consumption. Thus, as shown in Figure 2, electrolyte is recirculated through the cell from an external reservoir by pumps. In certain syntheses the reservoir will act as an external reactor when relatively slow homogeneous chemical or catalytic reaction is involved.

(iv) When electrolyte is recirculated within the electrolytic reactor, caused by *e.g.* a gas lift effect, or by thermal buoyancy forces, generated by electrolysis within the cell compartment of the reactor. This design of cell is adopted *e.g.* when a relatively large volume is required for an homogeneous chemical reaction, as in the case of the production of sodium chlorate by the electrolysis of brine (discussed in chapter 8).

4.1.4   Electrical Supply of Power

The steady supply of electrical power to a cell can generally be by three alternative methods;

1   at a constant electrode potential;
2   at a constant current; and
3   at a constant cell voltage.

A major cost factor in the installation of electrochemical processes is that of the power supply for cell operation. The supply of electrical energy to the cells using transformer-rectifier units is considerably cheaper than the use of potentiostats, which offer the means of controlling the electrode potential. The cost differential can amount to a factor of 5–10. Thus although potentiostatic control can, in principle, offer the advantage of higher yields and current efficiencies, in large scale operation this is generally outweighed by the higher cost and lack of availability of larger power rating units. An additional factor is the practical limitations of relying on the potential probing of an electrode in an 'industrial' electrolyte where surface characteristics may change due to corrosion or contamination, where there may be changes in the probe position, *i.e.*, changes in *IR* resistance, or where probe disconnection or blocking may occur.

In the case of continuous reactors there is no argument for the use of potentiostats as, apart from start up and shut down, the reactor is essentially at the steady state. Therefore, the cheaper power supplies, that output a constant voltage source, are used in practical electrolysis. Although in practice the overall current may be fixed during operation there is inevitably a distribution in both the electrode potential and current density in the reactor.

4.1.5   The Distribution of Power in Electrolysers

Industrial electrochemical reactor systems will generally be comprised of
several cell banks. The cell banks are units, which contain many individual cells. The electrodes in these cells will generally be large and thus
their electrical resistivity is a key factor which determines the way power is
fed into the unit. If the electrical resistivity of the material is relatively
high, say in the case of carbon, then it is generally not practical to supply
current at a peripheral point to each electrode in a monopolar configuration (see Figure 3), because of high voltage losses in the material and also
because of the resultant uneven potential distribution. Increasing the
thickness of the electrodes may not be cost effective, either the current
path must be appropriately short or the electrode conductivity must be
effectively increased by the use of conducting backing plates or metallic
inserts.

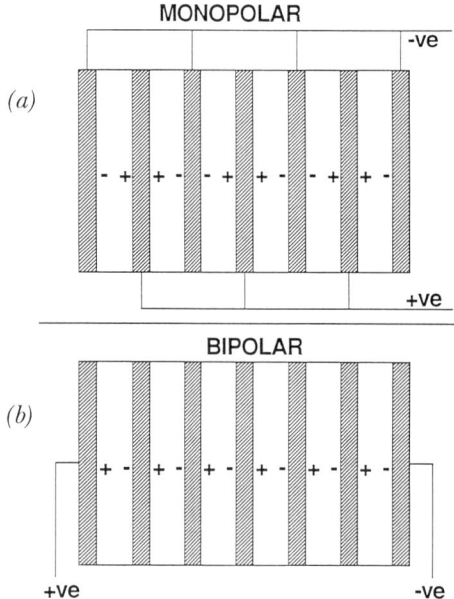

Figure 3   (a) *Monopolar and* (b) *bipolar connection of electrodes*

The alternative to a monopolar connection is a bipolar arrangement
(see Figure 3). This is a common technique whereby the electrical connection is made only to the end electrodes in a cell stack. Providing that
the end 'feeder' electrodes and connections are designed to minimise the
IR loss, this design generally gives a uniform distribution of the current

density. In both arrangements the power supplied to the cells is theoretically the same.

A disadvantage with the bipolar connection is that of current leakage or 'bypass' around the cells, which is due to the hydraulic connections between individual cell units forming alternative electrical paths for the flow of current around the electrodes. This causes a reduction in the Faradaic efficiency of the cell bank, an increase in the power used and a mal-distribution in the current density. A suitable design of the hydraulic manifolding and/or the use of plastic extensions to the electrodes can generally minimise the effect of current by-pass.

4.1.6   Current Distribution

The aspect of current distribution can have a major effect on the performance of electrochemical processes. A detailed analysis is outside the scope of this book although it is important to give a qualitative view of the effect due to its significance in the design of electrochemical cells for industrial processes. The analysis of current distribution can be divided into three general areas; batch electrolysers, flow electrolysers and three-dimensional electrodes.

With a batch electrolyser the concentration of reaction components may, as first approximation, be considered to be everywhere uniform except at the region close to the electrode. Ideally with a parallel plate unit, when the resistance between the electrodes is independent of position the current flow path is uniform between opposite faces of the electrodes, *i.e.*, the current is uniform. Factors which cause a variation in the resistance between opposite positions of the electrodes and thus a mal-distribution in the current include:

- uneven electrode structures
- non-planar or non-parallel electrode arrangements
- finite resistivity of the electrodes
- generation of a second phase, *e.g.*, gas bubbles or surface deposits at the electrode

Thus in electrolyser design for batch electrolytes the electrodes should completely fill the cross-section of the cell chamber. This, however, is not always possible in for example electrowinning cells where the growth of an electrodeposit is required.

In the case of cells with flowing electrolytes the unit is, by necessitity, in some way open to the flow of solution and this therefore provides an additional electrical flow path at some peripheral position(s) on the electrodes. Thus as the current will tend to flow along the path of least resistance at these peripheral positions, it will be higher there. As in the case of the bipolar connected electrodes, good cell design can minimise this effect. Thus in cells where the electrodes are not flush with the walls or do not fill the complete cross-section, there will be a non-uniform distribution of current density (see Figure 4).

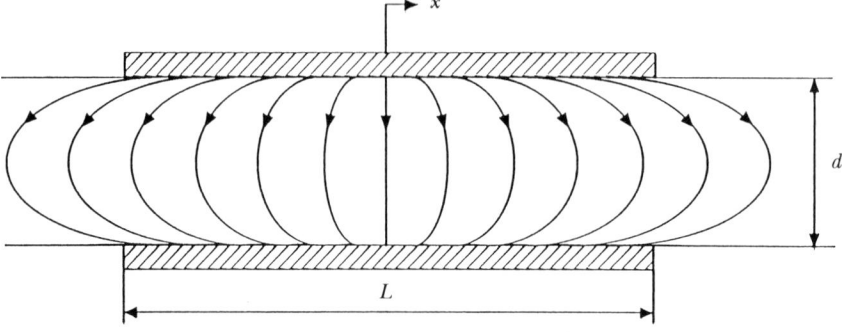

Figure 4   *Non-uniform current distribution in parallel plate electrolysers*

Early analyses of current distribution considered the so-called 'primary distribution' *i.e.* the current distribution as determined by the Laplace equation in the absence of polarisation or electrode kinetics. Such distributions generally give a poor representation of actual behaviour and thus the influence of polarisation and mass transport should realistically be included to determine the so-called secondary and tertiary distribution.

4.1.6.1 Secondary and Tertiary Distributions. Secondary distributions in current density include the influence of activation over-potential but neglect mass transport limitations. Typical distributions are characterised in terms of the Wagner number

$$\text{Wa} = \frac{\kappa \, d\eta}{d \, dj}$$

where d is a characteristic dimension.

Secondary current distributions obtained[1] for a parallel plate configuration, of electrode length, L, and interelectrode gap, d, (Figure 4) are shown in Figure 5 for the condition $L \gg d$. This distribution of current density is expressed as the difference between a local value j and the value as x/L approaches infinity, j_∞. From this, it is apparent that small interelectrode gaps, high electrolyte conductivity and rapid polarisation all improve current distribution.

Tertiary current and potential distributions correspond to situations

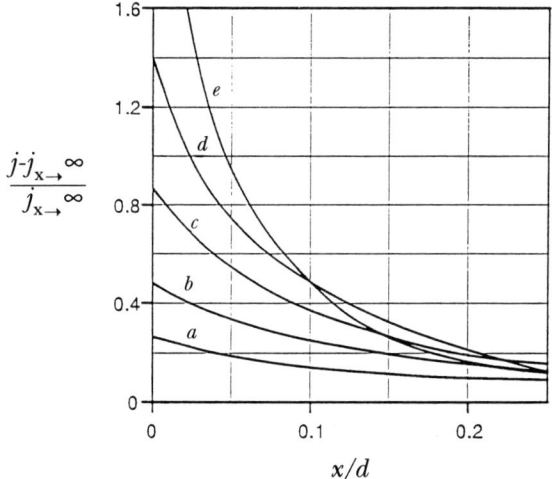

$$\frac{j - j_{x \to \infty}}{j_{x \to \infty}}$$

Figure 5 *Current distribution in parallel plate electrolysers; values of Wa; a=0.8; b = 0.4; c = 0.2; d<<L d = 0.1; e = 0*

when concentration over-potential or mass transport phenomena are important, *i.e.* when variations in the concentrations of reacting species are significant. To determine tertiary distributions essentially requires a reactor model, which adequately represents the processes occurring at the electrode surface; migration, diffusion and convection and the potential distributions in the bulk of solution as defined by the Laplace equation. This is outside the scope of this book.

A second factor with the use of flow cells is that as the electrolyte moves across the electrode(s) there will be a depletion in reactants and an accumulation of products. As a result there will be changes in the level of electrode polarisation, due to kinetic and mass transfer effects, and thus the values of electrode potential. Therefore these so called secondary and tertiary distributions will interact with the resistive (primary) distributions to determine the overall current distribution in the reactor.

4.2 CELL DESIGN

The design of the electrochemical reactor is largely determined by the appropriate operating conditions, the nature of the electrolyte medium, whether molten salt, aqueous or organic solvent, the form of the reactant or product, whether gas, liquid or solid and the type of electrode material and configuration. The emphasis in this section is for cells operating at relatively low temperatures in primarily the aqueous phase. An overall summary of the range of electrochemical cells is given in Table 2.

Table 2   *The range of electrochemical cell designs*

Electrode type	Geometry	Variants
Two-dimensional	Plate, mesh, expanded metal — in tank — in flow cells — stacked	 Moving sheet, wire Vibrating, reciprocating Flowing liquid metal
	Stacked discs	Rotating
	Cylinders — in tank —annular flow	Rotating in tank Rotating with annular flow
Three-dimensional	Packed bed — spheres, cylinders — coarse granular — fibre	Fluidised bed — circulating bed — moving bed — slurry — rotating bed — trickle flow
	Porous electrodes — cloth — reticulated — multilayer mesh — sintered — foam	 Spiral wound

4.2.1   Design Concepts

A good cell design should ideally satisfy a number of requirements, including:

- High productivity
- Good mass transport
- Good temperature control
- Low electrical resistance
- Ease of operation
- Safety in operation
- Provision for cell separators
- Ability to deal with gaseous products and reactants
- Minimum cost

Many of these requirements are satisfied in two general designs, based on either a tank electrolyser concept or a flow electrolyser concept. Tank electrolysers are conceptually the simplest of the two types. In most cases the electrodes, either in the form of sheet, mesh or gauze, are immersed vertically in the tank, arranged as alternate anode and cathodes. This design has the advantage of being robust and allowing inspection of the cell contents and the electrodes. Electrowinning and electrorefining cells are the classic examples of tank electrolysers where the increase in size of the cathodes, and their eventual removal and replacement, makes operation in any other form difficult. The interelectrode gap in tank electrolysers is generally as small as possible but is governed by practical limitations such as the requirement to separate gases in for example water electrolysers. The electrolyte in tank cells is generally not flowing and mass transport can thus be limited, although electrogenerated gas bubbles can cause a significant amount of electrolyte agitation. The tank electrolyser concept is restricted to processes, when there is no penalty paid for the relatively poor mass transport characteristic. This latter factor, together with low space-time yield, has tended to favour the adoption of flow cells using parallel-plate, three-dimensional and rotating electrodes in a large number of applications.

Flow electrolysers adopted in industry are frequently based on the parallel plate arrangement. There are practical (and economic) factors which limit the size of individual electrodes to areas of a few square metres, or less and thus electrochemical reactors are modular in design. The common design, shown schematically in Figure 6, uses vertically mounted electrodes in a plate and frame configuration mounted on a mechanical press. The multicell module, in a monopolar form, comprises of alternating anodes and cathodes separated electrically by the plastic cell frame or spacers, with suitable gasketed sealing. Provision for the use of membranes or diaphragms is usually made. This is in effect the closed compact version of the tank electrolyser. Electrolyte is pumped through each cell compartment to provide good mass transport and to facilitate gas removal from the electrolyte. The interelectrode gap can vary between approximately 1–20 mm depending upon the application and is determined,

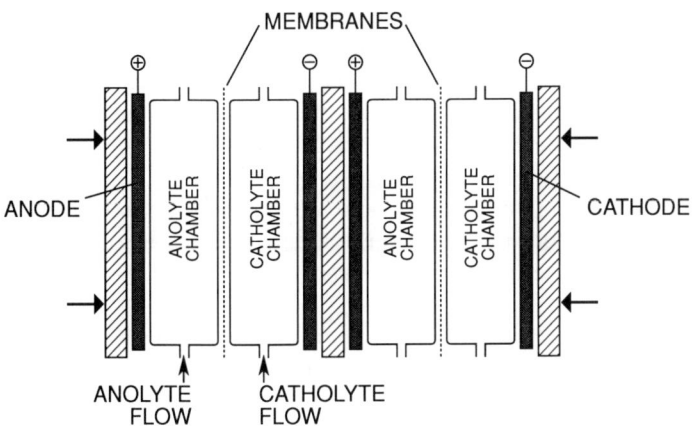

Figure 6 *Schematic of a parallel plate module*

largely by the conductivity of the electrolyte, and thus the *IR* losses, and also by the pressure losses and the requirements for manifolding of the external pipework for fluid flow.

In the cell design it is important to ensure a uniform distribution of electrolyte across the face of the electrode, and between the cells, by using suitably designed flow entry ports and manifolding. The manifolding for flow may be external to the cell or built into the frame as internal flow ports.

When a further increase in productivity, or scale-up, is required this is achieved by networking the cell modules, both hydraulically and electrically, into parallel or series combinations. This allows flexible operation as individual modules can be isolated to allow for maintenance, and electrode and separator replacement. Overall the reactor scale-up is in three stages:

(i)   Increasing individual electrode size from say 0.1 m² up to 2 m².
(ii)   Increasing the number of cell units in each module *e.g.* 10 to 100 cells in one module.
(iii) Increasing the number of modules.

Crucial factors in stage *(i)* include changes in the mass transport, potential distribution and gas voidage distributions on increasing the electrode length, as well as the electrolyte flow distribution. The parallel plate unit frequently contains turbulence promoters, typically plastic meshes, in the electrolyte channels to increase mass transport above that which could be achieved with open channels with similar volumetric flowrates. This improves the mixing in the cells and also provides for a more uniform distribution in mass transfer rate, which otherwise would be affected by hydrodynamic entrance effects.

In designs where a high electrolyte residence time is required the flow of electrolyte through the cells can be arranged to be in a series as opposed to a parallel flow arrangement (see Figure 7). Alternatively the flow can follow a serpentine path across the face of the electrode *via* an arrangement of baffles. In addition the cells can be operated with high surface area three-dimensional electrodes by filling the electrolyte chamber with suitable porous or particulate material.

There is a diversity of electrochemical cell designs which have been used in practice, but the selection of a particular concept is governed to a large extent by several general aspects:

* *Constructional simplicity.* The need to keep material and manufacturing costs to the lowest level is an obvious requirement.
* *Process operation.* The smooth running of the cell during operation should accommodate convenient feed addition and product removal. The inspection of the cell components, maintenance and replacement should be relatively easy. Some degree of versatility in

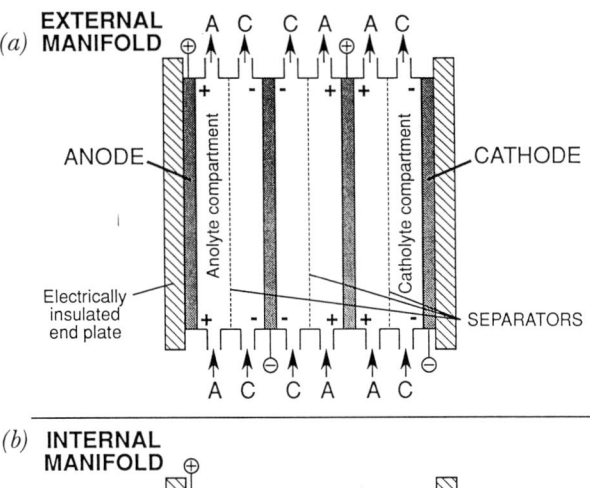

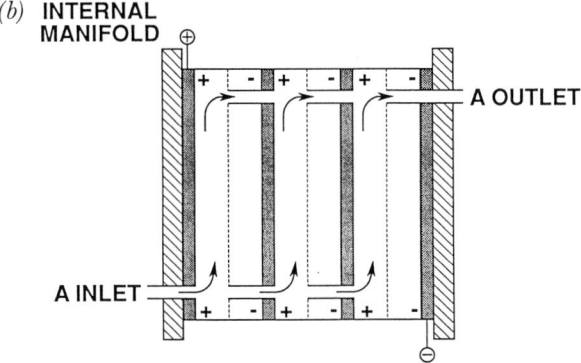

Figure 7 *Types of flow in cells:* (a) *parallel;* (b) *series*

operation may be required if there is a change in the product, or an operating condition.

- *Operating cost.* The conditions of operation should be selected which minimise the cost of material replacement, the cost of cell electrical energy and the cost of operating cell ancillary equipment such as pumps and stirrers, and heat exchangers.

- *Reactor engineering.* The cell operating parameters of current density, mass transport, temperature and pressure may need to be carefully selected to optimise yield, conversion and efficiency.

All the above features are essentially interactive and are largely determined by an economic optimisation of the particular process. Where market forces may necessitate a change in the product range then the final design of cell may be compromised by this factor.

4.3 ELECTROCHEMICAL REACTOR DESIGNS

4.3.1 Parallel Plate Electrolysers

A variety of reactor designs based on the concept of the parallel plate flow unit have evolved,[2] some of which are listed in Table 3. These units have been designed by leading companies in the area of electrosynthesis, and often in order to meet the requirements of specific reactions. All adopt bipolar electrical connection and external hydraulic connection, except the DuPont unit which uses an internal hydraulic connection.

Table 3 *Parallel plate flow electrolysers*

	Electrode area/cell (m^2)	Electrode shape	Cathodes	Anodes
Monsanto EHD	0.64	Plate	Pb	Pb (1%Ag)
Asahi EHD	1.09	Plate	Pb (Sb)	Pb (Sb)
DuPont ESE	0.84	Expanded metal	Steel	DSA
BASF Pilot Plant	0.40	Plate	DSA	PbO_2
Ionics Chemomat	0.9	Plate	Stainless steel, Ni, Hastelloy C steel	Pt/Ti, DSA

The Monsanto and the Asahi EHD design were developed specifically for the electrosynthesis of adiponitrile and both used membranes. The hydraulic flow path for the catholyte in the Monsanto electrolyser is in a series of 36 discrete vertical channels, formed by the set of internal polypropylene spacers. Flow is parallel along these channels. The anolyte flows through a channel formed by a polyethylene fabric, welded within the frame. The design uses spacers to prevent the contact of the membrane with the electrodes. The Asahi design uses electrolyte frames which give an internal multiple flow path (serpentine flow) across the face of the electrodes as shown schematically in Figure 8.

This arrangement gives a longer flow path in comparison to that of the Monsanto design and with typical electrolyte velocities, of between 1–2 m s^{-1}, used in both designs, a much lower flowrate. The Asahi reactor achieves higher conversions than the Monsanto reactor when operated at the same current density. The DuPont (ESE) design shown schematically in Figure 9, uses expanded metal electrodes rather than plate electrodes.

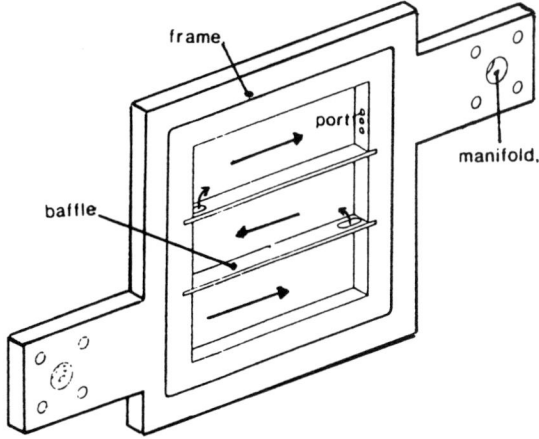

Figure 8   *Serpentine flow path of the Asahi cell*

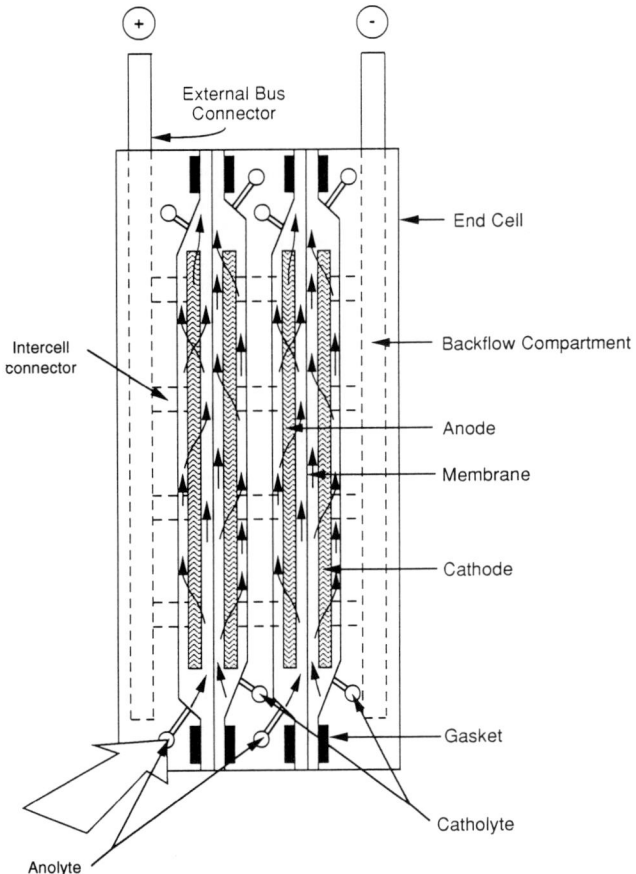

Figure 9   *DuPont ESE electrolyser*

This increases the area for electrolysis by approximately 80% in comparison to the flat plate electrode. The spacing between the membrane and the electrodes is maintained by plastic netting. The electrolyte flow in each cell is from bottom to top and is in parallel.

4.3.2   General Purpose Flow Electrolysers

The general purpose flow electrolysers currently available (see Table 4) are used in a wide range of applications and can be adapted to operate with a wide range of electrode materials. Among the variations in the design concept are the use of either internal or external flow manifolding, series or parallel flow, and bipolar or monopolar electrical connections. The design concepts of three commercially available cells are shown in Figure 10.

Table 4   *General purpose flow electrolysers*

	ICI	*Electrocell AB*	*DEM (electrocatalytic)*
Construction	Filter press bipolar (filter press monopolar)	Filter press, monopolar (bipolar possible)	Filter press mono- or bi-polar
Projected electrode area/m^2	0.01–0.2 (0.0064–0.21)	0.001–0.4 m^2	0.05–1.0 m^2
Electrode geometries	Plate Extended surface (Plate, blade, extended surface)	Plate, extendes surface	Plate dished electrode
Interelectrode gap/mm	1–10 depending on cell configuration	0.5–4.0	1–4
Separators	As required	As required	As required
Liquor distribution	Internal	Internal	External
Electrolyte flow velocity /m s^{-1}	0–0.2	0–0.6	0–2
Electrode materials	Graphite, lead dioxide and glassy carbon, metals/metal alloys and coated electrodes available on request	Metals, coated metal *etc.* wide range available	Metals, coated metal *etc.*
Frames	PTFE, PVDF or other plastic materials on request [a]	PTFE, PVDF, PP, other thermoplastic[a]	PTFE, PVDF, PVC[a]
Gaskets ('0' types)	EPDM. Other materials possible on request	Fluorinated silicone nitrile rubbers *etc.*	Fluorinated, silicone nitrile rubbers
Turbulence promoters	As required, can be provided in a variety of plastic mesh	As required	As required

[a] PTFE, poly(tetrafluoroethylene); PVDF, poly(vinylidene difluoride); PP, polypropylene; PVC, poly(vinyl chloride).

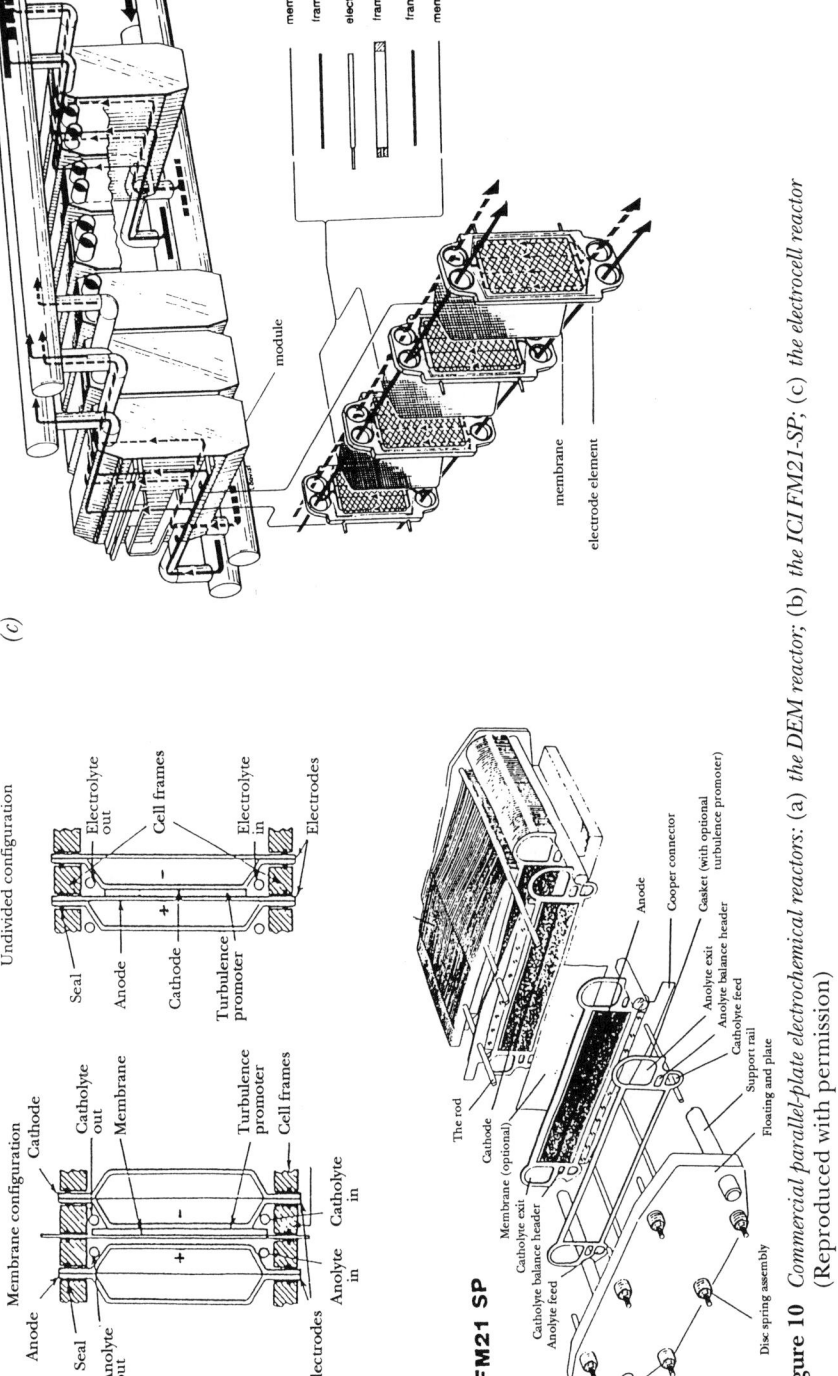

Figure 10 *Commercial parallel-plate electrochemical reactors: (a) the DEM reactor; (b) the ICI FM21-SP; (c) the electrocell reactor (Reproduced with permission)*

The DEM, dished electrode membrane cell, uses shaped electrodes to provide a low interelectrode gap, whilst allowing adequate space in the cell frame for pipe manifolding. The DEM unit can be operated in a divided or undivided configuration with mono- or bi-polar electrode connections. In the latter case the use of long flexible piping for electrolyte feed keeps bypass current to a minimum. As with all these 'parallel' plate electrode designs, good mass transport is achieved by using relatively high linear electrolyte velocities and/or turbulence promotors. The design has applications in organic and inorganic synthesis and effluent treatment.

The ICI FM21 SP cells [Figure 10(b)] arose from ICIs design experience with chlor-alkali membrane electrolysers, and is thus custom-made for reactions which involve the electrochemical generation of gases. The design has been used for both organic synthesis *e.g.* Kolbe synthesis, and for inorganic synthesis, *e.g.* oxidation of dinitrogen tetroxide to dinitrogen pentoxide in anhydrous nitric acid, discussed in chapter 8. The unique lantern blade electrode design of the ICI FM design gives an extended surface area, low effective interelectrode gap, and good gas release. This reactor contains no cell frames, and the electrolyte compartments are formed by the sealing gaskets. The company offers the cell technology in a wide range of sizes, from small laboratory cells through pilot scale up to full scale plant modules (see Table 4).

The electrocell unit [Figure 10(c)] is a multi-purpose unit available in a wide range of sizes for scale-up. It adopts an internal manifolding arrangement for electrolyte flow and usually monopolar electrical connections to the electrodes. It has been used in a number of electrosyntheses and effluent treatment applications, which are discussed further in this book.

The availability of general purpose flow electrolysers has cut out much of the development costs previously associated with electrochemical processes. Indeed the availability of units from bench scale-up to plant scale is a valuable assest for potential users of electrochemistry.

4.3.3   Other Reactor Designs

There are many reactor designs described in the open literature, of which a number have been successful at the commercial level. These can be categorised as follows:

1   Monopolar planar-electrode tank
2   Bipolar thin film
3   Spiral wound
4   Rotating electrode
5   Three-dimensional electrode

4.3.3.1 Thin Film Bipolar Cells. Generally when the electrolyte conductivity is low it is advantageous to use a small interelectrode gap in order to minimise *IR* losses. Thin film bipolar units (as shown in Figure 11), with interelectrode spaces of 0.1–1 mm, have been designed to satisfy this requirement. These units are an assembly of parallel, planar electrodes separated by insulating spacers and installed in an appropriate vessel. Electrolyte flow between electrodes is either by natural convection or forced circulation. Extreme simplicity of construction is offered, a stack of alternating electrodes and spacers, without the complex hydraulic system of a plate-and-frame design. Electrical connection is only made to the two end 'feeder' electrodes.

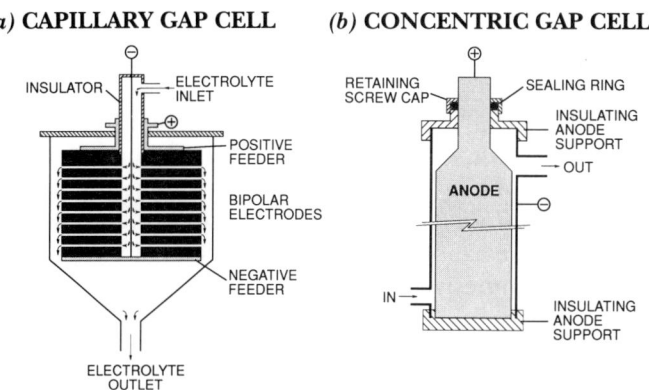

Figure 11 *Thin film bipolar electrolysers*

In general, bipolar thin film electrolysers are the preferred design for undivided cell processes, whereas plate-and-frame electrolysers are more commonly used in processes requiring a separator. The advantages of undivided cells over divided cells include:

- A single electrolyte system;
- Elimination of the *IR* drop of one electrolyte;
- Simple cell construction and low cost;
- Higher production per unit cell volume, *i.e.* higher space time yield;
- Lower power usage.

A key feature of the bipolar disc stack is that a single planar electrode serves as both the anode and cathode. When different materials are needed for the anode and cathode, a variety of schemes have been proposed for production of composite electrodes *e.g.* appropriate coatings of different materials onto the opposite faces of the electrodes and fusion of two dissimiliar materials. Disadvantages of the bipolar disc stack are the non-uniform distribution of flow, and the possibility of excessive current bypass around the edges of the electrodes.

The cell design used in the Monsanto process for the production of adiponitrile could also be classified as a thin film bipolar electrolyser. With an interelectrode gap of ≤ 2 mm, a large bank of approximately 100 bipolar electrodes are fitted into a cylindrical vessel. Further details of this are discussed in chapter 8.

4.3.3.2 Spiral Wound (Swiss Roll) Cell. The swiss-roll cell,[4] shown schematically in Figure 12 is a novel high surface-area device designed on the principle of a spiral-wound unit. A sandwich of alternating 'anode–spacer–membrane–spacer–cathode' is rolled around a cylindrical core, which is the current feeder for one electrode. The cell housing is designed to give electrical contact only to the opposite electrode. Electrolyte flows axially along the roll. The electrodes used in the cell must be flexible and are usually foils, which, due to limited electrical conductivity along the length of the roll, restricts the current density to approximately 10–50 A m^{-2}. The cartridge-orientated design means that for maintenance the unit must be withdrawn from its holder. The swiss roll cell has been proposed for effluent treatment applications and has seen commercial application in organic electrosynthesis.

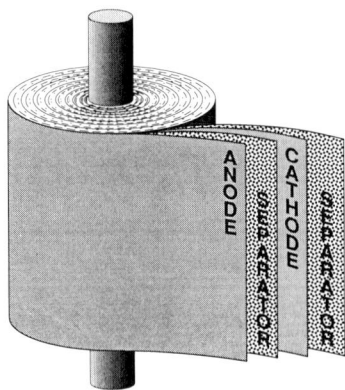

Figure 12 *The 'swiss roll' reactor*

4.3.3.3 Rotating Electrode Cells. By rotating an electrode, mass transport can be enhanced without significantly affecting the residence time of electrolyte in the cell. Two principle configurations have been employed, these are rotating discs and rotating cylinders as shown in Figure 13(*a*) and (*b*).

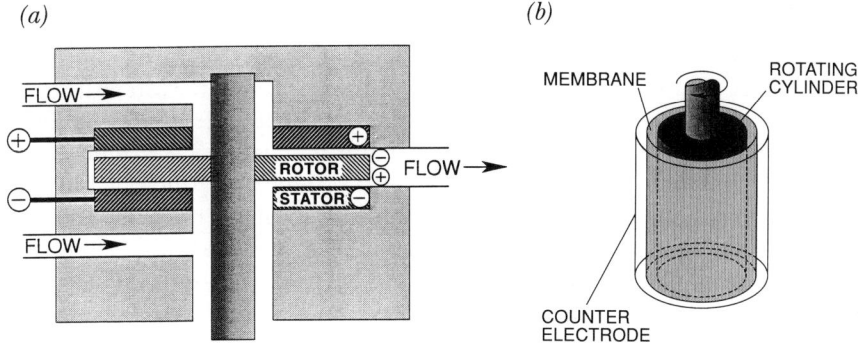

Figure 13 *Cells with rotating electrodes:* (a) *pump cell with rotating disc;* (b) *rotating cylinder*

The rotating cylinder cell in the form of the Udupa cell[5] is used extensively in India for the manufacture of a range of organic chemicals on a small scale, *e.g. p*-aminophenol and *o*-toluidine. Units are essentially batch tank electrolysers fitted with a number of rotating electrode elements and appropriate anode/cathode separators.

The rotating cylinder design is also used for the recovery of metals from effluents in which it operates in a continuous mode with axial flow of electrolyte in a annular channel formed between the inner, rotating electrode and the surrounding membrane.

One design of rotating disc electrode cell, the Cumberland Engineering cell,[6] utilises the rotation of a stack of parallel bipolar disc electrodes. It was devised to cope with electrolyses where the formation of insoluble precipitates, notably magnesium hydroxide during the production of hypochlorite from sea water, cause operational problems. The rotation of the electrode is used to remove the precipitate from the electrode and the cell. The electrical connection to the rotating electrodes is in a bipolar mode using stationary feeder electrodes embedded in the reactor walls. The electrodes are fitted with co-planar annular discs made from an electrically non-conducting material. These are fitted close to the inside and the outside of the electrode stack to limit current bypassing around the intermediate electrodes. Electrolyte feed is to the centre of the stack and electrolyte is pumped radially outwards. The pump cell[7] [Figure 13(*a*)] is similar in principle to the Cumberland cell and has been used commercially in metal powder production.

4.3.4 Three-dimensional Electrodes

The term three-dimensional electrode is used to describe electrodes in which the electroactivity is distributed in three dimensions by using suitable porous structures. Three-dimensional electrodes are used in cell designs where a high surface area per unit volume is required, thus providing acceptable space time yields when the reactant concentrations and/or the current densities are low. These electrodes are available as fixed bed, porous bed, fluidised, slurry and circulating beds. The major area of application is in waste water and effluent treatment, although notably a fixed bed electrode design using lead shot was used in the Nalco process for the manufacture of tetraalkyl lead. Perhaps this technology will re-emerge when organic electrosynthesis in non-aqueous solvents becomes more prominent, due to the application of sacrificial anodes.

The high surface area in three-dimensional electrodes is generally achieved using either particulates, porous or fibrous media. Depending upon the material, surface areas of the order of $1000-10\,000$ m^2 m^{-3} are possible, which is at least an order of magnitude better than for parallel-plate units. With sufficient flow of electrolyte through this media, good mass transfer rates can usually be achieved. There are a number of practical considerations in the implementation of three-dimensional electrodes. Pressure losses are significantly greater than for open channels, and mal-distribution of flow is likely due to channelling. The electrodes may not tolerate significant amounts of gas evolution, and may be susceptible to fouling by undissolved particulates, precipitation, *etc.* Operation at a high current density is desirable in order to increase the space time yield, however, this is generally limited by the rate of mass transport achieved which determines the maximum (limiting) current density of operation. Obviously a high current efficiency is desirable, not only to minimise energy usage but also to reduce problems associated with secondary reactions, *e.g.* precipitation and gas evolution. The operation and maintenance of three-dimensional electrodes should be relatively easy.

4.3.4.1 Electrode Configurations and Potential Distribution. In three-dimensional electrodes current flows in both electrolyte and electrode phases and their respective conductivities determine the associated distribution of electrode potential or reaction rate. An adverse effect of relatively high electrolyte resistance is that it can prevent the penetration of current into the pores and thus can stop the reaction from fully capitalising on the available surface area. With electrodes consisting of relatively poor conducting material, a potential drop in the electrode, in the direction of current flow, occurs which affects the current distribution. However, it is the potential availability of such high specific areas which has caused great interest in the use of three-dimensional electrodes.

One problem with porous electrodes, identical to that experienced with porous catalysts, is that if the reaction rate is large compared to diffusion then reacting species are consumed mainly near the boundary of the structure and the bed is under utilised. For electrolytic reactors made in the form of a packed bed rather than a porous matrix (with closed pores) this is not too great a problem when a flowing electrolyte is used.

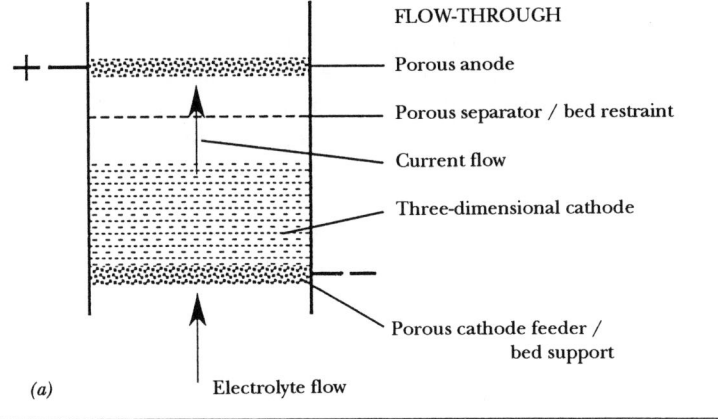

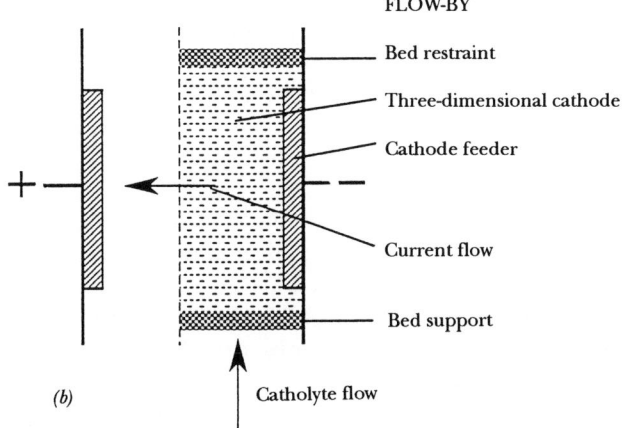

Figure 14 *Configurations of three-dimensional electrodes:* (a) *flow-through cell;* (b) *flow-by cell*

In practice three-dimensional electrodes operate (see Figure 14) either in a flow-through configuration, where the electrolyte flow and the overall flow of current are parallel or in a flow-by configuration where the electrolyte and current flow are perpendicular.

The first category *(a)* is generally limited in terms of scale-up due mainly to large changes in electrode potential in the structure, high cell voltages and low conversion. The potential distribution can be such that either the electrode potential is too high at certain points in the electrode and secondary reaction occurs, or electrode potential is too low and the electrode does not polarise. In many applications three-dimensional electrodes will operate at the maximum rate *i.e.* at the limiting current. For acceptable operation of the particulate bed, this distribution of potential must correspond to a suitable range[8] in which a limiting current plateau is identified. Generally the thickness of the bed is limited to a few centimetres especially if reasonable conversions of reactant are required.

The flow-by electrode design offers the means of independently varying the length of the electrodes in the direction of fluid flow and the thickness of the electrode bed in the direction of current flow. Thus in the case of a limiting current operation the design can in principle ensure an almost uniform limiting current operation, subject to restrictions and limitations in fluid flow and mass transport. These two factors are interrelated and the selection of the suitable electrolyte velocity is important for several reasons; a high value will give the following characteristics:

(a) a high rate of mass transport at the surface of the electrode to promote high limiting currents
(b) discourage variations in electrolyte solution pH
(c) assist in the removal of electrogenerated gases, which would affect the electrolyte conductivity
(d) increase the energy consumption required in electrolyte pumping because of the increase pressure drop in the bed
(e) give a relatively low degree of conversion of reactant
(f) improve the turbulent mixing in the bed and reduce the extent of electrolyte channelling or poor flow distribution

Thus overall a compromise on the value of electrolyte flow-rate has to be made taking on board the specific requirement of the extent of reaction. Ideally for maximum effectiveness, limiting current operation should be maintained throughout the bed. However, with certain reactions this is not desirable due, for example, to the occurrence of secondary reactions.

For limiting current operation the local distribution of electrode potential across the bed is essentially determined from Ohm's Law. For more general cases the calculation of electrode potential and current distribution is more complicated. The magnitude of the current and potential distributions exhibited by these three-dimensional, particulate electrodes can be appreciated by considering a limiting condition of operation, *i.e.* when conversions are low. Hence variations in concentration do not significantly affect the local reaction rate which is given, for Tafel kinetics, by:

$$\frac{j_m}{j_T} = \frac{2\theta}{\delta'} \tan\left(\frac{\theta x}{d} - \Psi\right) + \frac{e'}{\delta'} \tag{1}$$

where θ and Ψ are given by

$$\tan(\theta) = \frac{2\delta'\theta}{4\theta^2 - e'(\delta' - e')}, \qquad \tan(\Psi) = \frac{e'}{2\theta}$$

and

$$\delta' = dj_T\beta\left(\frac{1}{\kappa_s} + \frac{1}{\kappa_m}\right), \qquad e' = \frac{dj_T\beta}{\kappa_s}$$

where d is the electrode thickness, j_T is the total density based on the diaphragm cross-sectional area and, j_m, is the current density in the electrode phase. β is the polarisation parameter $= \alpha\, n\, F/RT$.

The current distribution (local reaction rate) is the change in metal phase current and is obtained from dj_m/dx. The current distribution determined by eqn. (5) enables a preliminary electrode design to achieve a 'uniform' distribution of reaction rate. This is important to fully utilise the electrode area and to achieve some control of the reaction selectivity. Typical current distributions are given in Figure 15 for the case when $\kappa_m \gg \kappa_s$. Clearly the parameter δ controls the distribution, with low values (<1) giving a more uniform distribution. Typically for a high electrode conductivity the current density is higher near the separator (or anode) and lower at the current feeder.

Generally in three-dimensional electrodes, when electrode conductivity is much greater than that of the electrolyte, good relatively uniform current distribution is obtained with a thin electrode, low current density and high solution conductivity. Three-dimensional electrodes are normally used when only low current densities are possible, and is to their advantage. They would generally not be used for current densities greater than approximately 200–400 A m^{-2} (based on the superficial cross-sectional area).

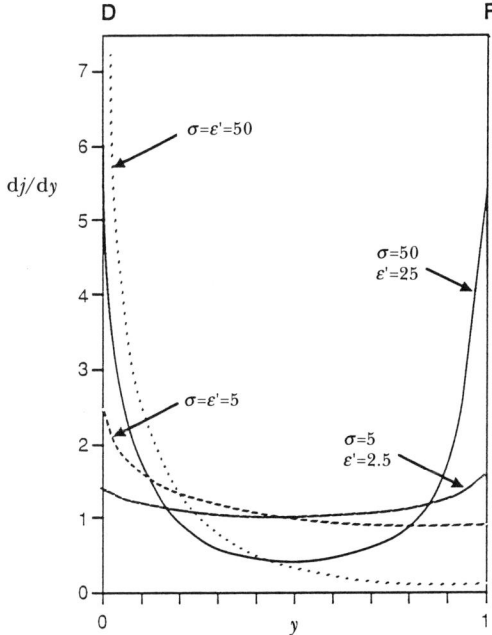

Figure 15 *Current distributions in a three-dimensional electrode under activation control;* $y = x/d$, $F = feeder\ electrode$

4.3.4.2 Moving Particulate Electrodes. Three-dimensional electrodes can be divided into two general categories, depending upon whether they are stationary beds or moving beds. The former are classified as either porous, where the electrode material is in the form of a continuous matrix, or as packed where the bed is constructed from 'loosely' contacting material. The moving beds cover a range of designs where the particulate electrode itself is in motion – either by mechanical means or hydraulically due to electrolyte flow. These designs include:

1   fluidised beds
2   circulating or moving beds
3   slurries
4   rotating and tumbling beds
5   vibrated beds
6   pulsed beds

Applications of these electrode designs are discussed in chapters 6 and 7.

An interesting design of three-dimensional electrodes is the fluidised bed (FBE), shown in Figure 16 as a typical rectangular configuration. The fluidised bed consists of an active particulate material which is freely supported by the buoyancy forces of an upward flowing electrolyte solution. The fluid flow into the bed is introduced through a porous distributor plate to maintain a uniform fluidisation, which is characterised by a random movement of particles throughout the bed. A variable in the operation of the fluidised electrode is the % expansion of the bed above its static height, which is determined by the liquid velocity. The greater the velocity, the greater will be the expansion. A high expansion of the bed limits the electrical contact between individual particles and can detrimentally affect the electrochemical reaction characteristics and thus bed expansion is limited in practice to *ca.* 20–40%.

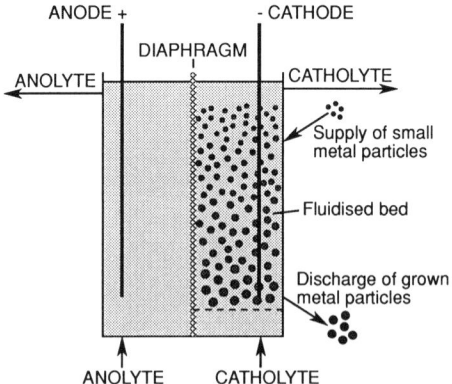

Figure 16 *Fluidised-bed electrode cell: rectangular configuration*

The attraction of the fluidised bed electrode (FBE) lies in the high active-surface area per unit volume, high mass transfer rate and the low pressure drop compared to fixed bed electrodes operating with the same flowrates.

The major disadvantage of the FBE for electrosynthesis and other applications is that it has an uneven and fluctuating potential distribution, which varies from cathodic to anodic potentials. This makes the selective transformation of species difficult to achieve. This phenomena is due to charge transfer processes occurring in the bed, where contact between adjacent particulate solid electrodes is not continuous. This has some attraction for many metal deposition systems, giving a certain degree of electrorefining capability.

An adaptation of the FBE concept is the bipolar fluidised bed configuration.[10] Typically, a bed of carbon particles is fluidised between two feeder electrodes. The bipolar design can offer an electrode cleaning effect as a result of the surface potentials changing continuously from anodic to cathodic values. This concept has been tested in several electrolyses *e.g.* anodic hypochlorite generation from sea water. However, a potential problem is that of the increased erosion and corrosion of the electrode material in this operating environment.

An adaptation of the fluidised bed electrode is the circulating bed electrode design in which particulate movement is in two regions — one where particles are hydraulically transported up an inclined surface and the other where the particles are falling, in a loosely packed region, to the bottom of the cell. Particles at the bottom of the cell meet a suitably directed electrolyte flow, which transports them into the rising layer. The electrochemically active region is generally towards the counter electrode. The moving electrode bed design[8] uses a solid separator between the rising layer and the 'falling packed' bed to restrict the electrochemical activity to the latter region. The major area of application of these designs is in metal electrodeposition in which the metal can be recovered continuously on the growing particulate material. The choice of bed material is somewhat limited, generally to 'uniform' particulate, because of the requirement of an even motion throughout the bed.

4.3.5 Reactor Design For Multiphase Reactions

Many established electrochemical reactor designs deal with processes in which the product is formed as a second phase, notably the electrochemical generation of gases such as chlorine, hydrogen and oxygen. The research and development associated with these processes has had a major impact overall on electrode and membrane materials and also on cell design. The concept of the 'zero cell gap', in which electrodes effectively sandwich the membrane material, has been exploited to minimise cell voltage requirements. This required further membrane development to facilitate good gas release away from the surface of the membrane contacted by the electrodes.

As a multiphase reactor in which the reactant is initially present in the non electrolyte phase, the electrochemical unit must couple together an interphase mass transfer process with the electrochemical reaction and any associated chemical reactions. This mass transfer process must ideally not be rate limiting or at least be high.

4.3.5.1 Gas–Liquid Reactions. Probably the simplest of procedures for carrying out electrochemical reactions involving gaseous reactants is to feed the gas directly into the cell as a dispersed two phase mixture or to sparge the gas directly into the cell. In this it is at least possible to consider the use of a general purpose flow electrolyser. Such a strategy has been carried out[11] in the electrosynthesis of epichlorohydrin (ECH) from allyl chloride (AC) using an electrosyn cell described previously. The reaction system for this process is shown schematicaly in Figure 17. The cell was operated in a divided configuration using a Nafion® cation-exchange membrane as the separator. The anode reaction generates chlorine for the conversion of AC to DCH (dichlorohydrin) [eqn. (2)]. The coproduced caustic is used in the dehydrochlorination step to convert DCH to

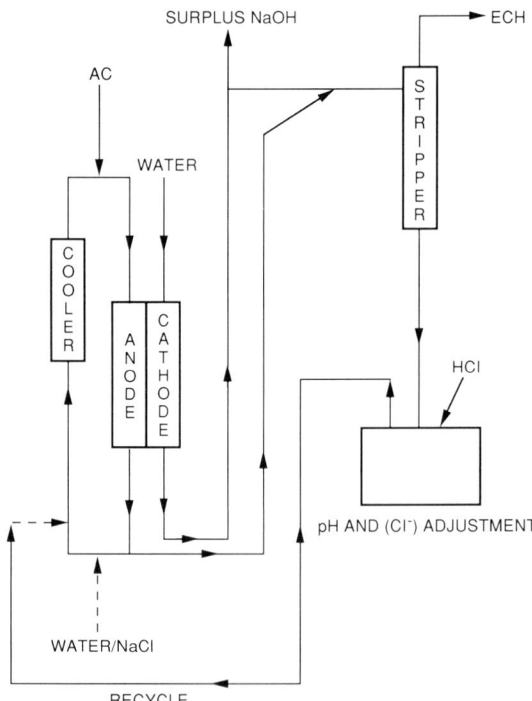

Figure 17 *Reaction system for epichlorohydrin production*

ECH. Streams from the anolyte and catholyte are contacted in a steam distillation column where reaction takes place and product is separated.

$$CH_2=CH-CH_2Cl + Cl_2 + H_2O \rightarrow CH_2(OH)-CHCl-CH_2Cl + Cl^- + H^+ \quad (2)$$
$$(AC) \qquad\qquad (DCH)$$

An interesting reactor design has been developed for the electrosynthesis of hexafluoropropylene oxide by Hoescht.[12]

$$CF_3-CF=CF_2 + H_2O \rightarrow CF_2OCF-CF_3 + 2\ e^- + 2\ H^+ \quad (3)$$

The electrolyte environment for the oxidation of the olefin is particularly aggressive, containing acetic acid, nitric acid and complexing fluorides. The anode material used is lead dioxide, which gradually corrodes during electrolysis, and together with the fact that no substrate was found to be stable in the electrolysis conditions meant that a thick, mechanically strong coating free from pin holes is used. It was also necessary to operate the process reactor at elevated pressure to increase the space time yield and current density. These factors resulted in a cylindrical configuration, shown schematically in Figure 18. The lead dioxide anode is formed on the inner wall of a stainless steel tube which is water cooled. The cylindrical configuration avoids critical zones at edges of the anodes and deformation of the substrate. A cylindrical cation exchange membrane surrounds the cathode to prevent the reduction of nitrates, the products of which would interfere with the epoxidation reaction.

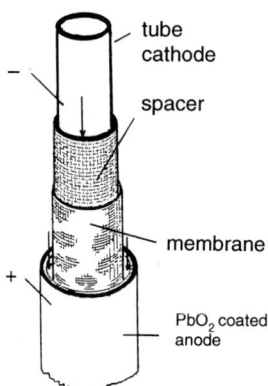

Figure 18 *Cylindrical cell design for hexafluoropropylene oxide production*

A recent reactor design is based on the sieve plate concept shown in Figure 19. The device gives, in principle, a means of carrying out the gas absorption in a sieve plate absorption tower simultaneously with the elec-

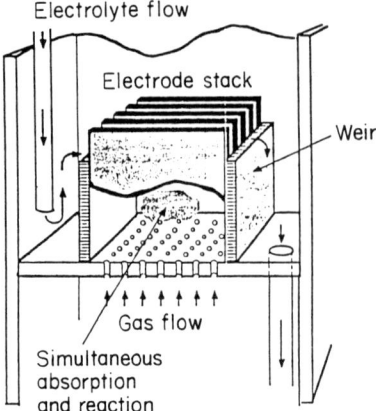

Figure 19 *Sieve plate electrochemical reactor*

trochemical reaction. By using parallel plate bipolar electrodes, uniform distribution of the electrode potential can in principle be ensured, which is not the case with other gas absorption electrochemical systems. A characteristic of the sieve plate electrochemical reactor (SPER), in its cylindrical configuration, is that it is readily amenable to the use of high pressure operation. The SPER was also used in the production ethene oxide and propene oxide.[13]

4.3.5.2 Packed Bed Reactors. Packed columns are frequently used for the absorption of gases into liquids and in many applications the absorption is enhanced by the occurrence of an appropriate reaction between the dissolved gas and the solvent phase species. This solvent phase can contain reagents which have been generated by an electrochemical reaction. This procedure is used in several applications and notably in the desulfurisation of sulfur dioxide using electrogenerated bromine, discussed in more detail in chapter 7.

   Several processes have considered using the packed bed as a two-phase electrochemical reactor in which the packing is the active electrode material. Examples of this include the epoxidation of propylene, the removal of chlorine by reduction, the oxidation of SO_2 directly on a carbon bed electrode and the production of hydrogen peroxide, discussed in chapter 8. In the latter case peroxide is produced by oxygen reduction on carbon electrodes and the cell is operated in a trickle flow regime.

4.3.5.3 Liquid–Liquid Dispersions. Packed bed electrochemical reactors have also been used for electrolysis in 'three-phase media', in for example the electrooxidation of benzene to *p*-benzoquinone[14] using trickle flow. A major limitation of the system is the poor solubility of the benzene in the aqueous electrolyte solution, which is a common problem associated with many other organic syntheses. There are several procedures, which can be adopted in this category of synthesis, to improve the performance and include:

(i)   electrosynthesis of micelle solubilized organic substances (using *e.g.* McKee salts)
(ii) phase-transfer assisted reactions
(iii) highly intense mass transfer in two-phase flow to maintain a near saturated solution of the electrolyte

A low solubility of reactant in the electrolyte phase has an advantage in indirect or mediated synthesis when the organic reactant can interfere with the electrode redox reactions.[15] In these cases the reaction takes place in a two-phase chemical reactor external to the cell and ideally should realise almost complete conversion of the organic in the aqueous phase before the electrolyte is separated and recycled to the cell for mediator regeneration (see chapter 2).

4.3.5.4 Solid Polymer Electrolytes. Cells which are based on solid polymer electrolytes, SPE, use ion-exchange material, typically in the form of a solid membrane, as the electrolyte. This polymer acts as the ion conductor and the electrode separator when sandwiched between two (porous) electrodes (see Figure 20). Several applications of this technology have been proposed in inorganic electrochemistry and include water electrolysis and the generation of ozone, which are discussed further in chapter 8. A major attraction of the SPE system is the ability to effectively electrolyse fluids or gases which have negligible conductivity. The technology is also finding some applications in organic electrochemistry where both cation- and anion-exchange materials are being used.[16] An example is the methoxylation of dimethylformamide (DMF) on a porous carbon anode. This reaction can operate at almost 100% current efficiency. The cell itself can operate in a flow-through mode due to the high rate of the electroosmotic transfer of organic species through the ion-exchange membrane.

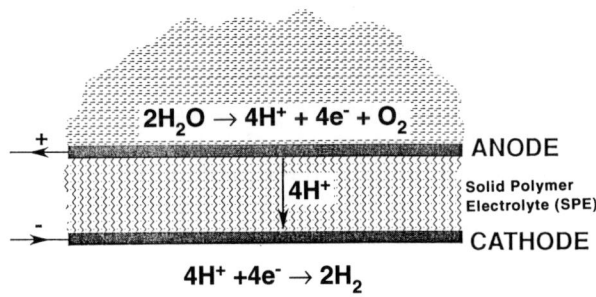

Figure 20 *Solid polymer electrolyte (SPE) cell for water electrolysis*

4.4   Mass Transport and Reactor Design

In many applications of electrochemistry, particularly in effluent treatment the concentration of the active species is very low. Thus, even with high surface area electrodes, the reactor can generally be operated near the mass transport limited current to try and maximise the performance in terms of the reaction rate, and thus to minimise the cost of the cells and associated power supplies. It is therefore appropriate to consider the mass transport characteristics of typical electrochemical reactors.

Correlations of mass-transfer behaviour relevent to electrochemical reactors are available in many references (*e.g.* ref. 17). Among the correlations of interest are those for rotating cylinder electrodes, parallel plate channels with and without turbulence promotors and particulate electrodes. The use of turbulence promotors is one of several ways to enhance the rate of mass transport to electrodes. The turbulence promotor acts essentially to disturb the flow of electrolyte at the electrode surface and causes localised eddy movement and mixing which intensifies the transport of material to the surface. Other techniques, which are used to enhance mass transport, are given in Table 5.

Table 5   *Methods of mass transport enhancement*

A *Electrode movement*	**B** *Electrolyte movement*
Rotation	Turbulent flow
Fluidisation	Gas sparging
Vibration	Inert fluidised beds
Reciprocation	Ultrasonics
Vortex flow	Pulsation
Centrifugal with fluid motion	Vortex flow
	Turbulence promoters
	— by surface roughening, extended surface, *etc.*
	— by baffles inert mesh or particulates

One of the significant benefits of turbulent promoters is that, at a macro level, they induce a uniformity of flow over an electrode in that eddy movement regularly disrupts the boundary layer development. This therefore suppresses the variation in mass transfer along the principle direction of the reactor. However, there are a number of factors against their use:

•   they cause a partial coverage of the electrode and thus a reduction in the available electrode area;

- they are susceptible to small particulate material which may cause a blockage or lodge at active regions on the electrode and disturb the electrochemistry;
- they can intensify corrosion or erosion where they contact the electrode or membrane surface.

Often with reactors of specified geometry the correlations of mass-transfer correlations simplify to an expression of the type:

$$k_1 A = k\,u^x \quad (\text{m}^3\ \text{s}^{-1}) \tag{4}$$

where k is a constant which depends upon the particular geometry, x is the velocity exponent which depends on the flow conditions and the geometry.

The area, A, is included here because for several systems it is not possible, or desirable, to experimentally determine these two factors separately. It is informative to consider the order of magnitude of the mass-transfer coefficient in electrochemical reactors to aid in rule of thumb estimations of performance. For a parallel plate reactor with turbulent flow:

$$St = 0.0278\ \text{Re}^{-0.125}Sc^{-0.79} \tag{5}$$

where $St = k_1/D$

For a system with a typical Schmidt number of 3000, this gives:

$$k_1/u = 0.0278\ \text{Re}^{-0.125}(3000)^{-0.79} \tag{6}$$

For a viscosity of approximately 10^6 m^2 s^{-1} and with a narrow channel where the hydraulic mean diameter is given approximately by $2\,d$, where d is the interelectrode gap, this gives:

$$k_1 = 8.96 \times 10^{-6}\ u^{0.875}\ d^{-0.125}$$

For an interelectrode gap of say 10 mm the value of the mass-transfer coefficient varies from 1.1×10^{-6} to 2.9×10^{-5} m s^{-1} in the velocity range $0.05 - 2.0$ m s^{-1}. This defines the broad range of moderate to good mass-transfer coefficients.

Thus with a value of mass-transfer coefficient of the order of 2×10^{-5} m s^{-1} then the mass transfer limited current density is given approximately by:

$$j = 2\,n\,C \tag{7}$$

where C is the concentration of the active species (mol m^{-3}).

Thus for example at a concentration of 1.0 mol m^{-3} (*ca.* 100 ppm) the mass tranfer limiting current for a two electron transfer process is *ca.* $j = 4$ A m^{-2}. Thus clearly to reduce the concentration of the species to the

1 ppm level at significant rates of reaction requires a significant enhance-ment of the mass-transfer coeffient and/or an increase in the active sur-face area.

4.5   ELECTROCHEMICAL REACTOR ANALYSIS

It is appropriate in this chapter on electrochemical reactor design to con-sider a brief example — in this case a batch recycle reactor. For further information on the subject of modelling of electrochemical processes a number of reference are available (*e.g.* refs. 8, 18 and 19). A typical batch recycle reactor with a plug flow reactor (PFR) is illustrated (also see Figure 2). Reactant A flows through the channel (see Figure 21), formed between the electrodes, at a constant flowrate v. There is no mixing or dispersion of reactant along the reactor only a gradual decrease in reac-tant concentration due to electrochemical reaction. At any plane, a dis-tance x along the reactor, the composition is assumed uniform, except at regions close to the electrodes where mass transport takes place. From a steady state material balance on a differential length of reaction, dx, the change in concentration C_A is related to the local current density in that section, j_x, according to

$$-v\,\mathrm{d}C_A = \frac{\sigma j_x\,\mathrm{d}x}{n\mathrm{F}} \tag{8}$$

where σ is the electrode area per unit length.

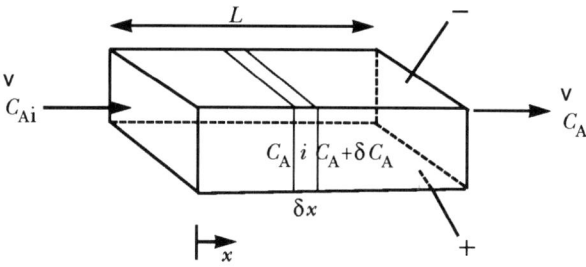

Parallel plate

Figure 21 *Plug flow reactor*

By defining the time a unit section of fluid takes to progress from inlet to outlet as the resistance time $\tau = L/u$ and noting that $\sigma = aS$, where S is the cross-sectional area in the direction of flow (assumed constant) eqn. (8) becomes:

$$\frac{-\mathrm{d}C_A}{\mathrm{d}\,\tau} = \frac{a\,j_x}{n\,F} \tag{9}$$

The above equation applies to a steady state operation of a PFR, whilst in the recirculating batch system inlet and exit concentrations vary with time. Fortunately this need not be treated rigorously, but rather a pseudo steady state approximation can be adopted in which the variation of concentration in the reactor with time are ignored. With batch recirculation this approximation is more accurate with a small volume of reactor in comparison to the volume of the mixing tank.

The instantaneous material balance over the mixer by vessel is:

$$v\,C_{A2} = vC_{A1} + V_m\,\frac{\mathrm{d}C_{A1}}{\mathrm{d}t} \tag{10}$$

which relates the outlet and inlet concentrations for the reactor.

As an example of a reaction system consider the case of the deposition of cadmium in the presence of dissolved oxygen and iron(III) ions. Cadmium deposition will only occur at values of potential where both oxygen reduction and iron(III) ion reduction, to iron(II) ion, occur at mass transport controlled rates. The analysis assumes that cadmium deposition is also at the mass transport limiting current density, which often represents the best case in terms of maximum reaction rate.

For the case of the recycle batch system with a plug flow reactor, in the pseudo steady state, in which the current density is given generally by:

$$j/nF = k_1\,C_i \tag{11}$$

eqns. (9), (10) and (11) give:

$$\tau_m\,\mathrm{d}C_{A1}/\mathrm{d}t + C_{A1} = C_{A1}\,\exp(-k_1 a\tau_R) \tag{12}$$

Integration of this gives the equation for the variation in reactant concentration as:

$$C_{A2} = C_{AO}\,\exp\{-t/\tau_m[1-\exp(-k_1 a\tau_R)]\} \tag{13}$$

Note that in this expression in term, $a\tau_R$, is identical to $k_1 A/v$. The current density for this reaction is therefore readily determined from eqn. (11). If the reactor is operated at a fixed electrode potential, such that other secondary loss reactions are time independent, then this effecively maximises the performance of the system. Figure 22 shows a typical performance of this system in terms of the variation of current efficiency with concentration. When a very large holding tank is used the system performance becomes identical to that of a closed batch reactor (see ref. 8).

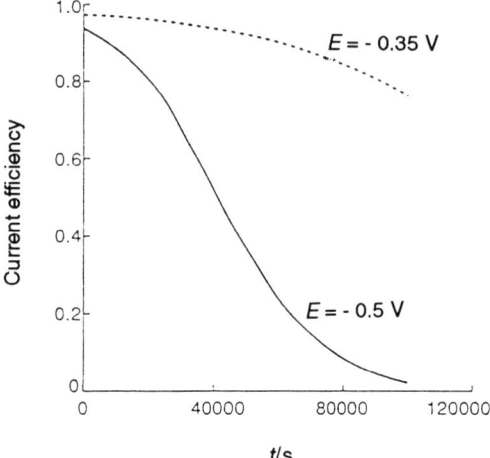

Figure 22 *Influence of oxygen reduction on the performance of a metal deposition reaction in a batch recycle reactor*

4.6   ENERGY TRANSFER IN ELECTROLYTIC REACTORS

Although in electrochemical processes the major use of direct electrical energy is in the cell bank(s), energy is also consumed in the pumping of electrolyte solutions and in controlling the temperature at the required operating level. To determine the appropriate conditions of operation, *e.g.* temperature, it is necessary to carry out an energy balance on the system and determine the ancillary equipment size, which will meet that temperature requirement. In a batch recirculation flow system there are two major pieces of equipment, the recirculation pump(s) and the in line heat exchanger(s), in addition to the cells and holding vessels (and gas–liquid separators if required) — see ch. 2, Fig. 3. All these units contribute to the net exchange in energy of the system during operation.

4.6.1   Pumping Requirements and Heat Transfer

The circulation pump in the reactor system serves to set the residence time of electrolyte in the cells and also the mass transfer rate. The lower the flowrate the smaller will be the mass-transfer coefficient generally, and the larger will be the conversion of the reactants through the cells. There will also be a greater rise in the electrolyte temperature in the cell in most cases. An increase in the flowrate will cause an increase in the mechanical energy in pumping. This energy is largely dissipated as heat and thus serves to increase the temperature of the electrolyte. This therefore puts a greater duty on the heat exchanger which is used to maintain the temperature within prescribed limits. The determination of the cooling duty can be obtained by performing an overall energy balance on the system as depicted in ch. 2, Fig. 3.

Overall the control of the operating temperature is generally carried out in three ways:

1   in an external heat exchanger as discussed in the case of a flow electrolyser;
2   in the reactor itself with the use of appropriate heat-transfer medium. For example in tank cells used for fluorine generation, using cooling coils or the reactor walls as appropriate cooling channels, or by adding additional cell frames to cool the electrolyte through the electrode surface.
3   by natural convection from the surface of the reactor.

All three methods are used in industry and examples are cited in this book.

REFERENCES

1   C. Wagner, *J. Electrochem. Soc.*, 1951, **98**, 133.
2   D.E. Danly, 'Emerging Opportunities for Electroorganic Chemistry', Marcel Dekker, New York, 1984.
3   D. Degner, 'Technique of Electroorganic Synthesis Part III, Scale-up and Engineering Aspects'. ed. N.L. Weinberg and B.V. Tilak, J. Wiley and Sons, 1982, ch. VI, p.283.
4   P.M. Robertson and N. Ibl, *J. App. Electrochem.*, 1977, **7**, 323.
5   H.V.K. Udupa, *A.I.Chem.E. Symp. Ser.*, 1979, **75**, No. 185, 26.
6   B.B. Greaves (to Cumberland Engineering Co.), Can. Pat., 966806 April 29, 1975.
7   D. Pletcher and F.C. Walsh., 'Industrial Electrochemistry', Chapman and Hall, 2nd edn., 1990.
8   K. Scott, 'Electrochemical Reaction Engineering', Academic Press, 1991.
9   J.S. Newman and C.W. Tobias, *J. Electrochem. Soc.*, 1962, **109**, 1183.
10   F. Goodridge and A.R. Wright, 'Comprehensive Treatise of Electrochemistry', ed. E. Yeager, J.O'M. Bockris, B.E. Conway and S. Sarangapani, Plenum Press, NY, 1983, vol. 6, ch. 6.
11   B.E. Bongenaar-Schlenter and L. Van Raam, *Chem. Eng. Res. Dev.*, 1991, **69**, 71.
12   S. Dapperheld and H. Millauer, 'Electro-synthesis, From Laboratory, to Pilot, To Plant', The Electrosynthesis Co. Inc., New York, 1990, ch. 5, p.115.
13   C.F. Odouza and K. Scott, *'Trans. Int.Chem. Eng.'*, 1995, **73**, 330.
14   C. Oloman and P. Reilly, *J. Electrochem. Soc.*, 1987, **134**, 859.

15 T. Tzedakis and A. Savall, *Chem. Eng. Sci.*, 1991, **46**, 2269.
16 K.H. Simmrock, J. Jorissen, R. Fabiunke and R. Gregel 127th meeting of The Electrochemical Soc., 18–23 Oct., 1987, in *Proc. Symp. Electrochem. Eng. Chlor-Alkali, Chlorate Ind.*, 1988, 383.
17 I. Rousar, K. Micka, and A. Kimla., 'Electrochemical Engineering', Elsevier, Amsterdam, 1986, vols. 1 and 2.
18 F. Goodridge and K. Scott, 'Electrochemical Process Engineering', Plenum Press, NY, 1995.
19 K. Scott, *J. Chem. Technol. Biotechnol.*, 1992, **54**, 257.

CHAPTER 5

Electrochemical Membrane Separation Processes

In this chapter the use of membranes in electrochemical separations is discussed. This primarily is in the use of ion-exchange membranes in electrodialysis and the other related use of ionic transfer through ion-exchange membranes. The use of electroosmosis and electrophoresis are discussed and also applications where electrochemistry is used to enhance other membrane processes such as micro- and ultra-filtration.

5.1 ELECTRODIALYSIS

The main applications of electrodialysis (ED) are for the concentration of electrolyte solutions, and for the diluting, or de-ionising, of solutions. The latter application has over the years been (see Table 1) the dominant technique for the desalination of brackish water. Electrodialysis is also used extensively for desalting and concentrating sea water for salt production. In principle, the technique has many potential applications for the removal or recovery of ionic species. Other applications, which have been commercially adopted, are in the food and dairy industries, pharmaceutical industries, metal-plating industry for effluent treatment, pulp and paper industries and chemical regeneration from salt solutions. Generally the process of electrodialysis is used to perform several operations:

Table 1 *Application areas of electrodialysis*

Production of potable water by desalination
Recovery of water and valuable products from industrial effluents
Removal of salts and acids in food processing
Desalting of pharmaceutical solutions
Production of salts from sea water
Production of acids and bases with bipolar membranes from salt water waste streams

- The separation of salts, acids, and bases from aqueous solutions
- The separation of ionic compounds from neutral molecules
- The separation of monovalent ions from ions with multiple charges
- The introduction of ionic moeities to generate new species

Electrodialysis competes with other separation processes, such as reverse osmosis, ion exchange, dialysis *etc.*, in many applications and can offer in many cases several significant advantages such as:

- High selectivity for charged components
- Lower energy and investment costs
- Continuous operation
- High product recovery rates
- Minimal change of feed water constituents due to chemical or thermal degradation
- No chemical regenerates or significant feed water pretreatment
- Proven membrane life

Electrodialysis is a method that uses a direct electrical current to transport ions through sheets of ion-selective membranes,[1, 2] which are either, anionic — permeable to anions and impermeable to cations, or cationic — permeable to cations but impermeable to anions.

These membranes are effectively impermeable to the hydraulic transport of water and other solvents. In practical applications the membranes are arranged alternatively in a stack, as shown in Figure 1, between anodes and cathodes and are separated by thin plastic spacers, some 0.5–2.0 mm in thickness. Two types of spaces are used in practice:

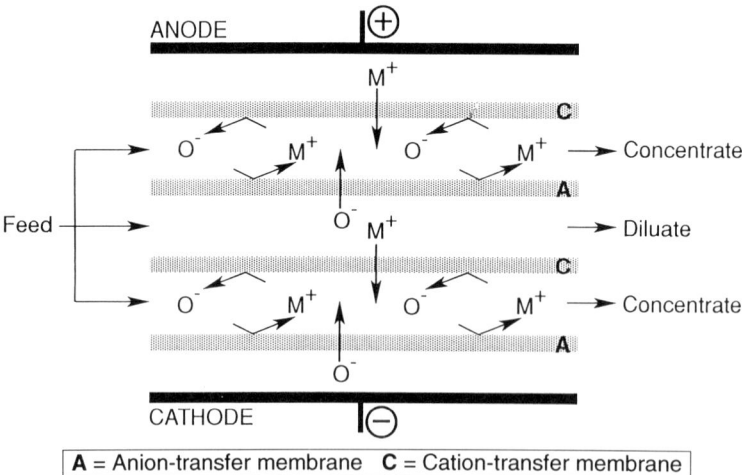

Figure 1 *Operation of an electrodialysis cell*

1   sheet flow spacers, woven or expanded sheet or plastic.
2   tortuous flow spacer made from thermoplastic sheet. The electrolyte flow path taken with this is serpentine, in channels *ca.* 10 mm wide.

The application of a direct electrical current causes anions to move towards the anode, passing through the anion-exchange membrane and are then stopped by the cation-exchange membrane. The cations behave in a similar manner, but move in the opposite direction and are retained by the anion-exchange membrane. Thus in time one stream will become ion enriched and the other will become ion depleted, forming the concentrate and diluate streams respectively. The combination of one concentrate and one diluate compartment is known as a cell pair. In an ED cell stack more than 300 cell pairs can be employed. The surface area of each membrane is of order of 0.5–2.0 m^2 and current densities are in the range of 200–2000 A m^{-2}.

In addition to the concentrate and diluate streams in the cell, there are two other streams, the electrode rinse streams adjacent to the anode and the cathode. The complete ED unit is usually mounted in a filter press, arrangement in which the concentrate and dilute streams are distributed by internal manifolding.

During practical operation an accumulation or depletion of solute(s) takes place at the surface of any membrane because of the permselectivity of the membrane. Thus a concentration variation is set up in the vicinity of the membrane in which the diffusion of the solute is a major factor determining performance. This phenomena is known as polarisation. In general an increase in polarisation decreases the efficiency of separation of the process as the flux of the more permeable species decreases. The effect of polarisation is to increase the cell voltage and also to instigate transport of other species. These species are typically H^+ ions, in cation-exchange membranes and OH^- ions in anion-exchange membranes, which lead to changes in the pH of the electrolyte solutions, because of the difference in mobilities of the anions and cations involved.

The membranes in ED stacks must be handled easily in stack assembly and be resistant to osmotic swelling and be impermeable to water pressure. The membrane must also be resistant to fouling or poisoning. Fouling can be caused by any material which settles on the membrane surface and de-activates the ion diffusion capability. This may be from organic macromolecules present in the water or by precipitation of colloids or calcium salts on the membrane surface.

Procedures, which are adopted to minimise the effect of fouling, include: pre-treatment of the feed, acidification of the concentrate streams and polarity reversal.

5.2   ION-SELECTIVE MEMBRANES IN SALT REGENERATION, RECYCLING AND EFFLUENT TREATMENT

The prime function of ED is the separation and thus concentration of a salt or ionic solution. Typically this may mean the removal of a contaminating ionic species from a product solution in, for example, the removal of salt from a dyestuff solution. The method is being widely applied for purification, recovery and recycling processes. [3] Asahi Chemical Co., have applied ED to the recovery of nitric acid from the waste water of an acrylic fibre manufacturing facility. [4] The technology can also be applied to the recovery of other mineral acids notably sulfuric acid. Ion-exchange membranes can also be used to 'split' salts and instigate 'reactions' in cell compartments, which form new products or recycle acid and base primary constituents.

5.2.1   Electrohydrolysis

Electrohydrolysis is a process by which an aqueous stream of salt, *e.g.* of sodium sulfate, is used to regenerate sulfuric acid and caustic soda by membrane ion exchange. The process can be operated in a three-compartment cell, shown in Figure 2, with the sodium sulfate stream fed to the central compartment. The sodium ions transfer across the cation-exchange membrane and combine with the hydroxide ion generated at the cathode. The sulfate ions transfer across the anion-exchange mem-

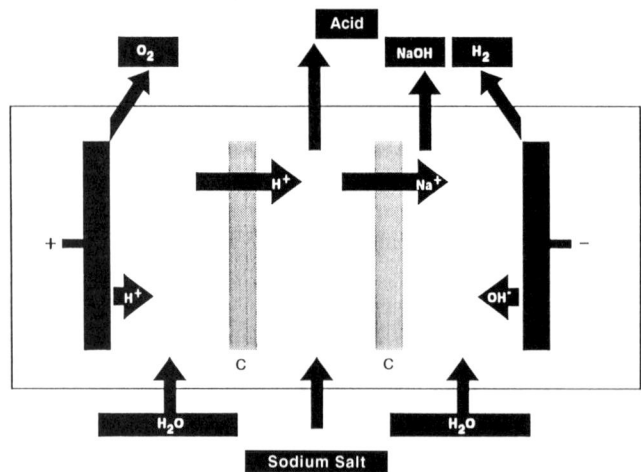

C - Cation Membrane

Figure 2   *Three-compartmental cell for electrohydrolysis*

brane and combine with the hydrogen ions, formed anodically, to produce the sulfuric acid. This process suffers from low current efficiency at
higher concentrations of sulfuric acid and relatively poor selectivity of
anion-exchange membranes.

An alternative process based on a two-compartment cell has been used[5]
to carry out this function (see Figure 3). In comparison to the two-compartment cell this new process gives a lower cell voltage, avoids the weak-
ness of the anion-exchange membranes and gives high current efficiencies
at sulfuric acid product concentrations of 15% and greater. Although the
sulfuric acid contains sodium sulfate, this is removed by crystallisation.
Importantly the catholyte product, typically 20% w/w NaOH is pure ($< 10^{-5}$
w/w sulfate).

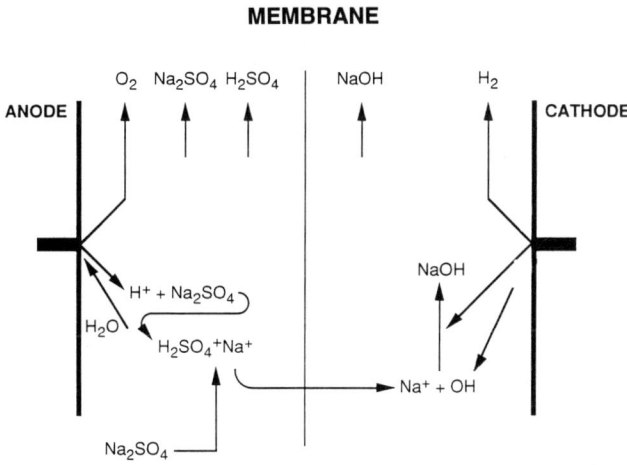

Figure 3 *Two-compartment cell for electrohydrolysis*

An alternative process for splitting sodium sulfate is in the generation
of ammonium sulfate and sodium hydroxide.[6] This is achieved in a three-
compartmental cell, in which Na_2SO_4 is fed to the central compartment
and SO_4^{2-} ions migrate through an anion-exchange membrane to the
anode. Ammonia is added continuously to this compartment to maintain a
constant pH of 1.5 and form the ammonium sulfate, which is to be used as
fertiliser. Maintaining a low pH is important to ensure stability of the DSA
anode, which otherwise, along with alternative materials (stainless steel,
Ni, Pt graphite), would corrode at high pH in ammonia solutions. The cell
uses Neosepta AMH anion-exchange and Nafion 902 cation-exchange
membranes and a Ni cathode to produce caustic and ammonium sulfate
solutions of concentrations greater than 30 and 40%, respectively. Current

efficiencies of the order of 90% are achieved. The use of the salt splitting technique can be applied widely to many inorganic and organic species. Three-compartment cells may be used in which two cation (or anion) membranes are introduced. An application of this method is shown in Figure 4 in the recovery of citric acid from sodium citrate.

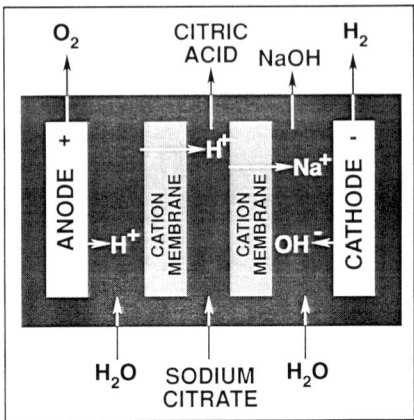

Figure 4   *Three-compartmental cell with two cation-exchange membranes for citric acid recovery*

5.2.2   Treatment of Plating Bath Rinse Waters and Waste Streams

The use of ED in recovery of metals from dilute waste liquors has increased rapidly, particularly in the USA. A main area of application is in the processing of rinse waters from the electroplating industry. Using ED, complete recycling of the water and metal ions can in principle be achieved (see Figure 5). In a typical plating line the plated metal parts are transferred from the plating bath to a still-rinse and then to the regular rinse bath. Owing to the relatively high salt concentration in the plating bath there is a considerable drag-out of salts from the plating bath to the still-rinse. The concentration in the still-rinse therefore increases rapidly and the drag-out of metal from the still-rinse becomes significant. Further treatment of the rinse solution is needed and by using ED the salts are removed and concentrated and recycled directly to the plating bath. The diluate is fed back to the still-rinse keeping the salt concentration in the still-rinse to a low level.[6] The ED process is being used successfully on plating baths for metals such as Au, Pt, Ni, Cu, Ag, Pd, Cd, Zn and Sn/Pb.[7] ED is used in the treatment of process solutions from chromium plating baths. Dilute rinsewater solutions can be concentrated by the removal of the CrO_7^{2-} ions across an anion-exchange membrane

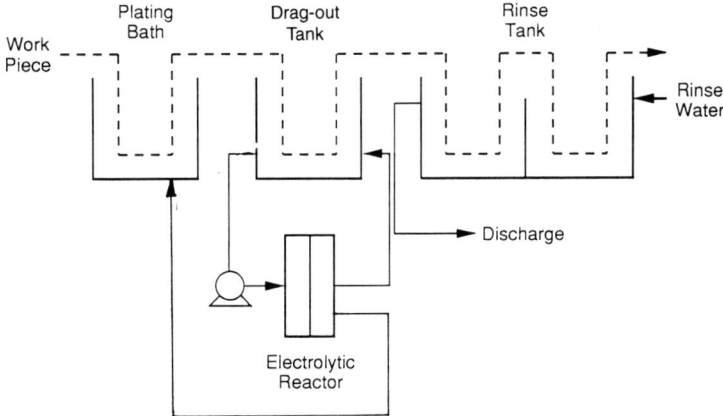

Figure 5   *Application of electrodialysis in plating bath rinse lines*

into the chromic acid concentrate. ED is also used as a method of purification of plating solutions by the removal of contaminant ions, *e.g.* Cu which are picked up during the plating operation (see chapter 7).

Although the costs of operating ED systems will clearly depend upon the application a perhaps not too untypical case is the recovery of chromium. Here the cost for 1.5 kg CrO_3 per day has been reported [8] as US $30 000 (1980 prices) at an operating cost of $1.9–2.0 per kg CrO_3.

There are many other examples of the use of electrodialysis in the treatment of waste streams for recycling and effluent remediation. Examples of these include:[9-13]

1   The separation of organic acids from Kraft Black liquor.
2   The recovery of ammonium sulfate (and some sulfite) from a waste water emanating from a plant to produce *p*-aminophenol.
3   The recovery and reuse of sodium hydroxide from industrial effluents typically from ion-exchange resin generation, pulp and paper, textile and various washing industries.
4   Barrier anolyte liquor catholyte process (BALC), which combines anion- and cation-exchange membranes with a neutral membrane for the processing of pulping spent liquors, for the renovating of cooking liquors and pulping chemicals and to recover lignin, carbohydrate and other wood derived chemicals.
5   The concentration and recycling of a dilute waste phosphoric acid solution produced in the generation of phosphoric acid from phosphate rock. This traditional application of ED was able to concentrate reagent grade acid by a factor of 2 –3, up to concentration of 1.0 mol dm^{-3} with energy consumptions of 1.73–2.5 kWh kg^{-1} of P_2O_5.

5.3   BIPOLAR MEMBRANES

A relatively new process for salt splitting is based on the use of bipolar
membranes. Bipolar membranes are polymeric materials composed of
two homopolar ion-exchange membranes; one cationic and one anionic.
When placed in an electrochemical cell, with the cationic layer in contact
with catholyte, current is carried by protons moving through this layer
and by hydroxide ions moving through the opposite anionic layer (see
Figure 6). Because of this property bipolar membranes are often referred
to as water-splitting membranes. The process is electrodialytic in nature,
merely involving the change in the concentration of ions already present
in the solution. For efficient operation the membrane should have a good
water permeability and efficient, low resistant, transport of the hydrogen
and the hydroxide ions.

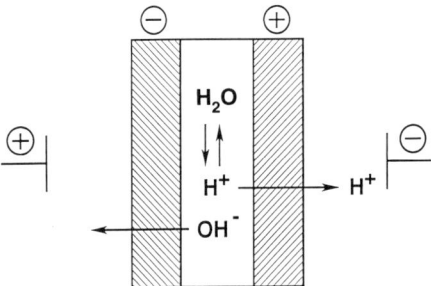

Figure 6   *A bipolar membrane*

Bipolar membranes are used with either two- or three-compartment
cells. The basic three-compartment cell, shown in Figure 7, consists of
one cation exchange, one anion exchange, and one bipolar membrane.
In operation, positive and negative ions migrate through the respective
monopolar membranes and concentrate in compartments on opposite
sides of the bipolar membranes. Water diffuses through these layers to an
interfacial region where it dissociates into the constituent hydrogen and
hydroxide ions, these ions diffuse back into the adjacent compartments,
in opposite directions, to produce alkali and acid solutions.

A typical example of the use of the bipolar membranes is in the treatment of concentrated salt solutions, such as Na_2SO_4 from the chemical
industry to produce H_2SO_4 and NaOH. The basic concept of the process
is shown schematically in Figure 7. The theoretical potential to achieve
the water splitting capability is 0.83 V at 25 °C. The actual potential drop
across a bipolar membrane is quite close to this being in the range 0.9–
1.1 V for current densities between 500–1500 A m^{-2}, which is the general
region of practical interest. The values of the membrane potential drop

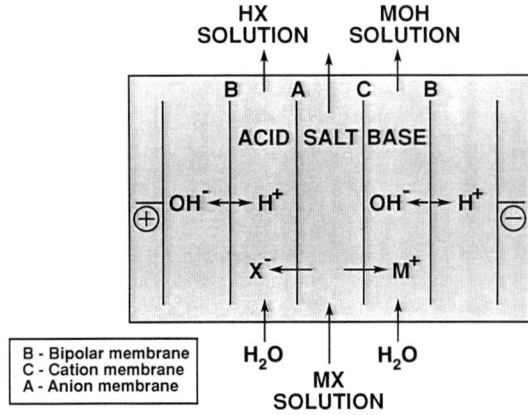

Figure 7 *Three-compartment cell for bipolar membrane operation*

equate to theoretical energy consumptions of the order of 600–700 kWh per tonne of NaOH.

Bipolar membranes can be used with various combinations of anion- and cation-exchange membranes. For example in an anion/bipolar membrane configuration, the anions move through the anion-exchange membrane and combine with the H⁺ ions arising from the bipolar membrane to form the acid product. This type of cell is useful for converting salts of weak bases (*e.g.* ammonium nitrate) to a salt–base mixture and a relatively pure acid. The performance of the two-compartment cells can be enhanced by the introduction of a third chamber as shown in Figure 8.

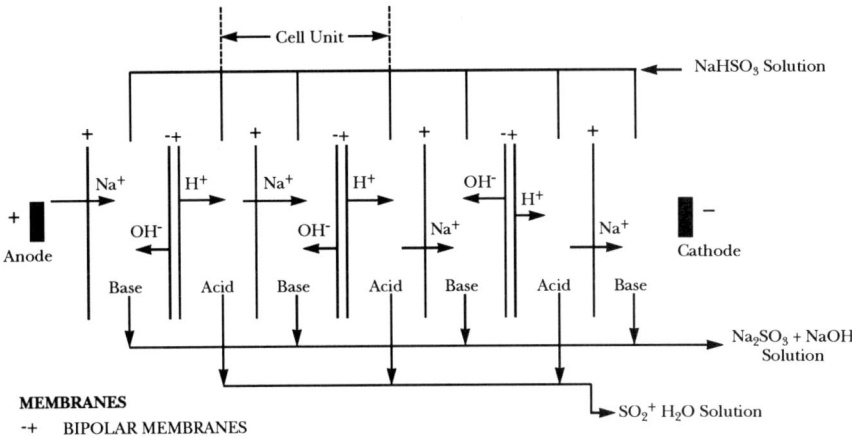

Figure 8 *Alternative operation of bipolar membranes with two cation-exchange membranes*

Table 2 *Bipolar membrane applications*

Salt splitting		Acid and base purification	
KF	Recovery of fluoride scrubber solutions	HCl	Mild steel pickling
KF/KNO$_3$	Reclamation of waste pickling acid	H$_2$SO$_4$	Lead acid batteries
KCl	Production of potassium hydroxide from minerals	HNO$_3$	Metals processing
NaNO$_3$	Battery waste processing	SO$_2$	Flue gas desulfurisation
NaF	Aluminium potliner reclamation	CO$_3$	Conversion of sodium carbonate to caustic
NaCl	Ion-exchange resin regeneration	NaOH	Ni–Cd batteries
Na$_2$SO$_4$	Rayon processing and pulp and paper processing; waste sodium sulfate conversion	KOH	Alkali scrubbing
NaRCO$_2$H	Organic acid purification and concentration (acetic, formic, citric)	NH$_3$	Catalyst processing; pigment processing
NaCNH$_2$RCO$_2$H	Amino acid production		

The multichamber, cation cell uses two cation exchange membranes. In operation the salt solution is first fed to the chamber between the two cation-exchange membranes and then passes to the acid compartment. This gives a salt–acid stream with a higher concentration of acid than the standard two-compartment cell.

There are only a few known suppliers of bipolar membranes and of plant for this technology, *e.g.* Allied Signals 'Aquatec' system, and Stantech, although other ion-exchange membrane manufacturers and research organisations have developed bipolar membranes. The potential applications of bipolar membranes are numerous (see Table 2) in recycling or effluent control and include the following:[14, 15]

1   *Regeneration of the spent pickling liquors* used in stainless steel manufacture (HF, HNO$_3$, KOH recovery). This process is operated at Washington steel, Pennsylvania with a capacity of 6000 m^3 yr^{-1} of liquor. The process takes a filtrate, containing KF and KNO$_3$ to generate a mixed acid stream and a KOH stream from the bipolar cell stack. The KOH stream is used in the neutralisation step to generate the metal hydroxides from the waste acid coming from the pickling bath. The acid product generated in this process is returned to the pickling bath.

2   *Gas desulfurisation* — The application of bipolar membranes for the desulfurisation of flue gas (FGD) is illustrated in Figure 9. In the absorption part of the FGD installation the following reactions take place:

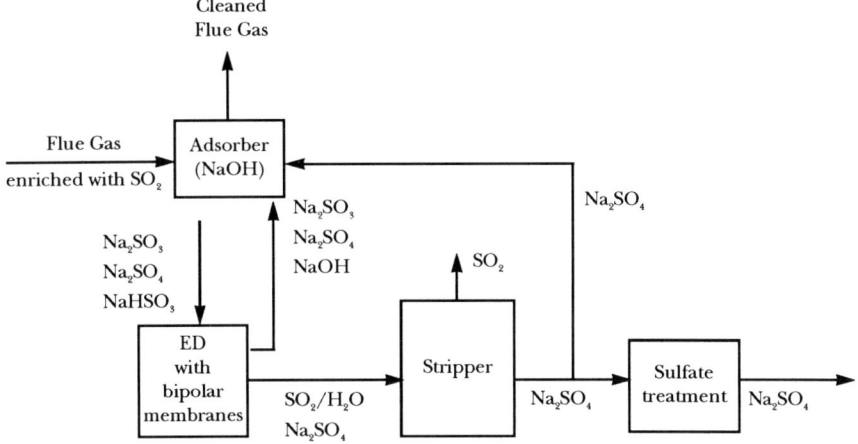

Figure 9   *Application of bipolar membranes in flue gas desulfurisation (FGD)*

$$SO_2 + 2\, NaOH \rightarrow Na_2SO_3 + H_2O \tag{1}$$

$$Na_2SO_3 + H_2O + SO_2 \rightarrow 2\, NaH\,SO_3 \tag{2}$$

$$Na_2SO_3 + 1/2\, O_2 \rightarrow Na_2SO_4 \tag{3}$$

The products resulting from absorption will be transformed in the bipolar ED-stack, in which cation-exchange membranes and bipolar membranes are arranged alternately, into sulfurous acid, sodium sulfite and sodium hydroxide. The following reactions occur in the two sides of the membrane cell:

Acid side:

$$NaHSO_3 + H^+ \rightarrow Na^+ + H_2SO_3 \tag{4}$$

$$Na_2SO_3 + 2\, H^+ \rightarrow 2\, Na^+ + H_2SO_3 \tag{5}$$

$$H_2SO_3 \leftrightarrow H_2O + SO_2 \tag{6}$$

Base side:

$$NaHSO_3 + OH^- + Na^+ \rightarrow Na_2SO_3 + H_2O \tag{7}$$

$$Na^+ + OH^- \rightarrow NaOH \tag{8}$$

By heating with vapour in the stripper, SO_2 can be recovered from the Na_2SO_4 solution and recycled.

3   *The recovery of ethylenediaminetetraacetic acid (EDTA)* — Waste water
     streams from some electroless plating baths consist of low quantities
     of unused formaldehyde and Cu^{2+} ions and relatively large quanti-
     ties of ethylenediaminetetra acetic acid (EDTA) and sodium sulfate.
     In the alkaline solution, the EDTA exists as a readily soluble sodium
     salt. By reducing the pH down to 1.7 the sodium salt of the EDTA is
     transformed into a less soluble acid. With electrodialysis the EDTA
     can be transported into the ED cell with the bipolar membrane,
     where H^+ ions are produced, and then precipitated in an external
     circuit as the less soluble acid. The precipitated free EDTA can then
     be recycled without any further treatment. The solution, which is
     left, consists essentially of sulfuric acid and formic acid and can be
     neutralised by the sodium hydroxide, produced by the bipolar mem-
     brane.

4   *The decarbonating of fruit juice* — Standard electrodialysis is a process
     used for the decarbonation of fruit juice. The electrodialytic water
     dissociation now competes with electrodialysis and has the advan-
     tage that for the decarbonation of fruit juice, NaOH is no longer
     necessary, and as a by-product citric acid is produced.

5   *The splitting of amino acids* — Amino acids have amphoteric proper-
     ties. At a fixed pH-value (isoelectric point) amino acids exist as neu-
     tral components. If the pH-value of the solution is lower than the iso-
     electric point of the appropriate acid it is positively charged and will
     be transported as a cation. If the pH-value of the solution is higher
     than the isoelectrical point, the amino acids are present as anions.
     This property can be used for the separation of amino acids by ion-
     selective membranes when the pH-value in the feed solution is
     between the respective isoelectrical points of the several amino acids.
     With a relatively simple regulation of feed pH, bipolar membranes
     can also could be used, and thus avoid the addition of chemicals.

5.3.1   Characteristics of Bipolar Membranes

In the application of bipolar membranes they should ideally possess the
following characteristics:

—   low membrane resistance (low voltage drop across the membrane);
—   high degree of water dissociation;
—   high rate of water diffusion;
—   high selectivity;
—   long life time with high chemical stability; and
—   acceptable price.

   For the bipolar membrane technology to be effective it should ideally
give current efficiencies of the order of 80% at current densities around

1000 A m^{-2}. One of the key features in operation is to minimise the formation of insoluble impurities in the cell, which could result in contamination of the membranes. This may necessitate careful pre-treatment of the feed prior to bipolar ED, for example the removal of multivalent metal ions using ion-exchange resins, or precipitation and filtration of metal ions such as Fe, Cr or Ni. The final product concentrations, which are possible using bipolar membranes, at acceptable efficiencies, depend on the acid or base to be regenerated. Strong acids such as sulfuric can only be produced directly at concentrations of 5–15 wt%, whilst organic acids with low dissociation constants (*i.e.* citric and lactic) can be produced at concentrations up to 30 wt%.

As already discussed there are several ways in which the cells can be arranged to achieve acceptable performance. In addition there are several operating techniques, which can be use to get the best out of the application, and these include:

5.3.1.1 Product Batching. The permselectivities of the membranes decrease as the concentration of product increases. Thus a process operating at the steady state, in a feed and bleed mode, and generating product at the highest concentration is operating at its least efficient. By product batching the bipolar membrane 'water splitter' can operate at lower average concentrations and therefore at higher concentrations.

5.3.1.2 Product Removal or Stripping. For example the removal of volatile species such as ammonia can enable lower concentrations of product to be maintained with higher water splitting efficiency.

5.3.1.3 Use of Supporting Electrolyte. The process efficiency of ED is influenced by the competitive transport of hydrogen and metal ions. Thus by introducing a supporting electrolyte, which raises the concentration of the metal ions, the transport of the latter species is favoured to a greater extent and thus efficiency improves. This strategy was adopted in the Soxal process for flue gas desulfurisation shown in Figure 9.

The economics of the use of bipolar membranes for many of the applications cited appear to be favourable with payback times of two years or less for applications such as ammonium nitrate and ammonium chloride recovery and in the recovery of the metal constituents of Ni/Cd batteries. There are many companies which supply electrodialysis modules units and associated membranes for a range of applications (see chapter 3). Several of these companies also supply bipolar water-splitting membrane modules. Module sizes range from relatively small bench scale units to full size plant. One interesting feature of at least one company's portfolio (Electrocell AB) is the use of cell modules, which can function as either electrodialysis units or as electrochemical reactors. This is a positive point for any company investing in electrochemical hardware, where if there is a decline of one marked in for example the electrosynthesis of a chemical intermediate the capital invested can be recovered in a salt separation or recycling process.

5.4   OTHER MEMBRANE PROCESSES

The application of an electrical field can under specific circumstances offer a number of advantages for membrane separations. It provides an additional process control variable (*i.e.* the applied potential) and can lead to enhancement of process rate, effectiveness and efficiency and thus can lead to a smaller plant and reduced overall running costs.

5.4.1   Electrokinetic Separations

Electrokinetics is used to describe the movement of charged particles and water molecules in an applied d.c. electric field. There are two main application areas, electroosmosis and electrophoresis, one being the converse of the other. That is to say that the movement of charged particles is always in the opposite sense to the electroosmotic flow of water. In general, any two-phase interface (solid–liquid, liquid–liquid) is charged and the electrokinetic phenomena are related to the distribution of charge at surfaces. Owing to this charge the particle is surrounded by a cloud of liquid containing ions with the opposite charge. This cloud is the diffuse double layer which serves to maintain electroneutrality in the system. In low ionic strength media the thickness of the diffuse double layer can be extensive (100 nm for a 10^{-5} mol dm^{-3} 1 : 1 electrolyte). In electroosmosis the flow velocity of water in a pore is given by the Helmoltz– Smoluchowski equation

$$u_f = \frac{eE\zeta}{4\pi\mu} \tag{9}$$

where, ζ, is the zeta potential, which is potential at the outer Helmholtz plane, the locus of the closest approach of hydrated ions to the solid–liquid interface, E is the field strength and e is the permittivity.

The zeta potential can be defined as the work necesary to take a unit positive charge from the bulk liquid phase up to the absorbed rigid layer, and is generally < 0.1 V. From eqn. (9) and from the application of Ohm's Law the volume of liquid displaced per second through the pore is related to the current flowing through the capillary by:[14]

$$V = \frac{e\zeta I}{4\pi\mu\kappa} \tag{10}$$

The equations used to describe the electrokinetic effects of electroosmosis and electrophoresis are essentially the same. In electrophoresis the velocity of the particle, u_p, is determined by the electrophoretic mobility, U_p, and the electric field strength E,

$$u_p = U_p E \tag{11}$$

The electrophoretic mobility depend upon the thickness of the diffuse double layer, the viscosity of the surrounding fluid and the zeta potential, and is given approximately by:

$$U_p = 2e(\zeta)\kappa r/3\mu \tag{12}$$

where r, is the particle radius.

For further details of this subject the book by Hunter[17] should be consulted.

5.4.2 Electroosmosis

Electroosmosis (EO) is a membrane based separation process, driven by the application of an electrical potential gradient across the membrane. This potential facilitates the transport of mobile ions and liquid through the pores of the membrane. The mechanism involves the formation of an electrical double layer at the surface, due to the ability of the membrane to acquire a charge when immersed in an aqueous solution. Further into the solution, although only a relatively small distance from the surface, a mobile diffuse layer of opposite charge is also formed. Upon experiencing an electrical potential gradient, this mobile layer of ions and water moves through the pores. Applications of this process are mainly in de-watering and for the treatment of colloidal suspensions and sludges in effluent and waste streams.

5.4.2.1 Electrokinetic De-watering. The selection of equipment in industry for the removal, or separation, of water from a particulate species depends upon several factors including the particle size and the solid content. For particle sizes of less than 10 μm electrokinetic de-watering can offer an attractive alternative to thermal drying or mechanical de-watering processes. The process can be used to concentrate slurries, or sludges, from 1–40% solids content under the influence of a potential field. In this process, the pores of a non-conducting microporous membrane act as a support for a fine layer of particulate species. The surface then gains a small immobile charge matched by an excess of ions of equal and opposite charge in the adjacent solution. The application of a potential field across the membrane causes the solution, with an excess of mobile counter ions, to transport along the pores. In addition to this effect, the dispersed particles are transported away from the membrane surface by electrophoresis.

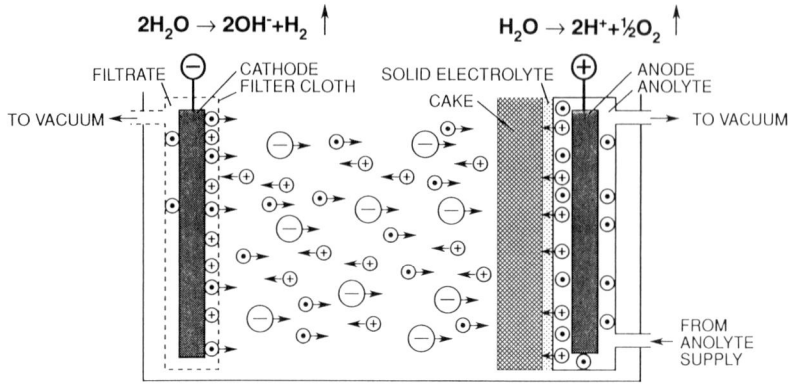

Figure 10 *Principle of operation of an electrofilter*

Units for electroosmosis have been under development at the UKAEA Harwell[18] for applications in radioactive waste processing and other areas. The Dorr–Oliver Company, USA, marketed an electrofilter for de-watering fine particulate suspensions of mineral materials. The principle of operation of the unit, shown in Figure 10, combines vacuum cake filtration with electroosmosis, electrophoresis, and electrophoretic deposition. There are three important elements to this device:

(i)   *Anode element.* This is composed of three parts, the anode (typically DSA), the surrounding electrolyte and a solid polymer electrolyte, which separates the anolyte from the media to be electrofiltered, whilst enabling the passage of the charge.

(ii) *Cathode element.* This is also composed of three parts, the cathode, the electrolyte and a filter medium which is in direct contact with the filter cloth. The filtrate which is drawn through the filter, under an applied vacuum, is the catholyte for the cathode reaction (hydrogen evolution).

(iii) *The slurry.* The application of the potential field across the slurry causes the solid and liquid to separate. The solid normally carries a negative charge and thus migrates towards the anode, where it is collected as filter cake and subsequently removed. In operation the formation of a cake at the cathode will reach an equilibrium thickness, when the velocity on the particles, due to drag, equals the particle electrophoretic mobility. The filtrate rate is determined by the applied field and the differential pressure applied across the filter cloth.

The energy consumption of electrokinetic de-watering is typically an order of magnitude lower than for thermal dewatering. The commercial unit was made possible by the introduction of cost-effective corrosion resistant DSA[R] anodes. This unit has an area of approximately 60 m^2 and has seen commercial service at a kaolin production plant.

5.4.3 Electrophoresis

Electrophoresis is also a separation process driven by an electrical potential. The physical process is more commonly associated with electrophoretic painting, where small particles of paint resin form a double (charged) layer. Upon application of a potential, the particles are attracted to an appropriate electrode (the article to be painted), forming an electrical circuit.[19] Electrophoresis has also been used for the formation of coatings of metal oxides, carbides, borides and alloys onto metal substrates. Electrophoresis has several applications in de-watering and thickening, which includes a proposed method for the removal or the concentration of colloidal particles (usually negatively charged in biological systems and in polluted waters), referred to as forced flow electrophoresis. By using microfiltration membranes, which are permeable to water and ionic species, but not the colloidal particles, a solution free of colloidal material is produced on simultaneous application of an applied pressure and an electrical potential.

A small scale application of electrophoresis is in the separation of biochemical species from natural sources using the Biostream electrophoretic separator, from CJB Developments.

5.5 ELECTROCHEMICALLY ENHANCED MICRO- AND ULTRA-FILTRATION

Electrofiltration is a term used to describe the use of an applied potential field across a micro- or ultra-filtration membrane to enhance production (permeation) rate. For example consider ultrafiltration which is a cross-flow membrane process for the separation of small, sub-micron particulate materials and macromolecules from aqueous solutions. The membrane materials, which can be used, are varied and can be polymer, metallic or ceramic. Ultrafiltration membranes operate through the application of a differential pressure between the feed and the permeate solution and the membrane permeability and selectivity depend upon the pore size and the pore size distribution. The application of an electrical potential field can in many cases be used to control the concentration polarisation in pressure driven membrane separations. An enhancement in the membrane flux can often thus be achieved due to the influence of the electrophoretic migration of the different species. With the use of suitable electronically conducting 'mineral' membranes this method of enhancing the rate of ultrafiltration can be realised.[20] These ceramic membranes, typically alumina, are coated with a metallic layer of nickel, nickel/tungsten alloy or with a RuO_2–TiO_2 mixture. The application of a potential field, particularly with the RuO_2–TiO_2 membranes, can produce a significant enhancement in the flux-rate. For example, an approximate doubling in the membrane flux-rate is achieved on the application of a potential field of 15 V.

The magnitude of the voltage applied depends on the design of the electroultrafilter and the conductivity of the solution feed.

REFERENCES

1   R. Rautenbach and R. Albrecht, 'Membrane Processes', J. Wiley and Sons Ltd, 1989.
2   M.C. Porter, 'Handbook of Industrial Membrane Technology', Noyes, 1990.
3   K. Scott, 'Membrane Separation Technology', S. T. I. Publications, Oxford, 1990.
4   'Ion Exchange Membrane and Its Applications', Technical Brochure, Asahi Chemical Industry Co. Ltd, Yurakucho Chiyoda, Tokyo, Japan.
5   A.D. Martin, *Inst. Chem. Eng. Symp. Ser.*, 1992, **127**, 153.
6   'ElecroCell AB Product Brochure', Taby, Sweden, 1993.
7   P.S. Cartwright, *Plating and Surface Finishing*, 1985, **7**, 28.
8   J.S. Lindstedt and W.G. Millman, *Plating and Surface Finishing*, 1982, **69**, 32.
9   T.R. Hanley, H.K. Chin and R.J. Urban, in 'Industrial Membrane Processes', ed. R.E. White and P.N. Pintauro, *Am. Inst. Chem. Eng. Sym. Ser.,* 1986, **248**, 121.
10   J.W. Rowe and P. Gregor, US Pat. 4 584 057, April 1986.
11   T.A. Davis, in 'Electrosynthesis From Laboratory, To Pilot, To Production', ed. J.D. Genders and D. Pletcher, The Electrosynthesis Co., 1990, ch. 15, p.239.
12   A.E. Simpson and G.A. Buckley '1st Symp. on Advances in Reverse Osmosis and Ultrafiltration', Toronto, 1988, 35–345.
13   G.A. Dubey, T.R. McElhinney and A.J. Wiley, *Tappi*, 1965, **48**, 95.
14   'Bipolar Membranes and its Application in Electrodialysis Process', Stantech GmbH, Technical Brochure, 1993.
15   K.N. Mani and F.P. Chlanda, ' Membrane Separation Processes, Proceedings of Int. Technol. Conf.', Brighton, UK, 1989, pp.235–252.
16   J.G. Sunderland, *J. App. Electrochem.*, 1987, **17**, 889.
17   R.J. Hunter, 'Foundations of Colloid Science', Oxford University Press, 1985, vol. 1.
18   A.D. Turner, 'Symp. on Future Ind. Prospects of Membrane Processes', Brussels, 1988.
19   C.F. Simpson and M. Whittaker, 'Electrophoretic Techniques', Academic Press, 1983.
20   C. Guizard, F. Legault, F. Indrissi, A. Larbot, L. Cot and C. Gavach, *J. Membrane Sci.*, 1989, **41**, 127.

CHAPTER 6

The Treatment of Industrial Process Streams and Effluents. Part I Metal Recovery by Electrodeposition

The use of electrochemical techniques for the treatment of a wide range of industrial effluents and process streams is increasing rapidly and applications include the removal and recovery of dissolved metals, the destruction of dissolved organics, and the treatment of waste gases. The methods are often used in conjunction with other methods because there may be several types of contaminating species present including suspended particulates, microorganisms, and dissolved organic and inorganic compounds. In many applications, specific technologies, although attractive, have limitations in their effectiveness unless they are coupled to other established or developing technologies. For example in the case of electrodeposition, when low metal ion concentrations are experienced, the efficiency and cost effectiveness start to fall and thus a method such as ion exchange, which can be more effective with lower metal ion concentrations, can be utilised to recover, concentrate and recycle the metal ions back to the electrodeposition. This example and other specific methods for recovery of dissolved metal ions are exclusively considered in this chapter.

6.1 RECOVERY AND RECYCLING OF DISSOLVED METALS

Pollution by metal ions arises in several industrial sectors and the resultant liquors may contain single uncomplexed metal ions, mixed metal ions in uncomplexed and complexed form and solutions also containing dissolved organics and other constituents. As well as a variation in the type of metal ion constituents, there may also be variations in electrolyte conductivity, overall ionic composition and pH. This, not surprisingly, has resulted in wide variations in waste management practices in industry. In most cases, however, the approach has been largely to consider the material as waste or effluent and to use the most convenient and 'economic' method of disposal. Methods for treatment of metal bearing solutions often involve one or more of the following steps:

(*i*)   Dilution, to lower the metal concentration below regulatory discharge limits. This may not satisfy total regulatory emission levels and could lead to high water costs.

(*ii*)   Mixing, of different process streams. This may tend to induce precipitation through pH moderation, and also is less controllable and potentially hazardous.

(*iii*) Temporary storage, followed by transportation to specialist waste disposal organisations.

(*iv*)   Chemical precipitation under controlled pH. The method is not very discriminatory between dissolved components, which may also adsorb onto the precipitate. The transportation and dumping of these materials is becoming increasingly more restricted and expensive. Although smelting of the precipitate is possible this can be beset by problems of inclusion or adsorption of other species derogatory to the smelting. Smelting is only feasible with hydroxide (or oxide) wastes, but the residual concentration level of the metal ions in solution is now approaching, or above, the legislative limits set on many metal ions in many countries.

The above scenario, in which a 'treatment step' is installed prior to the discharge point, is limited with regard to the flexibility and scope for material recycling and re-use. Precipitation is a return of the metal to its combined state to which large quantities of energy and materials have been previously expended to produce the metal or active metal form. What is ideally required are minimum waste strategies and/or point of source strategies designed into the process to re-use or recycle the components in their most useful and appropriate form. There are several incentives for the introduction of resource recovery and recycling:

(*i*)   increasing concern over environmental contamination in the face of decreasing discharge limits and reduced availability of disposal sites;

(*ii*)   the monetary or other value of the recovered material, which includes a reduction in supply and storage costs;

(*iii*) the increasing cost of process water and sewer discharge costs;

(*iv*)   a general reduction in the load of final effluent treatment.

Several methods have been researched and developed for the removal of dissolved metals in medium to low concentrations from waste waters including adsorption, ion exchange, extraction, precipitation, membrane separation and biochemical treatment (see Table 1). These techniques compete with one another, and can complement electrochemical methods, which are finding even wider practical applications. Notably the removal of the metal by electrodeposition, at source, is seen to be attractive due to the potential for a one-step clean method of metal recycle. The electrochemical approach will not always provide the primary solution to the problem, but can be part of an integrated waste management scheme.[1]

Table 1   *Alternative technologies for the removal of metal ion solutions*

Methods	Quality of treated effluent	Cost of effluent treatment (a)	Selectivity for individual metals	Feasibility of metals recovery for saleable product	Relative costs of metals recovery	Other factors
Precipitation *e.g.* as hydroxide or sulfide	Moderate good	Low–moderate (depends on soid–liquid separation)	None	Poor	High	Additional treatment to destroy complexing anions
Ion exchange	Very good	High	High or moderate (depends on pH *etc.*)	Good	Moderate	Resin may be affected by oxidising agents
Electrolysis	Moderate–high (suitable for closed loop operation)	High	High (with control of potential) depending upon standard potentials	Very good	Low (included in a)	Susceptible to other ions and organics
Reverse osmosis	Medium–high	Depends on desired effluent quality	None	Good for in-plant solution recycle	Low (included in a)	Restriction on pH of effluents
Evaporation	Minimal effluent	High	None	Good for in-plant solution recycle	Low	
Liquid–liquid extraction	Low–high (organic as well as residual metals in raffinate)	Moderate to high (depends on solvent loss)	High or moderate (depends on pH and extractant)	Good	Low for in-plant recycle otherwise moderate	Continuous countercurrent operation
Flotation	Moderate	Low–moderate	None	Poor	High	Relative
Cementation	Poor	Low	Poor	Good	Low	High effluent concentration
Electro-dialysis	Moderate	High	Poor	Good for recycling	High	Relatively high effluent concentration

6.2   PROCESS SOLUTIONS AND FINAL PRODUCT FORM

The recovery of metals from solution by electrodeposition can conveniently be divided into two areas depending upon the metal ion concentration to be treated.

(*i*)   In the metal ion concentration range 1–3 g dm^{-3} of solution feed, two-dimensional cathodes are generally used to maintain effluent concentration levels in the range 0.1–0.5 g dm^{-3}. Thus the removal of a metal from a process stream to a pre-determined concentration, enables the recycling of either solution or metal to the process.

(ii)   At approximate concentrations up to 100 mg dm^{-3} (100 ppm) of metal ion feed solution, three-dimensional electrodes are used preferably, especially if effluent levels of < 1 mg dm^{-3} (1 ppm) are required. In the case of certain precious metals, two-dimensional electrodes are used to treat lower concentration solutions when power requirements are not crucial. The removal of metal to very low concentrations enables the resulting spent liquor to be recycled, discharged, or passed to further treatment stages for other species.

Electrodeposition cells can operate continuously or in a batch mode, and can produce either compact metal for re-use or re-sale, powdered or flake metal, or concentrates for recycling. The metal may be produced as an alloy as an alternative to pure metal or occasionally as an insoluble metal compound (hydroxide, oxide and sulfide). The choice of the metal form depends largely on the use or destination of the material. It may be recycled directly within the process operation, or to another process, or reprocessed either in house or sold for refining or scrap.

6.3   THE INFLUENCE OF THE ELECTROCHEMICAL REACTION

There are approximately 32 metals that can be recovered effectively by electrodeposition from aqueous solutions, which is only partly indicated in the appropriate standard electrode potentials shown in Table 2. The simplest way of recovering the metal is by direct reduction of the metal cation:

$$M^{z+} + z\ e^- \rightarrow M \tag{1}$$

In certain cases the metal is not deposited from a free ion but from a complex with an inorganic or organic ligand *e.g.*

$$CuCl_3^{2-} + e^- \rightarrow Cu + 3\ Cl^- \tag{2}$$

$$CdEDTA^{2-} + 2\ e^- \rightarrow Cd + EDTA^{4-} \tag{3}$$

The chemistry of the system can have a significant bearing on the performance of the electrodeposition and the ability to recover the metal. The potential for an electrode reaction to occur at a desired rate depends on the thermodynamics and the kinetics of the system. The sign and magnitude of the equilibrium potential of species is a measure of the ability to accept or donate electrons. Redox systems with positive potentials will oxidise hydrogen to protons and increasing positive potentials correspond to increasing oxidation conditions. The converse is true for increasing negative potentials, reductants—electron donors will reduce protons to hydrogen. Thus increasing positive potentials equate to oxidation of metals to

Table 2   *The electromotive series*

Electrode reaction					Standard electrode potentials / V — $E°$, V at 25 °C
$S_2O_8^{2-}$	+	$2e^-$	$\rightarrow$	$2SO_4^{2-}$	+2.01
Co^{+++}	+	$2e^-$	$\rightarrow$	Co^{++}	1.82
H_2O_2	+	$2H^+ + 2e^-$	$\rightarrow$	$2H_2O$	+1.78
Au^+	+	e^-	$\rightarrow$	Au	+1.69
$2Cl$	+	$2e^-$	$\rightarrow$	$2Cl^-$	+1.36
$\frac{1}{2}O_2$	+	$2H^+ + 2e^-$	$\rightarrow$	H_2O	+1.23
Pd^{++}	+	$2e^-$	$\rightarrow$	Pd	+0.99
Ag^+	+	e^-	$\rightarrow$	Ag	+0.80
Fe^{+++}	+	e^-	$\rightarrow$	Fe^{++}	+0.77
Cu^+	+	e^-	$\rightarrow$	Cu	+0.52
Cu^{++}	+	$2e^-$	$\rightarrow$	Cu	+0.34
$2H^+$	+	$2e^-$	$\rightarrow$	H_2	+0.00
Pb^{++}	+	$2e^-$	$\rightarrow$	Pb	-0.13
Sn^{++}	+	$2e^-$	$\rightarrow$	Sn	-0.14
Mo^{++}	+	$2e^-$	$\rightarrow$	Mo	-0.20
Ni^{++}	+	$2e^-$	$\rightarrow$	Ni	-0.25
Co^{++}	+	$2e^-$	$\rightarrow$	Co	-0.28
Cd^{++}	+	$2e^-$	$\rightarrow$	Cd	-0.40
Cr^{+++}	+	e^-	$\rightarrow$	Cr^{++}	-0.41
Fe^{++}	+	$2e^-$	$\rightarrow$	Fe	-0.44
Zn^{++}	+	$2e^-$	$\rightarrow$	Zn	-0.76
Mn^{++}	+	$2e^-$	$\rightarrow$	Mn	-1.19
Al^{+++}	+	$3e^-$	$\rightarrow$	Al	-1.66
Mg^{++}	+	$2e^-$	$\rightarrow$	Mg	-2.36
Na^+	+	e^-	$\rightarrow$	Na	-2.71

solution species or to passivation (*e.g.* oxide) films, whilst increasing negative potentials equate to stable metal species. The ranking of potentials on the electrochemical series determines the relative oxidation (positive potentials) power and reduction power of species. Thus for two redox couples (denoted as 1 and 2) with values of standard potentials E_1 and E_2 the reduced form of couple 1 can be oxidised by the oxidised form of couple 2 when $E_2 > E_1$. Thus the thermodynamic driving force is positive and the reaction is spontaneous with species O_2 as the oxidant. Thus species such as Co^{II} are powerfull oxidants having quite large equilibrium potentials and species such as Cr^{III} are powerfull reductants.

   In electrochemical systems the feasibility of reactions occuring can be determined from the Nernst equation, which also takes into account the concentrations or activities of the electroactive species. In many systems pH plays a significant role and equilibria is commonly expressed in terms of potential–pH (Pourbaix) diagrams,[2] which incorporate both chemical and electrochemical (redox) reactions. A typical Pourbaix diagram is shown in Figure 1 for the case of Cu.

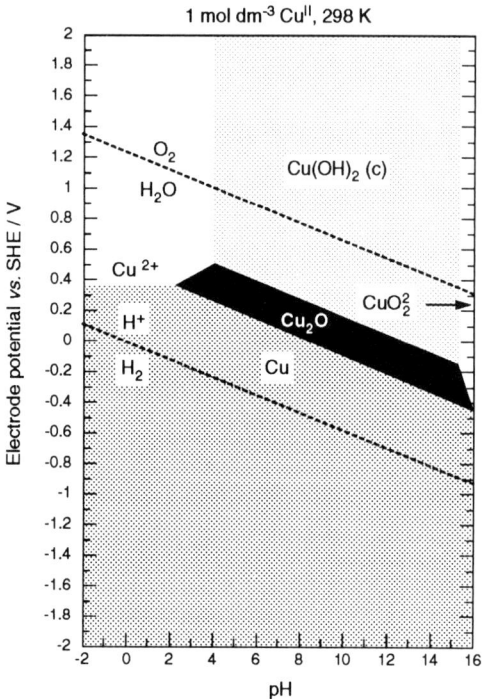

Figure 1   *Potential–pH (Pourbaix) diagram for copper*
        (With permission of G.H. Kelsall, course on Electrochemical Process
        Engineering, Newcastle-upon-Tyne, UK, 1994, July)

The broken lines are generally used to signify the equilibria between solution species and the solid lines define the equilibria between solid phases or between solid and solution phases. The diagram of Figure 1, which also shows the region of stability of water, show only the predominant phases under specified conditions of temperature, pressure and solution species activity. In the case of the copper system, with a value of E_o of 0.34 V, it can be seen that corrosion of the copper can only occur by the reduction of dissolved oxygen:

$$O_2 + 4\ H^+ + 4\ e^- \rightarrow 2\ H_2O \tag{4}$$

$$E_{O_2/H_2O} = 1.2291 - 0.0591\ pH + 0.0148\ \log(p_{O_2}) \tag{5}$$

and not by proton reduction,

$$2 \ H^+ + 2 \ e^- = H_2 \tag{6}$$

$$E_{H_2/H^+} = 0.0591 \ pH - 0.0296 \ \log(p_{H_2}) \tag{7}$$

In this redox system the reference potential under standard conditions of 0.1 MPa H_2 and unit proton activity is defined as zero.

The Nernst equation for hydrogen and oxygen define the bounds in which their oxidation and reduction respectively will produce water and conversely the region of potential in which oxygen evolution will occur (above the oxygen line) and hydrogen evolution will occur (below the hydrogen line).

Pourbaix diagrams give useful information on the behaviour of many systems relevent to environmental and synthesis applications, including metal deposition, from purely thermodynamic considerations. In practice electrolytic systems are driven by an over-potential and thus overall the performance of a system will be affected by the solution composition, temperature, over-potential and electrode material. Therefore in certain systems electrodeposition can be obtained contrary to thermodynamic predictions. For example, in the case of zinc, hydrogen evolution kinetics are very slow (see Figure 2), which enables the metal to be electrodeposited from aqueous solution even though the standard equilibrium potential is -0.763 V.

This particular system requires very pure solutions, free from impurities such as Fe^{II}, Cd^{II} *etc.*, which would otherwise co-deposit as their reversible potentials are higher (less negative) than that of zinc. These species would catalyse the hydrogen evolution reaction leading to an eventual zero efficiency of zinc deposition.

6.3.1   Complexation and Multiple Reaction Systems

If a metal ion is complexed, deposition occurs less readilly than from free ions *i.e.* there is a more negative reduction potential for the complexed metal ion. The shift in the potential can be as great as -0.6 to -0.8 V, and depends upon the stability constant of the complexed ion and the concentration of the ligand. It is therefore possible to alter the relative positions of metals in the electrochemical series by complexation. This feature is exploited in the plating industry, where conditions are chosen in which any impurities are selectively complexed, making their deposition potentials more cathodic than that of the metal being plated. This results in a improvement in the purity of the metal deposit.

The Nernst equation for the metal deposition–dissolution process in the presence of a metal complexing ligand Y has to be modified if some predominant complex species is formed by reaction such as:

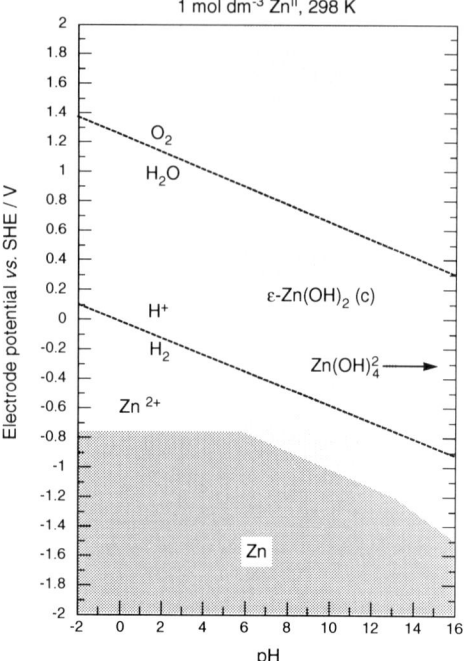

Figure 2   *Potential–pH (Pourbaix) diagram for zinc*

$$M^{n+} + Y^{z-} = MY^{(n-z)+} \tag{8}$$

The species Y may be neutral or charged. The predominant deposition (or dissolution) reaction is thought to be:

$$MY^{(n-z)+} + n\ e^- \longrightarrow M + Y^{z-} \tag{9}$$

which has a deposition potential given by:

$$E_{MY} = E_{M^{n+}/M} - (R\,T/n\,F)\ \ln(K_{MY}) + (R\,T/nF)\ \ln(a_{MY}/a_Y) \tag{10}$$

where

$$K_{MY} = MY^{(n-z)+}\ /M^{n+}Y^{z-} \tag{11}$$

A typical example is the complexation of zinc with cyanide where the standard reversible potential is shifted to more negative potentials and thus the metal is more active in dissolution.

The extent of complexation by a particular ligand depends upon the metal ion being complexed. For example in the case of EDTA the formation constant, K_f, has the following values for a Cd/ Fe system:

$$Fe^{II} \qquad \log K_f = 14.33$$
$$Cd^{II} \qquad \log K_f = 16.46$$
$$Fe^{III} \qquad \log K_f = 25.10$$

Thus it would appear that Fe^{III} ions form the stronger complex and undergo a more negative shift in deposition potential on complexation than does Cd^{II} or Fe^{II}. However, in solution the relative values of the stability constants are influenced by other solution conditions, notably pH. The value of pH influences the formation of metal hydroxides, the acid speciation of the ligand and the possible formation of secondary metal complexes.

6.3.1.1 Multiple Reaction Systems. At electrode potentials more cathodic than the equilibrium potential, the rate of metal deposition will depend upon the electrode kinetics of the particular reaction. Ideally the rate of deposition will be at the maximum Faradaic efficiency, up to values of the limiting current density. This frequently depends upon the relative over-potential for the reduction of protons to hydrogen gas which may be a competing reaction at the electrode surface (see Figure 3).

There are several other reactions that can lower the current efficiency of the metal deposition reaction:

(i)   oxygen reduction. This is more significant when dealing with low concentrations of the metal ions with well oxygenated solutions, especially those systems in which the anodic reaction is the evolution of oxygen gas. Here the concentration of oxygen can be super-saturated although the actual value depends significantly on the pH, the electrolyte composition and the temperature.

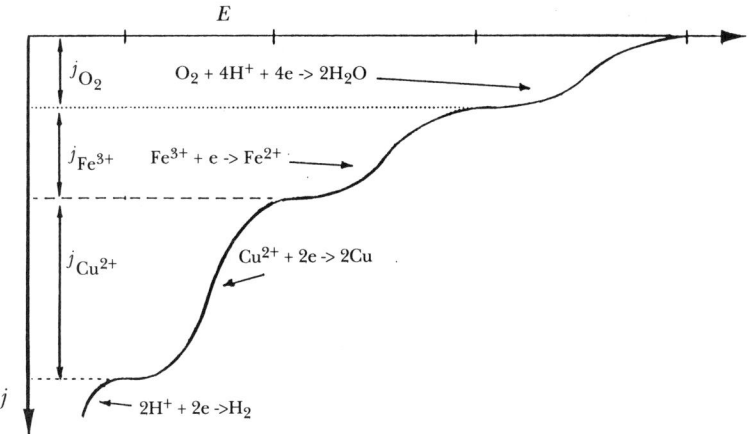

Figure 3   *Electrode potential characteristics for multiple reaction systems*

(ii)   the presence of other ionic species, metal or non-metal which may be cathodically active in the potential range of operation. For example Fe^{III} ions are reduced to Fe^{II} at potentials more anodic than for most metal ions and thus generally it will be under mass transport control. Other examples include the influence of thiosulfate ions in the recovery of silver from photographic fixing solutions and the formation of arsine.

(iii) the presence of redox species which may be chemically reactive with other dissolved ionic and solid species. For example metals which have been electrodeposited and which are not cathodically protected are free to corrode by reactions such as oxygen reduction or Fe^{III} ion reduction.

(iv)   the presence of other metal ions which also can be cathodically electrodeposited.

The performance of the system with a mixture of metal and other ions can be predicted from thermodynamics and the relative potentials of the species in the electrochemical series. This data can only indicate the possibility of selective metal recovery as in practice this is affected by the cathode substrate and the electrolyte composition. Thus it is usual to carry out experimental tests, to obtain appropriate current–voltage responses of the complete system. This data (polarisation curves), when obtained under well defined hydrodynamic conditions, enable the identification of conditions for selective (or otherwise) metal deposition in the absence of other competing reactions. A typical curve, which is representative of a system with a neutral electrolyte containing a mixture of Cu^{II}, Ni^{II} and Cd^{II} ions, is shown schematically in Figure 4. It is seen that selective Cu deposition is possible in the absence of hydrogen evolution, but that for selective Cd (and Ni) deposition the copper ion concentration will need to be reduced to very low values.

In electrodeposition systems of this type it is also advisable to carry out tests on the single metal ions in solutions and determine the influence of

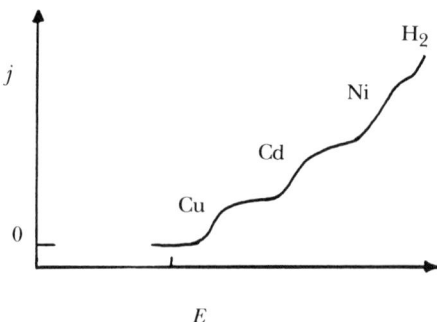

Figure 4   *Electrode potential characteristics for multiple metal deposition of Cu, Ni and Cd*

concentration in the kinetic range of metal deposition. In practice the system can be more complicated than suggested by the curves of Figure 4 due to the factors discussed above and also because of the following:

(i)   the nature and actual material form of the metal deposit will change during the course of the electrode reactions.

(ii) the evolution of hydrogen (and of oxygen) and reduction of oxygen can cause pH changes at the electrode surface, which can lead to changes in the chemical form of the deposit. This for example could be the formation of metal hydroxides *via* chemical reactions of the type:

$$M^{n+} + n\ OH^- \longrightarrow M(OH)_n \qquad (12)$$

      This clearly affects the quality of the metal deposit but could also lead to passivation of the cathode surface.

(iii) the possibility of the deposition of metal alloys.

(iv) a complex metal deposition reaction involving adsorbed intermediates.

6.3.1.2 Influence of the Anodic Reaction. In general the overall cell chemistry is influenced by the counter electrode reaction, and generally in the case of metal ion recovery a preferred reaction is the evolution of oxygen at a suitable inert surface;

$$e.g.\ at\ pH>7;\ 4\ OH^- - 4e^- \longrightarrow O_2 + 2\ H_2O \qquad (13)$$

   This reaction generally serves to cause the minimum in disruption to the cathode environment although clearly there will be changes in the pH of the solution. A drawback of this reaction is that most anode materials have relatively high oxygen over-potentials and thus result in relatively high energy consumptions. Other possible anode reactions are:

(i)   evolution of chlorine in a chloride electrolyte;

(ii) oxidation of solution species such as:

$$Fe^{2+} - e^- \longrightarrow Fe^{3+} \qquad (14)$$

$$SO_3^{2-} - 2\ e^- + H_2O \longrightarrow SO_4^{2-} + 2H^+ \qquad (15)$$

(iii) the reduction of hydrogen using porous gas diffusion anodes;

(iv) anodic dissolution of the anode metal.

   These reactions must not have a negative effect on the cathode reaction. In several of these cases, such as hydrogen and sulfite ion oxidation, there is a considerable reduction in the anode potential resulting in

improved energy consumptions. Some of these reactions, *e.g.* anode dissolution, are not compatible with metal recovery from dilute solutions and
would require the use of a cell separator. A suitable ion-exchange membrane can thus minimise problems in operation by the:

(a)   exclusion of aggressive electrolyte media, such as chloride, fluorobo-
      rate or nitrate ions;
(b)   exclusion of dissolved oxygen;
(c)   prevention of redox shuttle reactions;
(d)   provision of a useful anode reaction, in, *e.g.*, the anodic degradation
      or regeneration of printed circuit board etchants;
(e)   prevention of contact between non-adherent metal deposit and the
      anode.

6.4   SILVER RECOVERY FROM FIXING SOLUTIONS

During the processing of photographic media 'fixing' is used to remove
the unwanted silver halide produced during the development of the
image. The fixer solutions are usually based on aqueous ammonium thiosulfate and the following reaction typically occurs:

$$AgBr + n\,S_2O_3^{2-} = Ag(S_2O_3)_n^{(2n-1)-} + Br^- \tag{16}$$

with the predominant complex $n=2$.

Although the electrochemistry of practical photographic fixer solutions is complex, it is a favoured method for the recovery of the silver,
because it is a clean process, controllable, capable of dealing with a wide
range of solution compositions, offers a means of recycling fixer and also
gives a pure silver product. A simplified picture of the reactions, which
exert a significant effect on the electrorecovery of silver, has been presented by Walsh.[3] The required cathodic process is:

$$Ag(S_2O_3)_2^{3-} + e^- = Ag + 2\,(S_2O_3)^{2-} \tag{17}$$

At more negative potentials, the reduction of bisulfite ions may occur

$$6\,H^+ + HSO_3^- + 6\,e^- = HS^- + 3\,H_2O \tag{18}$$

or even the reduction of thiosulfate ions

$$S_2O_3^{2-} + 8\,H^+ + 8\,e^- = 2\,HS^- + 3\,H_2O \tag{19}$$

This may then lead to the formation of toxic hydrogen sulfide gas at
sufficiently high over-potentials.

The formation of silver sulfide may result from the reactions:

$$S^{2-} + 2\ Ag^+ = Ag_2S \tag{20}$$

$$HS^- + 2\ Ag(S_2O_3)_2^{3-} + 3\ H^+ \rightarrow Ag_2S + 4\ HS_2O_3^- \tag{21}$$

which will clearly have a detrimental effect on the electrodeposition in terms of:

(*i*)   a reduction in the current efficiency of silver recovery;
(*ii*)   contamination of the silver deposit;
(*iii*) contamination of the fixer solution by colloidal Ag_2S;
(*iv*)   catalytic decomposition of the fixer.

The anodic reactions, which may occur during electrodeposition, will depend on the value of the anode potential and the nature, and type, of electrode surface. The following major reactions are possible: oxidation of sulfite [eqn. (15)] may occur and at higher potentials, thiosulfate oxidation

$$2\ S_2O_3^{2-} = S_4O_6^{2-} + 2\ e^- \tag{22}$$

and the possible formation of polythionates *e.g.*

$$SO_3^{2-} + S_2O_3^{2-} = S_3O_6^{2-} + 2\ e^- \tag{23}$$

The above reactions of the sulfur species may be accompanied by reactions from material present in the developer carry-over, such as the anodic oxidation of hydroquinone to quinone, which may react further with sulfite or thiosulfate. In bleach fix solutions, used in colour film developing, the bleaching agent and its break-down products may instigate other reactions at the cathode, *e.g.* the reduction of $Fe^{III}EDTA-$ complex, which is readily reversed at the anode.

The cathodic behaviour of the reduction of silver in thiosulfate and bisulfite solutions has been studied by Weise *et al.*[4] and typical polarisation curves of the system are shown in Figure 5.

The reduction of bisulfite ions starts at potentials some 100 mV more negative than that for the reduction of silver, which imposes a limitation in the range of the deposition potentials of silver if current efficiencies approaching 100% are to be realised. This effect is intensified at low concentrations of silver. If high electrode potentials are used then, as well as a loss in current efficiency, the formation of Ag_2S may occur. This behaviour can limit the type of electrochemical reactor which can be used in practice–particulate electrodes, which can exhibit significant distributions in electrode potential, would have to be carefully designed. Cell designs, which have been used for the recovery of silver, include the cyclone cell and the rotating cylinder electrode and are discussed later in sections 6.5.1.1 and 6.5.1.3.

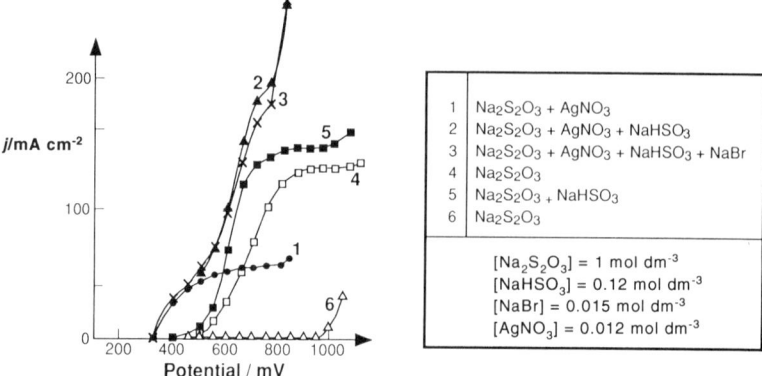

Figure 5   *Cathodic polarisation curves for the reduction of simulated spent fixing solutions*

6.5   APPLICATIONS OF METAL DEPOSITION SYSTEMS

In the treatment of dilute solutions of metal ions the attraction of a clean one-step recovery of the metal value is a compelling argument for its adoption. The parallel plate, tank electrolyser is widely used for the electrowinning of metals from high concentrations of ions, but is generally unsuitable for low concentrations of the order of 1 kmol m^{-3}. The major limiting factor is that the poor hydrodynamics and thus mass-transfer rates produce limiting currents, which are low and uneconomic. Thus for waste water applications considerable development into cell technology has occurred to improve the performance. Available cell designs can be put into two general categories (see Table 3):

(a)   direct metal recovery;
(b)   indirect recovery as a metal ion concentrate (concentrator cells).

Table 3   *Categories of cell designs for metal deposition*

Deposited metal	*Reusable anode*	*Metal on a combustible substrate*	*Metal ion concentrate*	*Metal flake or powder*	*Particulate metal*
Parallel plate	Chemelec cell in electroplating	Packed–porous cathodes	All three-dimensional *e.g.*	Rotating or vibrating cells	
— Recowin					
— Retec cell		— Retec cell	— HSA reactor		Moving beds
			— Environ-cell	RCE cells	Circulating beds
Concentric cylinder	Foam electrodes	— Enviro-cell	— ER cell	(batch or continuous)	Fluidised beds
— Cyclone cell				Goecomet reactor	
Continuous belt electrodes				— CEER cell	

In these categories both two- and three-dimensional electrodes are used and a range of reactors used for metal deposition is shown in Table 4, virtually all of which have seen some commercial application.

Table 4   *Commercial reactors for electrodeposition*

Reactor	Cathode	Method of product removal	Operating range (ppm)
Chemelec cell (BEWT Water Engineers Ltd)	Vertical mesh (or plate) in an electrolyte with fluidised glass beads	Discontinuous by manual scraping or re-use as anodes in electroplating	750
Concentric cell (Wilson Process Systems)	Inner surface of a cylindrical foil	Discontinuous – brass cathode may be furnace refined in the case of gold	100
Rotating electrode cell	Rotating cylindrical foil, or static cylindrical foil with rotating anode	Discontinuous by manual scraping or flexing	1–100
Eco-cell (Steetley Engineering Ltd)	Outer surface of a rotating cylinder (divided cell)	Continuous *via* powder formation	1–100
Reconwin (Ecotec)	Vertical plates, air agitation	Manual stripping of sheet	750
CEER cell (Finishing Services Ltd)	Vertical plate (divided cell)	Discontinuous metal powder or flake	> 10^3
Retec cell (Eltech)	Vertical metal (or carbon) foam electrodes in a tank	Discontinuous, cathode may be steel which may be chemically stripped or carbon (for precious metals) cathode may be re-used	0.1–100
Haraeus cell	Rotating Zn band	Continuous by manual peeling	1–100
SCADA Systems	Carbon fibre	Discontinuous by stripping	0.1–100
ER cell (ElectroCell AB)	Packed bed of carbon particles within a parallel plate and frame (divided cell)	Discontinuous *via* leaching	0.1–100
Enviro-cell (Deutsche Carbone)	Contoured packed bed of carbon (divided cell)	Discontinuous *via* leaching or vacuum removal of bed	0.1–100
FBE reactor (Billiton Research bv)	Fluidised bed of metal particles in a tube-and-shell type geometry (divided cell)	Continuous *via* with-drawal of grown particles	0.1–100
HSA reactor	Packed bed of carbon particles (divided cell)	Discontinuous by leaching	0.1–100
Porocell (EA Technology)	Carbon fibre high surface area	Incineration of carbon substrate	0.1–100
Geocomet	Rotating tubular or impact rod	Continuous as powder	1–100

The performance of reactors for waste water treatment is based on several criteria; energy consumption, space time yield (discussed in chapter 3) and the specific investment cost for a unit volume of solution treated. This can be quantified in the normalised space velocity, nsv, *i.e.* the volume of process solution whose concentration can be reduced ten-fold in a unit time in a unit volume of solution treated. For a batch reactor, operating with a mass transfer limited current:

$$nsv = 3600 \, I\text{CE} \, / \, (C_i - C_o) \, V_r \, n \, \text{F} \log \, (C_i/C_o) \, (\text{m}^3\text{m}^{-3}\text{h}^{-1}) \qquad (24)$$

where I is the total current, C_i and C_o are the initial and final concentrations of the active species and V_r is the active reactor solution volume.

6.5.1 Electrodeposition from Single Metal Ion Solutions

Plate and frame cells, in which the metal is deposited in a sealed chamber, are generally impractical for several reasons including difficulties in removing the electrodeposit. However, cell designs (*e.g.* Electro-cell) are used in conjunction with a packed or porous bed (of carbon) for some applications and are described in section 6.5.1.5.

6.5.1.1 Two-dimensional Electrodes. The simplest cells use vertical, plate or mesh, electrodes in tanks where turbulence is provided by using either inert fluidised beds as shown in Figure 6 (Chemelec Cell, BEWT Water Engineers Ltd) or air agitation (Reconwin cell), in conjunction with electrolyte pumping (see Figure 7).

In the Reconwin cell[5] a uniform curtain of fine air bubbles is directed across the face of the cathode to increase the mass-transfer coefficient in comparison to unagitated systems. The degree of enhancement in mass-transport rate is up to seven times the value for unagitated systems. Typically, at a concentration of 1 g dm^{-3}, the limiting current for say copper deposition is approximately 100 A m^{-2}. This equates approximately to a

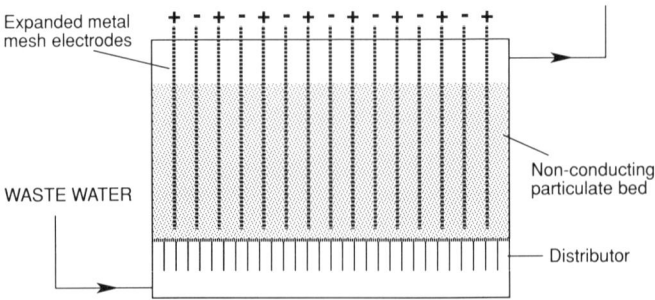

Figure 6   *The Chemelec cell using an inert fluidised bed*

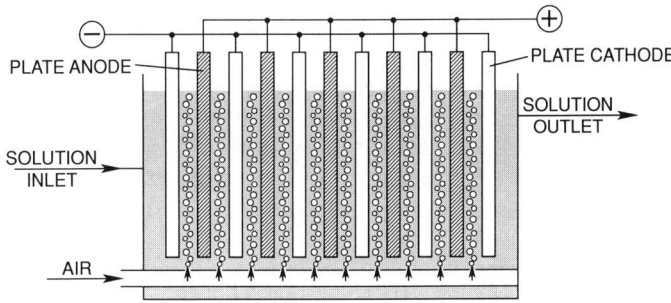

Figure 7 *The Reconwin cell with air agitation*

mass-transfer coefficient of the order of 4×10^{-5} m^3 s^{-1}. The cathode material is a permanent metal (stainless steel) blank of approximate size 0.6 m × 0.6 m, from which the electrodeposited metal is recovered as foil. The cell is used in an undivided configuration in which both sides of the cathode are active. To prevent the deposited metal enveloping the cathode sheet, which would make metal removal extremely difficult, a technique referred to as current shadowing is used to reduce the current density at the edge of the cathode sheets. This makes the metal readily peelable and eliminates the use of non-conductive edge strips, which often encourage the production of dendrites at the edge junction. The interelectrode gap in these cells is half that used in conventional electrowinning tank house cells, *i.e.* 2.5 cm, and serves to keep energy requirements to acceptable levels. The anode material used in the Reconwin cell is either a lead–calcium alloy, which has excellent stability characteristics in acid sulfate electrolytes, or more expensive DSA type coated materials. The cell is used in applications such as direct rinse water recovery and treatment and acid processing/etching bath treatment, recovering metals such as Cu, Cd, Ni and Zn. In practice the cell is operated with current proportional control in which the applied current is adjusted according to the on-going metal ion concentration, measured through suitable analysis. A disadvantage of the cell is that the electrolyte is saturated with oxygen from the air and thus cathodic reduction of this species could occur.

The Chemelec unit uses mesh or expanded metal electrodes from which deposited metal can be recovered by dissolution or by an electrorefining process. The cell benefits from the good mass transfer arising from the turbulence caused by the inert fluidised bed of ceramic (glass) beads. Electrolyte flows up the cell through a porous distributor, which serves to maintain a 'uniform' fluidisation. Typical applications are in the metal finishing, electronic, mining, photographic and electroplating industries to recover metals such as Cu, Ni, Zn, Cd, Rh, Pt, Sn, Au, Ag and brass.

Generally the performance of the Chemelec cell (and Reconwin cell) is not good enough to meet current metal ion discharge levels in a single step at acceptable costs and so the usual area of application is in the recycling of process solutions. For example removal of cadmium carried over into the first static rinse tank in plating bath operations maintains a low metal concentration in the tank and reduces considerably the carry-over of toxic metal into the running rinses. The cadmium cathodes are then recycled through the plant by stripping the metal off anodically in the plating bath itself. The performance of the cell keeps the concentration in the drag-out tank between 50–150 ppm Cd and the cell runs at up to 75% current efficiency.

The Chemelec cell comes as a standard module to accommodate a maximum of twelve double expanded mesh electrodes having a total cathode area of 3.3 m². The overall dimensions of the cell are 0.5 m × 0.6 m × 0.75 m. For smaller scale applications half-size and quarto-size modules are available. The major costs associated with the operation of the Chemelec cell are the electrical power and the cell capital. The electrolysis power required to recover 1.0 tonne of metal is of the order of 6 000–10 000 kWh. The capital cost of the cell depends on the scale of operation, but typically for the maintenance of the metal ion levels in static drag-out at around 100 ppm to effect recovery at 99% would cost (1993 prices) the following:

> £6 000-9 000   for 5 kg per week recovery
> £20 000   for 40 kg per week recovery
> £50 000   for 140 kg per week recovery
> £250 000   for 840 kg per week recovery (This is an application
> of the recovery of brass from cyanide chemistry.)

The Chemelec Cell is of considerable commercial interest in metal recovery and recycling and currently there are over 2 000 Chemelec systems in operation worldwide. The relative capital and running costs compare favourably with those of solvent extraction, ion exchange, reverse osmosis, evaporation and freezing.

6.5.1.2 Cylindrical Electrode Cells. The use of a concentric cathode cell[6] has been described for the recovery of precious metal on a small scale (Wilson process systems), from spent photographic solutions and from electroplating and refining wastes. The recovery of the metal is by scraping manually from the cathode or by furnace refining as in the case of gold, however, the design is not likely to find other applications.

6.5.1.3 Rotating Electrodes. Cells, which utilise rotation of the cathode, produce the metal deposit as a powder or a particulate. This enables the cell to be operated in a continuous mode when the powder is displaced freely from the electrode surface. There are two general devices using either cylinders or discs:
(a) Rotating cylinders — A rotating cylinder electrochemical reactor (RCER) for electrodeposition[7] can be applied as a single cell or as a cascade of cells (see Figure 8) on a common rotating shaft.

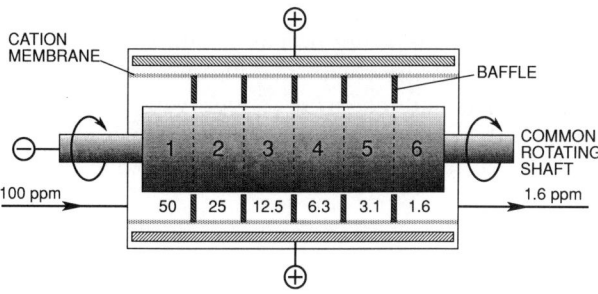

Figure 8   *The Eco-cell with cascade configuration*

This unit can operate on a continuous or batch basis, and notably amongst two-dimensional electrode cells it can reduce metal ion concentrations to the single ppm level. The cell is operated with the inner cathode cylinder rotating with tip velocities in the range 0.6–20 m s^{-1}. This results in the formation of Taylor vortices in the electrolyte flow that produce a high degree of agitation and high mass-transfer rates. The concentric anode is arranged to give an interelectrode gap of between 1–2 cm. The deposited metal is in the form of flakes or powder, which can readily be removed using an appropriate mechanical scraper. The loosely adherent surface deposit serves to increase the effective electrode area for electrodeposition and causes a significant enhancement in the rate of mass transport over and above that for smooth rotating cylinder electrodes. A typical mass-transfer correlation is:

$$k_1 = 0.079 \; u \; \mathrm{Re}^{-0.08} \; \mathrm{Sc}^{-0.644} \tag{25}$$

where u is the peripheral tip velocity and Re is the Reynolds number based on this velocity.

A reactor with a cathode of diameter 0.258 m and length 0.254 m (area 0.198 m^2), rotating at 750 min^{-1} and with an applied current of 500 A can reduce concentrations of Cu ions from 500 ppm to 4 ppm in a batch recirculation mode. In a single pass operation the reactor operates with energy consumptions of approximately 8.3 kWh kg^{-1}, with normalised space velocities of *ca.* 19 m^3 m^{-3} h^{-1} and with exit concentrations of 100–175 ppm. By using a cascade of rotating cylinder electrodes in series the final exit concentration can be reduced to a few ppm in continuous operation. Although the energy consumption appears high in comparison to plate electrowinning cells (2 kWh kg^{-1}) it should be remembered that there are much higher electrolyte concentrations used in the latter systems (a factor of 1000 or more higher).

A second device based on the movement of a cylindrical cathode is the impact rod or rotating tubular reactor.[8] This cell uses a set of rod cathodes positioned in the annular channel formed between concentric anodes and a set of cathode guide rails. Movement of rods serves to dislodge the metal during deposition. The rotating tubular bed reactor is used for gold recovery in the electroplating industry. This cell design is a variation on the particulate electrode design with mechanical bed agitation combined with forced electrolytic flow bringing about improved mass transfer as well as separation of deposited metal from the cathode substrate.

(b) Rotating disc electrodes — The rotating disc electrode in the form of the pump cell (see chapter 3) has been considered for the recovery of metal (as powder). In a particular design it consists of a set of discs mounted on a common rotating shaft. The cell has been used commercially in the treatment of photographic fixing solutions. The use of a rotating disc electrode for the selective removal of metals from mixed metal ion solutions using a disc cathode continuously coated with a thin film of mercury is discussed in section 6.5.2.

6.5.1.4 Three-dimensional Electrodes. Three-dimensional electrodes can reduce metal ion concentrations to 100 ppb and less. The applications of three-dimensional electrodes can be subdivided into three areas depending upon the subsequent form of the metal recovered:

(i)   where the metal is produced as a freely moving particulate form.
(ii)   where the metal ion is deposited into an electrode structure from which it is recovered by dissolutuion (chemical or electrochemical) *i.e.* concentrator cells.
(iii) where recovery is by incineration of the combustible substrate.

The fluidised-bed electrode (FBE) in the form of a cylindrical unit has been used commercially. The main characteristic of the FBE (see Figure 9) for electrodeposition is the ability to continuously produce a solid

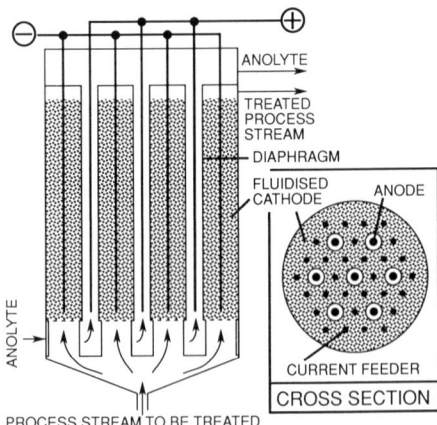

Figure 9   *Fluidised-bed electrode cell; cylindrical configuration*

metal deposit by the continuous addition and withdrawal of particulate substrate. Some of the main problems of early fluidised beds; particle agglomeration, metal deposition on the feeder and membranes, damage to membranes and difficulty in scale-up have been overcome in this cell design and technology developed by AKZO, which has been acquired by Billiton Research bv.[9] The design gives uniformity of fluidisation, uses cylindrical ceramic separators, and has improved potential distribution due to strategic positioning of the cathode feeders. The success of this cell owes much to the development of the ceramic diaphragm, which has high chemical and mechanical stability. The latter allows the anode compartment to be subjected to over-pressure and thus prevent contamination by permeation from the catholyte. The diaphragm has a low permeability and enables operation of the cell with a large height, thus enabling a high depletion of the metal ion per pass through the cell. The smooth surface of the diaphragm prevents adhesion of the particulate material and subsequent scaling. Applications of the FBE have been widely researched in the deposition of metals such as Ag, Cu, Cd, Co, Ni, Zn and Hg. The reported performance of the system has been compared economically against non-electrodeposition processes in the following applications:

(i)   copper removal from a chlorinated hydrocarbon waste stream, of composition 100 ppm Cu, 5–30 g dm^{-3} chloride ion, pH 0–1, chlorinated hydrocarbon 0-16 g dm^{-3} and suspended solids 0.5–5 g dm^{-3}. With a 1.2 m high, 0.35 m diameter unit the copper concentration was reduced to less than 1 ppm in a single pass with a current efficiency of 70%. Economic comparisons with a precipitation–filtration treatment, which formed $Cu(OH)_2$ sludge, for a 15 m^3 h^{-1} plant capacity, showed the FBE system to have both investment and operation costs some $200 000 per year lower (1978 costs).

(ii)   The recovery of Hg from a brine stream in a chlor-alkali plant of composition, 100 g dm^{-3} chloride ion, pH=1, 5 ppm Hg. The cell used a Cu bed to deposit the Hg as an amalgam, and reduced the Hg content to 50 ppb.

(iii)   The separation of Cu from electrolytes containing either Ni or As. At a Ni concentration of 50 g dm^{-3} the Cu concentration was reduced from 2000 ppm to 0.1 ppm and the Cu deposit was 99% pure. In the presence of As (5 g dm^{-3}) the Cu concentration can be reduced to < 0.1 ppm before the formation of As hydride occurs. In this latter case the reduction of the Cu concentration to 100 ppm gave a deposit which contained only 0.3% As.

(iv)   Purification of sulfuric acid solutions by the removal of Cu ions from a concentration of 300 to 5 ppm, at the rate of 1 kg h^{-1} of Cu metal. This operation replaced ion exchange and gave a reduction of $200 000 in the annual operating cost, mainly as a result of the reduction in the amount of acid and ammonia, used in the neutralisation of the ion-exchange solution.

(v)   Purification of a Zn electrolyte of composition, 800 ppm Cu, 800 ppm Cd and 150 g dm^{-3} Zn. In sequential FBE, the Cu concentration was reduced to < 0.1 ppm and the Cd to < 1 ppm and the electrodeposits were 99% Cu and 90% Cd, respectively. The current efficiencies of the deposition were 70 and 40%, respectively. The replacement of Zn dust cementation, for the purification of the electrolyte for a 150 000 tonne per annum Zn electrowinning factory, with the FBE was economically favourable. The investment in the FBE was $2×10^6 lower, and the net annual operating profit was $2.8×10^6 higher, than for the cementation process.

The use of fluidisation is not the only means to achieve continuous particle movement in attempts to achieve continuous metal deposition. Particle circulation or spouting, vortex flow and a moving bed have been used (see chapter 4). An alternative moving electrode is the tumbled bed, which consists of a rotating chamber (see Figure 10) that holds the appropriate particulate material. Through rotation of the chamber or drum the particulate substrate for metal deposition is turned over and thus agglomeration or fusing of the deposited metal is avoided. The operation is similar to that of a barrel electroplating unit.

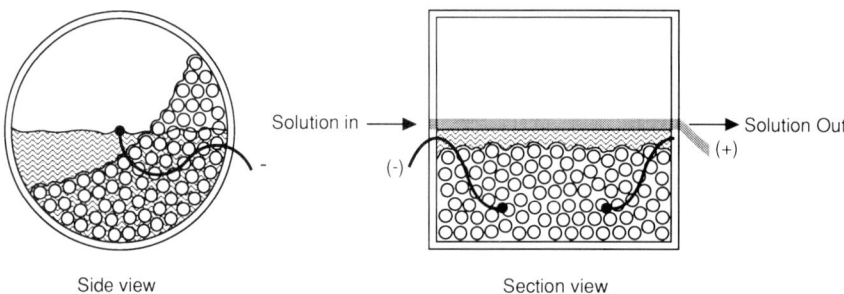

Side view   Section view

Figure 10   *The tumbling-bed electrode*

An adaptation of a particulate bed cell, which enables the production of metal onto a particulate substrate without 'fusing' the bed together, is the pulsed percolated porous electrode (pppe). This cell uses a regular pulse of upflowing electrolyte to unsettle a particulate bed during electrodeposition, preventing the metal deposit solidifying the bed into a fixed porous mass. The bed therefore can be considered to be operating intermittently between a fixed and fluidised state. An industrial prototype has been built in France by the Martineau Co. who specialise in the recovery of photographic fixing salts. The unit has also been tested on a cadmium effluent.[10]

6.5.1.5 Fixed- or Porous-bed Electrodes. One of the first commercial applications of three-dimensional electrodes in metal recovery was in the primary gold winning of dump leachings of low grade gold ores, or old mine tailings, with cyanide solutions. Recovery of the gold by carbon adsorption and electrowinning has led to the development of special cells (Zadra cells) in order to cope with the low concentration of gold in the eluate from active carbon stripping. Typically the electrolyte concentrations are between 0.5–2 g dm^{-3} gold, which is too dilute for direct electrowinning. The cells contain anodes of stainless steel and cathodes of steel wool, contained in pervious polypropylene baskets, which provide a large surface area to volume ratio for gold deposition. After sufficient deposition of the gold, onto steel wool, the cathodes are removed and smelted to produce gold bullion. Improvements in the original prototype cell have been made by using a stainless steel tube anode electrolyte distributor to eliminate channelling of the electrolyte flow. The application of a rectangular design is reported to be in current industrial use.

Early fixed- or porous-bed electrodes for electrodeposition processes used finely divided graphite or carbon chips, which act as both anode and cathode (see ref. 11). The feed liquor enters between the electrodes with about 99% of the flow passing through the cathode and copper is plated out onto the carbon surface. When the cathode becomes plugged with copper the cell polarity is reversed and the copper is anodically dissolved to yield a concentrated solution. This type of system was studied for the electrodeposition of copper as a substitute for liberator cells, in copper tank house bleed streams in South Africa. The Kennecott Copper Corporation (USA) developed a thin disposable particulate coke-bed cathode for the same purpose. As the dilute acidic leach liquor passed through the cathode, copper is electrodeposited in a thin layer forming a solid copper–coke sheet. This product is smelted in a furnace to separate the coke and then molten copper is fire refined to produce a saleable product. Economic assessment of this system was favourable.

Although carbon is a cheap and relatively stable substrate for cathodes in three-dimensional electrodes, metal electrodes are used in two-cell designs based on spiral wound modules.

1   The extended surface electrolysis (ESE) cell for removal of heavy metals from waste streams has been developed by DuPont. This is of sandwich construction consisting of a packed stainless steel mesh cathode, separator and a screen anode rolled up like a swiss roll. These cells can operate with fluid velocities of 1–10 cm s^{-1} and with apparent current densities of 10–200 mA cm^{-2} at the separator. In field trials a single ESE cell removed 75–90% of copper from a waste stream containing 5–100 ppm copper. This type of cell is a concentrator device for metal ions and in this application the copper was recovered from the cell by leaching with a H_2SO_4–H_2O_2 mixture.

2   The swiss-roll cell from ETH in Switzerland. Typical applications
      are: *(i)* the reduction in the Cu ion concentration from 5–0.05 mol
      m^{-3} in a 1 mol m^{-3} sulfuric solution with a 57% efficiency and with a
      cell voltage of less than 2.0 V; *(ii)* the recovery of Ag from a used
      fixer solution down to a concentration of 0.1 ppm; *(iii)* the removal
      of Hg from waste stream using a Cd electrodeposited cathode,
      which fixes the Hg as an amalgam; *(iv)* the treatment of zinc cyanide
      plating bath rinse waters which contain the $Zn(CN)_5^{2-}$ complex ion.
      In this application the anode reaction was used to oxidise the
      cyanide ion to assist the deposition of the Zn metal by freeing the
      divalent Zn ion.

6.5.1.6 Commercial Applications. In the commercial packed and porous
electrodes listed in Table 4, three use particulate carbon or carbon fibre
cathodes, and the fourth cell uses a reticulate foam of carbon (or metal).

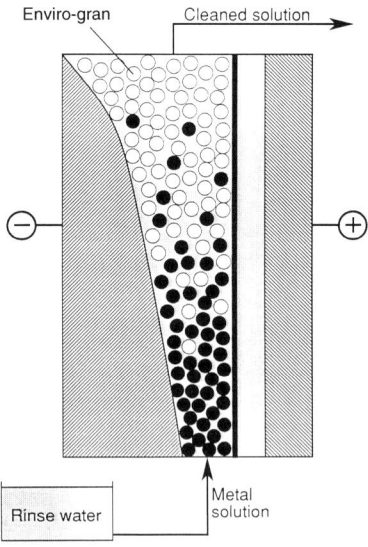

Figure 11 *Fixed-bed electrode; Enviro-cell*

A packed-bed cell called the Enviro-cell, (see Figure 11), developed by
Deutsche Carbone[12] uses a tapered cathode compartment filled with elec-
trographite granules, specially impregnated and screened to produce a
definite particle size. The profiled packed bed has been designed for systems where the residence time of electrolyte in the cathode is high, and
thus with low metal ion concentrations (or for single pass operation), the
conversion of ion concentration can be significant.

Table 5   *Applications and performance of the Enviro-cell* [a]

Application	Metal	Flow /m³ h	Conc.-in (ppm)	Conc.-out (ppm)	Energy cons. /kWh m⁻³	Anode size /m²
Chlor-alkali electrolysis	Hg	3.5	1	0.05	0.16	20
Production of cellulose acetate	Cu	20	20	1.9	0.08	40
Dye stuff production	Cu	6	400	2	4.0	90
Dye stuff production	Hg	2	4	0.05	2.5	15
Oxychlorin process (under construction)	Cu	22	6	0.5	0.15	40
Production of fabrication material for plastics (under construction)	Cu	3	50	0.5	1.8	26

[a]   K.J. Muller, *Dechema Monograph*, 1991, vol. 123, p. 199.

The Enviro-cell, has been applied to several industrial applications (*e.g.* plating and photographic) and typical performance data of the system is presented in Table 5 for both Cu and Hg recovery. The reactor has also been used for the cathodic treatment of effluent containing both Cu ions and HOX [chlorinated organics (*e.g.* epichlorohydrin and chloro-ethanes)]. Typically at pH 10 and a temperature of 48 °C the HOX con-centration is reduced from *ca.* 20–1 ppm and the Cu to 0.1–0.2 ppm.

6.5.1.7 Rectangular Parallel Plate Electrolysers. The Electrocell AB range of electrochemical reactors (see chapter 4) has been partially redesigned as the ER cell, a flow-by three-dimensional reactor. Active packing materials, which have been used, are porous graphite, reticulated or felt metals and carbons and regular or irregular packings. In a typical design of cell the bed is 1.0 m high, 0.25 m wide and 1.08 cm thick. The relatively narrow cell design is used to minimise channelling of electrolyte. Electrical con-tact to the cells is *via* a titanium grid.

The estimated performance[13] of the cells for a 1 m³ h⁻¹ flowrate of effluent is indicated in Table 6, for a single stack with between 5–10 cath-odes. Several metals, such as Cu, Zn and Ag, can be removed from solu-tion to below the 1 ppm level with reasonable current efficiencies and energy consumptions — cell voltage being typically in the range 2–3 V. These cell voltages are typical values experienced with electrolytes having

Table 6 *Performance of the ER cell*

Metal	Inlet conc. (ppm)	Solution Characteristics	pH	Particle size/mm	Flow-rate /dm³ min⁻¹	Cell voltage/ V	Cell current/ A	Outlet conc. (ppm)	Current efficiency (%)
Cu	67	CuSO₄/H₂SO₄ $\kappa = 0.030\ \Omega^{-1}\ cm^{-1}$	1.4	1.0–1.4	2.0	2.2	13	0.03	52
Cu	26	Complexed with quadrol $\kappa = 5.3 \times 10^{-3}\ \Omega^{-1}\ cm^{-1}$	2.1	2.0–3.15	1.0	2.3	4	0.6	35
Zn	44.6	ZnSO₄/K₂SO₄/H₂SO₄ $\kappa = 0.019\ \Omega^{-1}\ cm^{-1}$	5	2.0–3.15	1.0	3.2	8	0.44	27
Ag	910	Thiosulfate fixer solution	3.6	2.0–3.15	1.0	0.5–2.0	2	0.7	20

conductivities of 0.006–0.03 ohm⁻¹cm⁻¹. The metal loading of the electrode, which can be achieved before the cell pressure drop becomes too high, is up to 0.2 kg. Metal recovery from the electrode is achieved by current reversal or using a chemical etchant.

6.5.1.8 Carbon Fibre Electrodes. A recent development to increase the available surface area for metal deposition has been cathodes based on carbon fibres (see Figure 12) or woven cloth. It is claimed that a significant increase in mass-transfer rate with simultaneous controllable and uniform electrode potential over the entire electrode surface can be obtained.

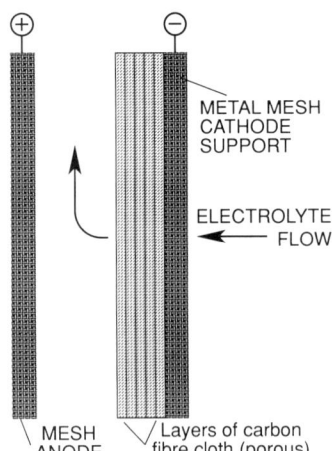

METAL MESH
CATHODE
SUPPORT

ELECTROLYTE
← FLOW

MESH
ANODE

Layers of carbon
fibre cloth (porous)

Figure 12 *HSA porous carbon reactor*

The carbon fibres are typically between 5–10 μm in length and 1 g of fibre has typically a surface area of 2.6×10^6 cm^2, providing an effective electrode surface area approximately 12 000 times the geometric cross sectional area. The carbon fibre electrode has an approximate interfibre distance of 290 microns and porosities up to 90–97%, depending upon the compression of the material and corresponding specific surface area. The metal is recovered from the bed by an electrorefining process. A feature of the system is the use of controllable power supplies[14] to fix the electrode potential of operation and minimise possible side reactions. Typical commercial applications of these fixed-bed electrodes are in the recovery of cadmium and copper from the drag-out and rinse waters of electroplating baths. The economics of this process are considered to be very favourable for most metals and a detailed analysis of the costs of Ag recovery is given[14] which is compared against a precipitation method using borohydride.

A relatively new carbon fibre electrode is the POROcell from EA Technology UK. This is a cylindrical cell configuration that operates in the flow-through mode using a suitably designed electrolyte flow distributor. The fibre cathodes are in the form of cartridges, which are readily removed after they become loaded with the metal. These are then furnace refined to recover the metal. The claimed advantages of carbon fibre cathodes are:

1   Current legislated discharge levels are achieved.
2   No sludge is generated.
3   Very low operating cost.
4   No consumable reagents required.
5   Easy replacement of cartridge cathodes.
6   Applicable to all plateable metals.
7   Eliminates conventional chemical precipitation.
8   Generates a saleable metal.
9   Is economically attractive.

6.5.1.9 Reticulated Vitreous Carbon. A promising high surface area material for three-dimensional electrodes is reticulated carbon (RVC), or metal. This material is used in the Retec cell, which is an undivided monopolar connected reactor utilising a bank of vertical foam cathodes (60 × 50 cm) with interspersed DSA anodes. Up to 50 electrodes can be used in the cell which has provision for electrolyte recirculation and also gas sparging to improve electrolyte mixing. With 50 foam cathodes the cell would typically operate at currents of *ca.* 500 A.

The specific surface area of RVC increases with the material grade (ppi), with a 100 ppi material giving values of specific mass-transfer coefficients (k_1a) of 0.07–0.23 s^{-1}. The material has a high porosity enabling, in principle, a large amount of deposition in the structure. It has been demonstrated,[15] using a 0.4 m long cathode, that the concentration of

Cu^{II} (in sodium sulfate, pH 2) can be reduced from 10 ppm to approximately 0.1 ppm in a single pass even in the presence of air saturated solutions. Scale-up estimates indicate that a 1 m × 1 m × 0.012 m sheet of RVC could remove 99% of the metal from an effluent flowing at a rate of 1 m^3 h^{-1}. The normalised space velocities of this system are quoted at 50–450 m^3 m^3 h^{-1} and the normalised volumetric power consumption is in the range 1.3–6.4 kW m^{-3}. However, RVC is relatively expensive and fragile which is likely to impose limitations in its use.

6.5.1.10 Metal Sludge Production. Recently a cell has been described that uses a porous graphite or carbon cathode operated in a flow-through mode much like the operation of a pre-coat filter.[16] In one application Cu is removed from waste water in the form of a sludge of metal (50–70%) containing carbon. The arrangement can deal with low metal ion concentrations and high current densities *e.g.* 1 ppm and 1000 A m^{-2}.

6.5.1.11 System Comparisons. Overall the performance of different three-dimensional electrodes is similar, as borne out in Table 7, which gives a brief comparison of the energy consumptions and final effluent concentra-

Table 7   *Comparison of the performance of metal recovery cells*

Type	Typical max anode current density A m^{-2} for a dilute (250ppm) conductive solution	Effective conc. limit (level where current efficiency is 25%) /ppm	Energy consumption kWh kg^{-1} Cu at 150 ppm Cu, 100 g dm^{-3} H_2SO_4
Electrodialysis		30	>20
Spiral wound cell	500	0.5	8–9 [a]
DuPont ESE cell	500	0.5	8–9 [a]
HSA	2000	0.1	7–8 [a]
2D Cathode (mesh)	50–100	10	3.5–4.5
Rotating cathode: coherent deposit	300	10	4.5–6.0
Rotating cathode: powder production	2000	<1	6.5–7.5
Fluidised: non-conducting particles	100	10	3.5–5.0
Fluidised-bed electrode: conducting particles	1500	<1	6.0–7.0
Particulate-bed electrode cell: conducting particles (recirculating bed cell)	1500	<1	5.0–6.5

[a] Includes energy for winning from concentrated solution.

tions achieved by the cells for Cu recovery. Like all sectors of the process industries there will usually be more than one manufacturers device that will do the job effectively. Clearly a limitation of the concentrator reactors is that they can essentially only recycle the metal as a more concentrated solution. Thus they are competing with procedures such as ion exchange which is discussed in section 6.5.5. However the use of cheap combustible carbon substrates is an attractive alternative in many applications.

6.5.2   Metal Separations from Mixed Metal Ion Solutions

Aqueous solutions containing mixtures of metal ions occur in many process industries. There are in certain cases limitations to selective electrodeposition from mixed metal ion solutions which will require the use and integration of other separation methods. With some mixed metal ion solutions the difference in equilibrium potentials for the reactions can indicate that selective electrodeposition is probable, *e.g.* with solutions of Zn or Ni and Cu in dilute H_2SO_4 solutions, with the fluidised-bed electrode. Another example is the use of a flow-by carbon felt electrode in the extraction of Cu, Pb and Hg in a batch recirculating reactor.[17] The felt was of fibre diameter 1.1×10^{-3} m, with a porosity of 0.91 and a specific area of 2200 m^2. With a supporting electrolyte of $NaNO_3$ at pH 2 the selective deposition of Cu from a lead–copper ion solution in a divided cell with an anion-exchange membrane is achieved. By fixing the initial electrode potential to a low value of -300 mV [*vs.* standard calomel electrode (SCE)] the deposition of Cu occurs without lead deposition. When the residual Cu concentration reaches the prescribed level the potential is increased to -600 mV to remove the Pb ion. Overall the system reduced the concentration of Cu from an initial 50 ppm value to 0.05 ppm and then reduced the concentration of Pb from 100 to 1 ppm (see Figure 13).

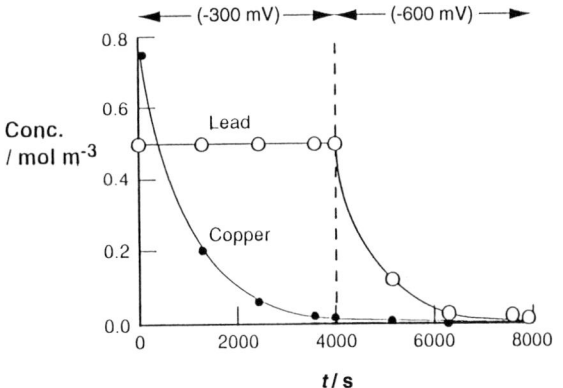

Figure 13 *Extraction of Cu and Pb with a carbon felt electrode*

6.5.2.1 The effect of the cathode substrate. As discussed in section 6.4 the difference in decomposition potentials for metal deposition also depends on the metal cathode as well as the solution composition and temperature. Although generally once a metal is electrodeposited the substrate will take on the characteristics of that deposit, there are a few instances where the influence of the substrate may be used to advantage. One example is the use of Hg, applied in the form of a rotating disc electrode, for the selective removal of metals from mixed metal ion solutions.[18] The disc cathode is vertical and is continuously coated with a thin film of Hg. The metal is extracted from the aqueous solution in the Hg film, which is then continuously removed form the disc surface. The Hg exits at the bottom of the cell without building up a pool. The electrowon metal is recovered from the metal amalgam in a second electrorefining cell equipped with a rotating anode. In this cell the metal is oxidised and recovered as a solid metal.

The cell has been used for the selective separation of Ni and Co. This is possible as although the difference in the reversible potentials is only 37 mV, the difference in measured deposition potential for Ni and Co is 170 mV. Figure 14 is a proposed process for the selective recovery of Cu, Ni and Co from an acid sulfate or chloride solution which also contains Fe. A feature of this scheme is that the metals are recovered in two stages:

(*i*)   the individual metal is first extracted at a high current density to a moderately low concentration *e.g.* Cu at 0.1 g dm^{-3};

(*ii*) to avoid the use of uneconomic low current densities the metal is then recovered together with a second metal with a deposition

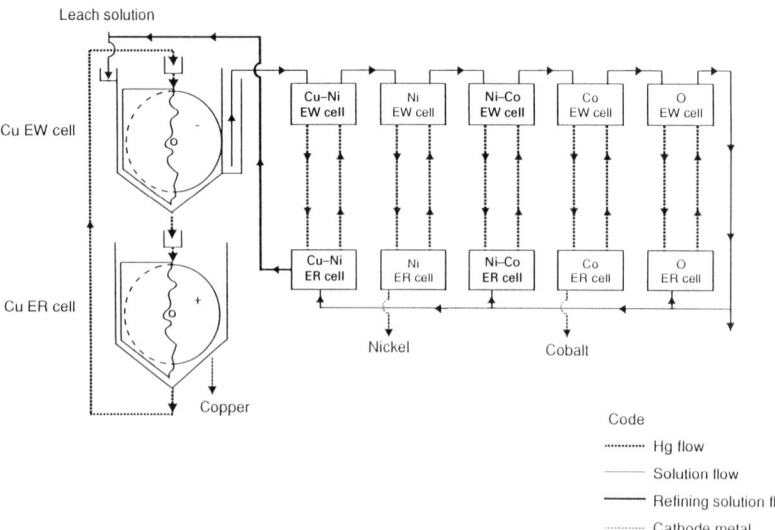

Figure 14 *Separation of Cu, Ni and Co on a Hg-coated disc electrode*

potential close to that of the primary metal of interest *e.g.* Cu–Ni. The second metal is then electrowon as a single metal.

Overall in this process there are six electrowinning–electrorefining stages for the three metals to be won. The process has also been demonstrated for the recovery of Cu, Ni, Co and Mn. The problem of residual Hg in the aqueous solution can be minimised by using a mercaptan ion-exchange resin to reduce the concentration to <2 ppb.

The use of Hg as a cathode is also applied in the form of coated metal wools for the removal of Hg from gold cyanide solutions formed during Au-bearing ore processing. Although there are technical advantages to the use of Hg generally its application as a cathode is not considered attractive for environmental reasons.

6.5.3   Combined Electrochemical Processes

In efforts to improve the process efficiency of the electrodeposition alternative anode reactions have sometimes been employed. Cells such as the Enviro-cell, the swiss-roll cell *etc.*, have been used in applications that combine electrodeposition and anodic oxidation, typically that of cyanide, and some organics. Other examples include the following:

6.5.3.1   Tin Recovery from Sludge.[19] In the electrodeposition of tin from a fluoroborate electrolyte, a sludge containing approximately 50% tin is formed. This sludge results from the anodic oxidation of Sn^{2+} ions and comprises a mixture of various Sn^{2+} species and some bath additives. The tin can be recovered by electrodeposition of a leachate of the sludge obtained by reacting with concentrated hydrochloric acid.

$$H_2SnO_3 + 6\ HCl \rightarrow H_2SnCl_6 + 3\ H_2O \qquad (26)$$

After dilution of the leachate to give a Sn^{IV} ion concentration of 50 g dm^{-3}, the tin can be electrodeposited onto steel plate in tank electrolysers. The electrorecovery process can be made 'continuous' by adding concentrated tin solution to the used electrolyte. The anode reaction in this process is chlorine gas evolution (on graphite electrodes) which is absorbed in sodium hydroxide solution to form sodium hypochlorite, which is used in another part of the plant. Tin is recovered as a 3 mm thick compact deposit with a current efficiency > 90%.

6.5.3.2 Treatment of Etching Solutions. Etching involves the dissolution of metal, typically copper, from the board by an etching–oxidation solution. Several etchants can be used including iron(III) chloride and copper(II) chloride. During operation, as the copper dissolves, the effectiveness of the etchant solution progressively falls due to the consumption of the etchant by the reaction, which in the case of a copper etchant is:

$$Cu + Cu^{2+} \rightarrow 2\ Cu^+ \tag{27}$$

The etchant must be regenerated in the case of copper(II) chloride or discarded in the case of iron(III) chloride. The traditional method of regeneration is by oxidation of the copper(I) ion, formed by dissolution, using hydrogen peroxide. This process produces an excess of etchant, which must be stored, and was typically disposed of by precipitation as copper oxide and subsequent landfill.

In the continuous electrolytic regeneration of copper(II) chloride etchant the copper(I) chloride is oxidised anodically in a cell (see Figure 15) while the cathode of the cell recovers the copper as a solid flake deposit. The cell is divided by a membrane (either PVC-based, or cationic-exchange) that limits the transport of copper ions, which are in fact complexed, probably mainly as $CuCl_3^{2-}$. To maintain the required rate of deposition of Cu the membrane transport rate is augmented by introducing the flow of etchant from the regeneration circuit.

An economic analysis of the process realised a two year payback on the capital investment. In addition the process offered the following advantages:

- improved quality of the printed circuit board
- virtual elimination of disposal costs
- the etching solution is at its optimum composition
- Cu is recovered in a high value form
- no additional hazardous chemicals need be handled

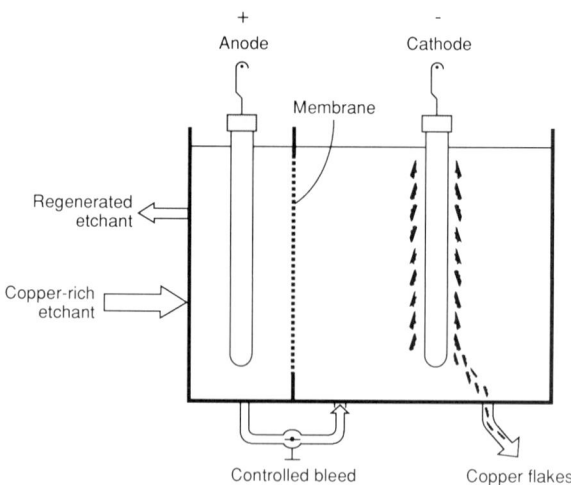

Figure 15 *Continuous electrolytic regeneration of copper(II) chloride etchant*

In the case of the alternative etchants, such as iron(III) chloride, continuous regeneration is also feasible.

6.5.3.3 Simultaneous Tin Electrodeposition and Cr^{III} Reduction.[20] A reactor with a Sn-coated Ti-mesh cathode immersed in an inert fluidised bed to enhance mass-transport rates has been used for the simultaneous electrodeposition of Sn^{II} with electroreduction of Cr^{III} to Cr^{II}. In the envisaged electrohydrometallurgical process for tin extraction by reductive decomposition of SnO_2 with Cr^{II}, both reduction processes can be telescoped to produce a saving in capital and cell component costs. The reactions in this process can be summarised as:

$$Cathode \qquad 4\ Cr\ Cl^{2+} + 4\ e^- \rightarrow 4\ Cr^{2+} + 4\ Cl^- \qquad (28)$$

$$2\ SnCl_4^{2-} + 4\ e^- \rightarrow 2\ Sn^0 + 8\ Cl^- \qquad (29)$$

$$Anode\ (e.g.) \quad 2\ H_2O \rightarrow O_2 + 4\ H^+ + 4\ e^- \qquad (30)$$

Reactor:
$$SnO_2 + 4\ H^+ + 4\ Cr^{2+} + 4\ Cl^- \rightarrow Sn^0 + 4\ CrCl^{2+} + 2\ H_2O \qquad (31)$$

with

$$Sn^0 + 2\ H^+ + 4\ Cl^- \rightarrow SnCl_4^{2-} + H_2 \qquad (32)$$

$$2\ Sn^0 + 4\ H^+ + O_2 + 8\ Cl^- \rightarrow SnCl_4^{2-} + 2\ H_2O \qquad (33)$$

The electrodeposition of tin when studied separately produced an adherent deposit with current efficiencies >90% at concentrations as low as 1 mol m^{-3}. The combined reduction process gave overall current efficiencies >90%.

6.5.3.4 The Use Of Hydrogen Anodes. The replacement of oxygen evolving anodes with hydrogen gas oxidation anodes is driven by the goal of reduced energy consumption. There are several technical areas of application which include their use in the electrowinning of zinc, the treatment of pickling solutions and copper(II) chloride etching solutions.

For example traditional zinc electrowinning uses Tainton (Pb with 1% Ag) anodes for the evolution of oxygen from sulfuric acid electrolyte. With a gas fed hydrogen anode the difference between the potential of the oxygen evolving anode and the hydrogen anode is 1.23 V. During polarisation of the electrode this potential difference actually exhibits a maximum value with increasing current density due to the superior electrocatalytic nature of the hydrogen anode. The working cell voltage is typically in the range 1.8–2.0 V constituting a significant (*ca.* 50%) saving in electrical energy. Other benefits that arise from the use of hydrogen anodes are a reduction in cooling requirements and a reduction in ventilation costs associated with the generation of acid mist from oxygen evolution. Clearly the use of hydrogen anodes must be balanced against the higher invest-ment costs, the cost of hydrogen and possible greater replacement costs of the anodes, which are susceptible to poisoning, allowing however for the recovery of the platinum electrocatalyst. An alternative cheaper electrocat-alyst material, tungsten carbide, is however also suitable for Teflon-bond-ed hydrogen anodes and is currently of some technical interest.

6.5.3.5 Combined Electrodeposition and Synthesis. An alternative to oxygen evolution is to use an anodic electrosynthesis reaction to form commer-cially valuable products. In this way the power and the invested capital for the cells for the individual processes are effectively halved. There are potentially many possible combinations of reactions, with interest shown in the coupling of zinc electrowinning with the oxidation of Mn^{II} to MnO_2 (E^o=1.21 V) and the coupling of copper and zinc electrowinning with the manufacture of sodium perchlorate from the oxidation of sodi-um chlorate. There are several key factors which will decide the possible success of this type of operation:

(i)   the current densities of the operation should, for parallel plate operation, be matched as closely as possible. The use of high surface electrodes may enable a suitable matching of overall currents of operation.

(ii)   the membrane must prove to be an effective barrier to components which can upset the operation of the opposing electrode reactions.

(iii) conditions of operation such as temperature, pressure and pH should be compatible.

In the case of perchlorate manufacture coupled to Zn or Cu elec-trowinning, operating temperatures are generally comparable and can be set in the range 30–60 °C. The current densities for efficient operation of the two electrode reactions are different, and thus the electrode areas are different, in the case of Cu electrowinning the cathode to anode ratio is between 5–3.

6.5.4   Integrated Separation Processes

Limitations in the effectiveness of metal deposition can arise from the presence of other ionic species, the presence of organic or neutral species and also when relevent ionic species are in low concentrations. To improve the efficiency a combined, or 'integrated', approach to metal recycling can be used which incorporates other chemical and physical separation processes in conjunction with electrodeposition. A simple example is a combination of ion exchange (resins or membranes) with electrodeposition to achieve selective metal ion separation.

The combined approach of electrodialysis and electrodeposition is realised in the process termed electro-electrodialysis (EED) shown schematically in Figure 16. In this, for example anions are transported (selectively) across an ion-exchange membrane from the catholyte to the anolyte. At the cathode appropriate cathodic reduction can take place. A similar performance is also possible with the use of supported liquid membranes containing suitable metal complexing agents which act as carriers.

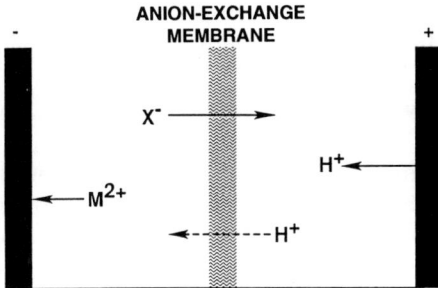

Figure 16 *Electro-electrodialysis*

A major factor in the application of integrated systems is the content of the liquors or waste waters to be treated *i.e.*

1   Is the solution a mixture of dissolved and suspended solids, metallic and other ions, organic species and other components such as oils?
2   Does the water contain a mixture of similar materials such as a range of metallic ions?

In the first of these areas a combination of separation methods is likely to be required to process the different materials. Incineration of these types of liquors can clearly be considered and energy consumption can be reduced by introducing stripping and evaporation prior to incineration. More traditional methods would involve the treatment in several stages with for example:

- solvent extraction to remove dissolved organics
- steam stripping to remove dissolved gases
- chemical treatment (oxidation) of cyanides, ammonia *etc.* and dissolution of metals
- biological treatment
- carbon adsorption
- reverse osmosis to reduce the hydraulic loading or for final dissolved ion and organic recovery.

6.5.4.1 Metal Ion Sources and Dissolution. Potential sources for metal include catalysts, ores, scrap, batteries, circuit boards *etc.*[21] Processing of this material will generally require a dissolution process. The technologies for metal recovery from spent catalysts is reviewed by Litz[22] with particular interest in the recovery of Pt group metals from automobile catalysts. Platinum group metals can be recovered from autocatalysts by leaching in HCl–HNO$_3$ mixtures and Ni can also be recovered from low grade spent catalysts by leaching in HCl. The recovery of Mo, V, Ni and Co from spent hydrodesulfurisation catalysts used in hydroprocessing and of Co/Mo, Ni/Mo and Ni used in edible oil hydrogenation is feasible. The process for recycling the catalysts comprises of pre-treatment by calcination, solubilisation of metals by acid, alkali and oxidative leaching, separation and recovery of the Mo, as high purity MoO$_3$, by solvent extraction (hydroprocessing catalyst only), and electrowinning of the Co and Ni as cathodes and metal salts.

The recovery of metals such as Pt and Pd from low grade ores can also be achieved by leaching in cyanide, chloride and bromide media. The use of chloride solutions for the leaching of sulfide is quoted as having beneficial effects on electrodeposition. In many cases the dissolution of metals from waste and ores can be accelerated by the use of anodic oxdiation. Processes include:

1   The electrodissolution of copper sulfide ore slurries.
2   Electrooxidation for the recovery of carbonaceous gold ores and molybdenum concentrates.
3   The anodic dissolution of molybdenum in ammoniacal solutions.
4   The electrolytic oxidation of arsenopyrites slurries.
5   Electrochlorination of sea nodules to recover copper, nickel and cobalt. The complex ores are treated by anodic dechlorination to recover metal. The addition of sodium sulfides improves metal recovery to greater than 80%.
6   Catalytic electrochemical dissolution oxidises and dissolves typical catalyst metals and contaminants. The process is seen as a replacement to the muiltistep treatment of for example Naphtha catalyst, involving roasting, acid leaching, caustic dissolution, reaction with H$_2$S or Cl$_2$ and calcination. The method uses anodically generated Ag^{2+} (or Ce^{4+}) ions, which are recycled to oxidise the hydrocarbons and carbonaceous material adhered to the catalyst surface. Sulfides are converted to aqueous sulfur species.
7   The recovery of bismuth in acidic media, using a niobium electrode based on electrochemical dissolution and electrodeposition.
8   The recovery of copper from scrap metal pickling waste water by electrodeposition has also been demonstrated.

Alternative electrochemical methods, to anodic dissolution and electrodeposition, for metal recovery are also available. These include an electroreductive stripping processes (combined electrochemical reduction and solvent extraction) for separation of europium, samarium and gadolinium. Electrodialysis has also been used to recover metal (rhodium) catalysts from tar byproducts formed during acetic acid and acetic anhydride production.

The ash residues from municipal solid waste incinerators often contain toxic metals such as Pb, Cd and Cr. These metals can be extracted and recovered by electroplating. Non-ferrous waste from Cu, brass and lead smelters are also amenable to electrochemical recovery. Leaching of the metal in sulfuric acid followed by electrodeposition is also feasible.

6.5.4.2 The Recycling of Lead Acid Batteries. The recovery of metal ion constituents along with other materials from disposable or spent batteries is a developing area for the application of electrodeposition. One example[23] of the recovery of lead from spent batteries is a commercial process which does not involve crushing of the batteries. The lead in the batteries is dissolved in a fluoroboric acid electrolyte and deposited on the cathode in tank cells using insoluble lead dioxide coated graphite anodes. Of the metal ion impurities present in the electrolyte formed by battery solution only trivalent Sb is found to co-deposit significantly with the Pb. By oxidising Sb^{III} to Sb^V, the amount of co-deposition becomes acceptably small for the process.

A second hydrometallurgical process[24] consists of crushing batteries and separating metal pieces from electrodic pastes. The sludge is then solubilized in a double stage reduction process (alkaline and acid, respectively) and the purified solution is submitted to electrowinning. The plant operation is simple, with high efficiency, high reaction rates, and negligble waste gas emission. Calcium sulfate is the only solid waste. The plastic scraps are recovered free of lead compounds and may be re-used, after granulation. The lead is recovered in two grades: hard lead, which is simply melted and cast, and electrolytic grade lead which is won from the sludge. Electrical energy requirements of the whole process are *ca.* 0.5 kWh kg^{-1}.

6.5.4.3 Scrap Metal. An alternative electrorefining cell has been described[25] in the production of high purity Co and Ni from super alloy scrap. The cell uses a double membrane (DMEC), *i.e.* two anion-exchange membranes positioned between the impure anode and the high purity cathode (see Figure 17). The function of the DMEC is to separate the impure anolyte from the pure catholyte, using a stream of spent catholyte in the central membrane compartment.

A second example has considered the use of a single-step process for the separation and recovery of the components of scrap galvanised steel. The technical and economic feasibility has been favourably evaluated.[26] The process involves the anodic dissolution of the zinc from the scrap in hot caustic solution, whilst the Zn is simultaneously electroplated.

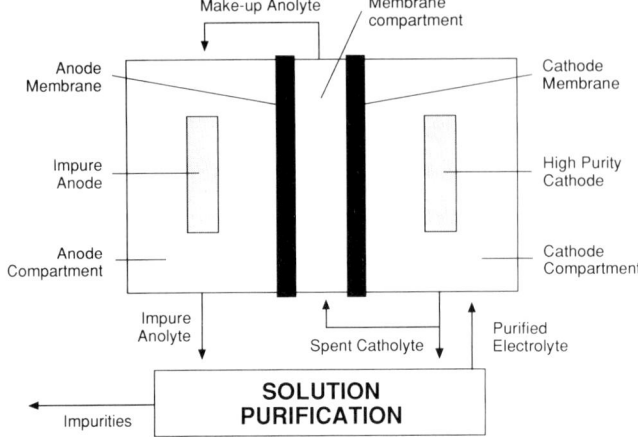

Figure 17 *Double membrane cell (DMEC) for super alloy scrap recovery*

The recovery of tin described in section 6.5.3 used chemical dissolution to facilitate the electrochemical recovery of tin. Recovery of tin from scrap can also be achieved by anodic dissolution in heated alkaline leaching solutions prior to electrodeposition.

6.5.4.4 Processing of Nickel–Cadmium and Alkaline Batteries. A process for recovery of the metal ion constituents from Ni–Cd batteries[27] is illustrated in Figure 18.

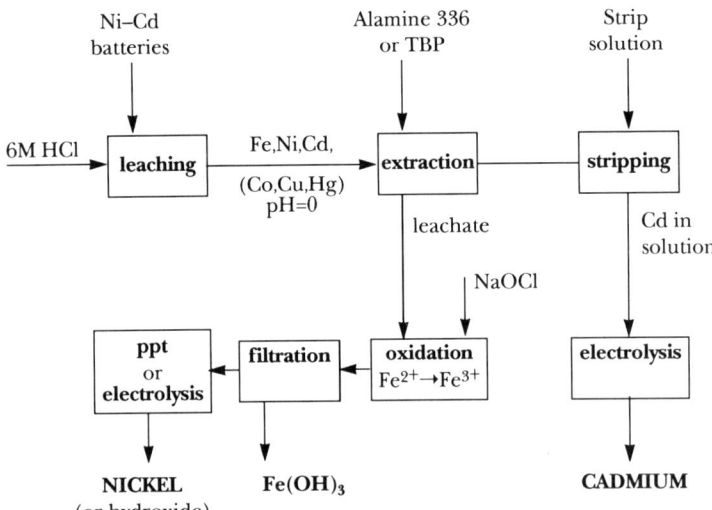

Figure 18 *A process for the recycling of Ni–Cd batteries*

In the process the Ni–Cd batteries are first shredded and leached of the metal components in concentrated hydrochloric acid. This leachate consists of a solution of the dissolved metals, mainly Fe, Ni, Cd with small amounts of Co, Cu and Hg at a pH *ca.* 0. The Cd is extracted with a commercial metal ion extractant, such as Alamine 336 (or TBP). Stripping of the Cd from the extractant produces an aqueous electrolyte solution from which cadmium is electrowon. After extraction of the cadmium, the leachate is contacted with sodium hypochlorite, at pH 4, to oxidise Fe^{II} to Fe^{III} and induce precipitation of iron(III) hydroxide. The iron(III) hydroxide precipitate is then filtered off and washed with dilute hydrochloric acid and contains less than 0.001 wt% of Cd with 0.1 wt% Ni and 0.25 wt% Co. The filtrate now contains mainly Ni which is recovered by electrodeposition (or by precipitation as the hydroxide). The electrodeposition steps of this scheme are likely to be carried out in two stages — using two-dimensional and then high surface area (HSA) electrolysers. The operation for 'eliminating' metal ion discharge may be backed up with ion exchange. The processing of 1000 kg of batteries will produce 200 kg of Ni metal and 159 kg of Cd metal and 500 kg of iron(II)–nickel scrap.

The processing of alkaline batteries can proceed in a similar way to that of the Ni–Cd battery. After leaching in 6 mol dm^{-3} HCl, the leachate containing Hg, Zn and Fe and Mn has the Hg and Zn removed by solvent extraction with Alamine 336. Both metals are relatively easily stripped from the solvent with NaCl solution and the strip solution is electrolysed to first electrodeposit mercury (< 1 ppm), and then zinc is removed by electrodeposition or by precipitation at pH 10. After extraction of Hg and Zn, the leachate is oxidised with hypochlorite at pH 4 to precipitate Mn and Fe. This precipitate may be used as a raw material for the winning of manganese dioxide.

An alternative process,[28] operated in China for the recovery of Ni and Cd from Ni–Cd batteries, is a multistep process involving roasting, leaching, filtration, precipitation and neutralisation. Electrodeposition is used to recover the Cd from the concentrated solutions which are produced as part of the processs strategy and also featured in the purification of the ammonium sulfate reagent solution.

6.5.5 Combined Electrodeposition and Ion Exchange

The integration of electrochemical recovery and ion-exchange systems to treat the various metal pollutants, at point of source or end of pipe, has already been mentioned as a potentially useful method, which is of some interest.

A basic description of an ion-exchange resin is a solid, insoluble acid or base capable of entering into a chemical reaction with aqueous based ions. The species can be anion exchangers, capable of exchanging anionic complexes such as chromate, or can be cation exchangers, capable of exchanging cations, for example simple metal ions. In practice the cation exchanger replaces the hydrogen ions in the matrix with the metal ions, such as Cu, in the solution. These are subsequently regenerated by recontacting with a fresh mineral acid stream. The important feature of this operation is that the metal ion concentration is drastically increased from what can be values of a few ppm to more the 30 000 ppm. The operation usually requires an excess of acid and so the recovered solution is acidic and frequently appropriate for direct treatment by electrodeposition. In addition the ion exchange operation can be used to regenerate the metal ion solution in a form more suitable for metal deposition by using a suitable anion in the regenerating acid. For example Cu ion solutions that contain nitrate or chloride, which cause some difficulties in electrodeposition, can be regenerated by the use of sulfuric acid.

In general ion exchangers have high surface areas and thus are not limited by mass transport to the same extent as electrodeposition processes. They are generally suitable for the treatment of large flows of dilute ion solutions as operating costs are based on the amount of pollutant whilst capital cost is based on the hydraulic loading. In general ion exchangers offer only modest selectivity for individual metal ions while chelating ion-exchange resins, *e.g.* iminodiacetate, have greater potential for metal ion separation as indicated in Table 8 [29] in terms of the selectivity coefficient. Chelating ion-exchange resins are also used for the recovery of metal ions which are complexed with agents such as quadrol, citrates and EDTA. It has been claimed that a significant advance in ion-exchange systems has been the introduction of the Recoflo system. This is a compact bed (of fine particle resin) between 15–60 cm high. These ion exchangers utilise

Table 8   *Characteristic selectivity coefficients of chelating ion-exchange resins*

Ion	pH=4	pH=2
Fe^{+++}		325 000
Cu^{++}	2 300	130 000
Hg^{++}		>43 000
Au^{+++}		>8 100
Ag^{+}		4 600
Ni^{++}	57	3 200
Cd^{++}	15	620
Fe^{++}		190
Mn^{++}		120
Zn^{++}	17	50
Al^{+++}		20
Mg^{++}		
Co^{++}	6.7	
Fe^{++}	4.0	
Mn^{++}	1.2	
Ca^{++}	1.0	1.0

primarily the ion-exchange sites at the surface of the resin and thus have modest loadings. The Recoflo system is marketed by Ecotec and is employed industrially in several sites in conjunction with the Reconwin electrowinning cell.

In broad terms the applications of the combined ion exchange–electrowinning approach fall into the following categories:

(i)   removal of contaminants from electrowinninng electrolytes;
(ii) pre-concentration of dilute metal streams followed by electrowinning;
(iii) selective recovery of particular metals from solution containing undesirable, unwanted or toxic metals or compounds;
(iv) conversion of chloride or nitrate metal ion solutions to sulfates.

6.5.5.1 Application in Metal Plating and Finishing. The combined use of electrodeposition and ion exchange can be used effectively to give appropriate levelling control in individual metal finishing and plating applications (see Figure 19).

Many metal finishing and plating operations often have more than one metal contaminant present in their waste water. Segregation of the different waste waters according to their metal contaminant is therefore important because:

(a)   the metals are easier to sell or re-use and are more valuable if they are relatively pure, and

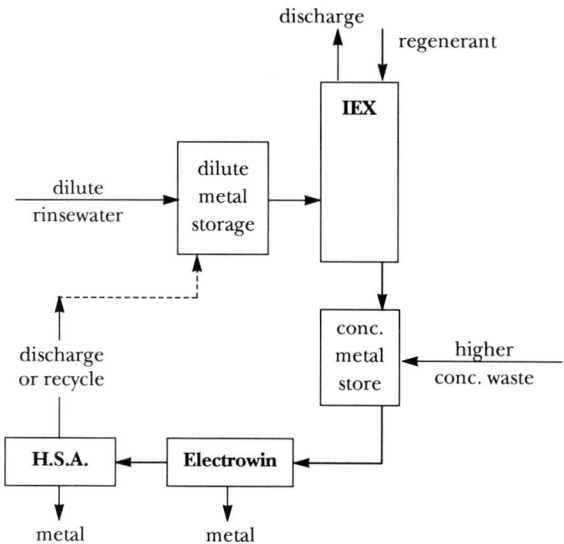

Figure 19 *Process for metal recovery using integrated ion exchange and electrodeposition*

(b)   combinations of metals may be difficult to recover because required
       process conditions (such as pH) may not be the same for different
       metals.

As well as segregation by metal type, streams should also be segregated
by concentration. Usually waste streams with high metal concentrations
(such as dumped metal plating baths) can be directly electrowon (see
Figure 19). Waste water streams with low metal concentrations (such as
rinse waters) usually are concentrated by ion exchange (IX), and the IX
regenerant is electrowon. Waste streams that fall between these two
extremes may be treated either way, depending on the waste water.

Overall each metal recovered will generally require a separate elec-
trowinner because of requirements of different bath conditions (e.g. pH
and temperature) and varying removal rates. For several metals electrodeposition becomes impractical below certain concentrations and thus the
cell effluent electrolyte can be recycled back to the ion-exchange system.
An alternative to IX for treating the cell effluent is to use a high surface
area electrode for further metal recovery. In the operation of integrated
processes of this type it is important to assess the likely build-up of contaminants due to recycling and the requirements of water, acid and alkali
balances.

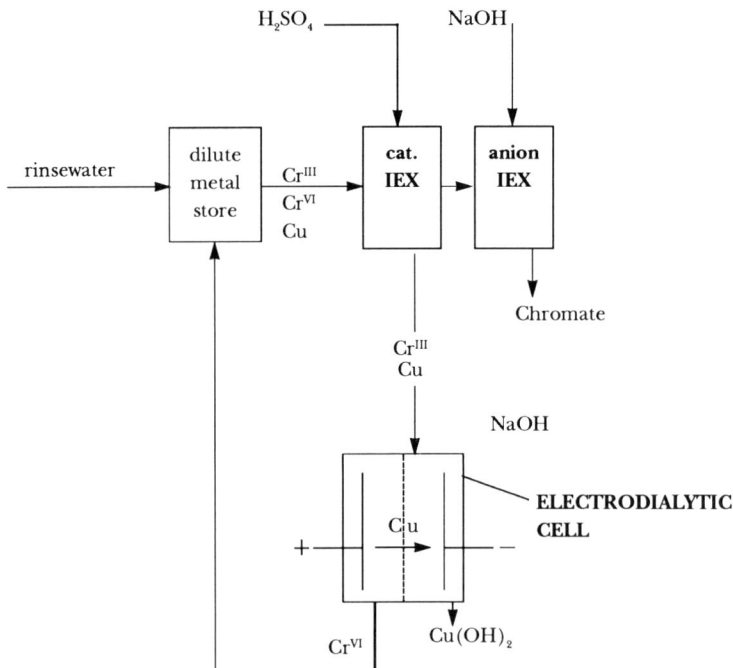

Figure 20 *Separation of chromium and copper ions using ion exchange and electrolysis*

Combined IX and electrodeposition (or electrolysis) can also be applied where a mixture of heavy metal such as Cu with dissolved homogeneous redox systems, such as Cr^{VI}/Cr^{III}, occurs in for example metal treating baths containing chromic acid (see chapter 7). In one example electrodialytic membrane separation is adopted to separate copper from the chrome(III) species (see Figure 20).

This stream is produced by the cation exchanger. The cation-exchange membrane permits transfer of copper from the treatment tank into the catholyte chamber, where copper hydroxide is formed in the alkaline solution. The trivalent chrome ion has a much lower electromotive potential than the Cu ion and its transfer across the membrane is less than 0.1%. The trivalent chrome is anodically oxidised to chromate and this solution is returned to the anion IX effluent for the recovery of chromate. The anion-exchange regenerant (primarily sodium chromate) is converted to the chromic acid ion in a similar electrodialytic recovery unit by transfer of sodium cations across a cation-exchange membrane and generation of H^+ ions at the anode.

6.5.5.2 Applications. Overall the combination of ion exchange and electrodeposition has several applications and the following are examples of the use of one system, the Recoflo–Reconwin combination.[5]

- Cu and Zn recovery from pickling baths and dilute rinse waters
- Cu and Ni recovery from printed circuit board, plating and etching operations
- Ni or Co recovery from plating or aluminium electrolytic colouring rinse waters
- Selective removal of Sn^{II} from Sn^{IV}
- Phosphoric acid recovery
- Gold cyanide recovery
- Uranium recovery

A typical situation is effluent containing a very low concentration of metal, for example copper at a concentration of 1.0 g dm^{-3}, which passes through a cation-exchange column to give a product solution containing less than 1.0 mg dm^{-3}. Elution of the cation exchanger with sulfuric acid then yields a metal–sulfate solution containing 10–40 g dm^{-3} of metal. This metal is then electroplated in the electrowinning cell until a metal ion concentration of 1.0 g dm^{-3} remains in solution. Simultaneously an equivalent amount of sulfuric acid is liberated. This can be either used in the cation exchanger, for subsequent regeneration or neutralised if electrodeposition is dependent upon pH. For example in the case of nickel it is not possible to electrowin below a pH *ca.* 2. The acid generated during electrowinning, must be continuously neutralised using sodium hydroxide and thus cannot be recycled to the cation exchanger. Recycling technology may consider the use of electrodialytic salt spitting to recover the acid from the sodium sulfate solution. If a chelating resin is employed in the ion exchanger, the spent electrolyte can be combined with the dilute effluent feed stream to the ion exchanger to recover the residual nickel. The chelating resin selectively recovers nickel, rejecting the sodium in this case.

6.5.6   Electrochemical Ion Exchange

An alternative ion-exchange system has been developed by AEA Harwell, in which an electrochemical potential is applied to the ion exchanger. The process referred to as electrochemical ion exchange, EIX, is a method of separating ionic species in which an electrochemical potential field is used to enhance the normal operation of standard ion exchange. In this enhanced ion-exchange operation, an ion-exchange material is attached to an electrode structure (typically platinised titanium)[30] using a suitable binder (see Figure 21). The adsorption and elution properties are controlled by an external applied potential. The principle of operation combines migration of absorbed ions within the ion-exchange media, specific ion interaction in the media, and local pH changes at the current feeder. The principle can be applied, without eluent chemicals, to concentrate feed solutions.

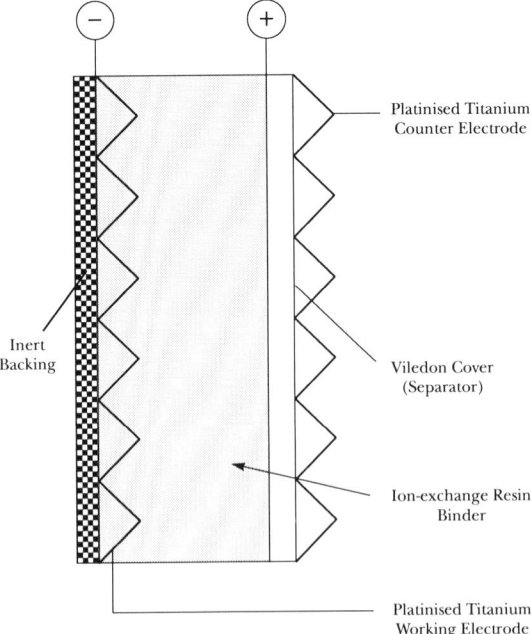

Figure 21 *Structure of an electrochemical ion-exchange electrode*

The ion-exchange materials used in the EIX electrodes are generally weak acid cation exchangers or basic anion exchangers. The resin activation process involves ionisation of the respective active group. The absorption and the elution reactions involve protonation and de-protonation, respectively, of the exchanger by the perturbation of pH caused by either the reduction or oxidation of water at the electrode. Thus the typical cycle for the case of cation exchange, using for example a carboxylic acid group, is:

(*i*)   deprotonation to give an overall negative local charge

$$R–COOH + OH^- -> R–COO^- + H_2O \tag{34}$$

(*ii*)   absorption of the cation to maintain a charge balance

$$R–COO^- + M^+ -> R–COO^-M^+ \tag{35}$$

(*iii*)   elution to re-protonate the active group

$$R–COO^-M^+ + H^+ -> R–COOH + M^+ \tag{36}$$

For anion-exchange processes the secondary or tertiary amine ion-exchange groups are activated initially by protons. Clearly in operation the application of an applied potential initiates ionic motion, which assists both the absorption and elution stages of the process. In the complete cell the counter electrode is a source of ions, provided through the electrolysis of water. For cation exchange, the oxidation of water supplies protons, which will decrease the pH of the external solution and also be absorbed as interfering ions, which will eventually de-activate the electrochemically activated site by recombination. The influence of the interfering ions on the current efficiency of the process will depend upon the pH of the feed solution and the buffer capacity .

The original application of EIX was in the nuclear industry to reduce the concentration of radionuclides. The suggested non-nuclear applications are in the removal of toxic metals, precious metals, nitrates and corrosive anions and in the de-ionisation of water. The technique has been demonstrated in the removal of chloride and sulfate in the presence of borate/boric acid and for the removal of Co from feeds, which contain concentrations of 100 ppb, to undetectable levels. A comparison was made between EIX, ED and EDIX for the removal of Co from a feed containing borate and lithium ions. EDIX is a procedure in which an ion-exchange material is used as the intermembrane packing in an ED unit. This resin, which has a higher conductivity than the feed stream increases the effective conductivity of the solution (thus reducing the cell voltage) and also increases the contact area with the feed. The performance in this application of EIX, with the same decontamination factor for Co, was

good; the size of the EIX cell is approximately five times smaller than that
of the EDIX cell and the ED cell.

6.5.7   Metal Ions Recovery by Complexation Ultrafiltration and Electrolysis

A process is under development for the recovery of metal ions from dilute
solutions by a combined process of complexation–ultrafiltration and electrolysis.[31] Ultrafiltration (UF) is a membrane separation process, in which
selectivity is linked to the size of the membrane's pores. UF is not capable
of concentrating uncomplexed metal ions as they are too small in size to
be retained by the membranes. To achieve separation of the metal ion it
is therefore necessary to increase the metal ion size artificially, by complexation with a macromolecule of sufficient molar mass. The macromol-
ecular complex thus formed is retained by the UF membrane and concentrated, while the solutes of low molecular weight pass through the
membrane as the permeate. The coupling of the UF separation with
extraction by electrolysis has been demonstrated for copper and cadmium.

6.6   CEMENTATION

Cementation is a method in which a second metal, C, in solid powder
form, displaces the dissolved metal ion, M^+, from aqueous solution by
virtue of the galvanic cell reactions:

$$M^+ + e^- \rightarrow M \tag{37}$$

$$C \rightarrow C^+ + e^- \tag{38}$$

The process takes place on the surface of the metal agent C, which is
anodically dissolved into the solution under open circuit conditions.
Typical metal cementation agents, which are used in several hydrometallurgical applications, are Fe and Zn. Typical reactors for cementations are
agitated tanks, rotating drums, oscillating chambers and fluidised beds,
all designed to achieve good mass transport and surface renewal.

Inefficiencies in the process, which result in dissolution of the metal
substrate without metal displacement, are due to the corrosion reaction:

$$C + n\,H^+ \rightarrow C^{n+} + n/2\,H_2 \tag{39}$$

and possible parasitic reactions if the substrate metal C forms redox
species on dissolution, *e.g.* the Fe^{II}/Fe^{III} couple.

A typical application is in the cementation of Cu from relatively dilute
solutions (5 g dm^{-3}) using Fe:

$$Fe + Cu^{2+} \rightarrow Fe^{2+} + Cu \tag{40}$$

A recent process[32] claimed to improve the efficiency of this cementation process uses a fluidised bed of Fe particles subjected to a pulsating magnetic field (Actimag, Switzerland). The pulsating magnetic field in conjunction with fluidisation gives increased rates of reaction through higher rates of mass transport and continuous exposure of the Fe surface. The process is operated continuously by the regular addition of the Fe particles to the fluidised bed. This process has been operated with a bed 3 m long and 0.25 m wide for the recovery of 99% of the copper from a 5 g dm^{-3} solution of copper sulfate (pH 3). The final solution contained *ca.* 100 ppm of copper and the solid copper product was typically 97% pure.

The advantage of cementation processes is the simplicity of operation and the use, as in the case of Fe, of relatively cheap reagents. A limitation of the process is that it is not technologically clean, *i.e.* no net removal of metal from solution (mol mol^{-1}). Final cemented metal ion concentrations are not low enough for the process to be used for effluent treatment. An interesting adaptation of cementation has seen the use of electrodeposition to continuously regenerate the metal reagent used in the cementation process. This regeneration takes place *in situ* with the cementation reaction and typically uses a fluidised bed as the substrate material. For example in using Zn for the cementation of Ag, a fluidised bed of carbon particles is used for the electrodeposition of the Zn, which is then used in the cementation of the Ag powder.[33] The circulation of particles within the fluidised bed enables the deposition of the Zn to take place in a relatively small region of the bed away from the bulk of the cementation reaction. The rate of the cementation reaction can largely be controlled by the current applied in the electrodeposition.

REFERENCES

1   B. Fleet, *Coll. Czech. Chem. Commun.*, 1988, **53**, 1107.
2   A.J. Bard, R. Parsons and J. Jordan 'Standard Potentials In Aqueous Solutions', Dekker, NY, 1986.
3   F.C. Walsh, 'Electrochemical Engineering', *Int. Chem. Eng. Symp. Ser.*, 1986, **98**, 139.
4   L. Weis, M. Giron, G. Valentin and A. Storck, *Int. Chem. Eng. Symp. Ser.*, 1986, **98**, 49.
5   C.J. Brown, 'Metal Recovery by Ion Exchange and Electrowinning, in Electrochemistry For A Cleaner Environment', ed. J.D. Genders and N.L. Weinberg, The Electrosynthesis Co., 1992, ch. 7.
6   D. Pletcher and F.C. Walsh, 'Industrial Electrochemistry', Chapman and Hall, 2nd edn., 1990.
7   D. Robinson and F. C. Walsh, *Hydrometall.*, 1991, **26**, 93; 115.

8   R. Kammel, NATO Conf. Ser. 6–10, Hydrometall. Process Fundam., 1984, 617.

9   C.M.S. Raats, H.F. Boon and G. van der Heiden, *Chem. Ind.*, 1978, July, 465.

10   G. Lacoste, 'Electrochemical Cell Design and Optimisation Procedures', Dechema Monograph, 1991, vol. 123, 411.

11   D.N. Bennion and J. Newman, *J. App. Electrochem.*, 1973, **2**, 122.

12   K. J. Muller, ref. 10, p.199.

13   D. Simonssen, *J. Appl. Electrochem.*, 1984, **14**, 595.

14   J. Farkas and G.D. Mitchell, *A. Int. Chem. Eng. Symp. Ser.*, 1985, **81**, 57.

15   D. Pletcher, I. Whyte, F.C. Walsh and J.P. Millington, *J. App. Electrochem.*, 1993, **23**, 82.

16   G. Huber, W. Gosele and W. Habermann, 'Elektrochem Stoffgewinnung: Grundlagen Verfahrenstech', Dechema Monograph, 1992, vol. 125, p.475.

17   A.M. Polcaro and S. Palmas, 'Electrochemical Engineering and The Environment', *Int.Chem. Eng. Symp. Ser.,* 1992, **127**, 85.

18   T.M. Morris, *Eng. Mining J.*, 1977, **178**, 86.

19   S.A.N. Sheya,T. Stefanowicz, T. Golik, S. Napieralska and M. Osinska, *Resources, Conservation and Recycling*, 1991, **6**, 61.

20   G.H. Kelsall and F. P. Gudyanga, ref. 10, p.167.

21   K. Scott, in preparation.

22   J.E. Litz, *Soc. of Mining Engineers of AIME*, 1986, 11.

23   M. Maja *et al.,* 'Electrochemical Engineering', *Int.Chem.Eng. Symp. Ser.*, 1986, **98**, 155.

24   U. Ducati, Italian Pat. 23 745 AA, 1981.

25   D. Redden and J.N. Greaves, *Hydrometall.*, 1992, **29**, 547.

26   F. J. Dudek, E. J. Daniels, Z. Nagy, S. Zaromb and R. M. Yonco, *Sep. Sci. Technol.*, 1990, **25**, 2109.

27   *Div. Technol. Soc. TNO Report 89-127/R.22/CAP, Recycling of Spent Batteries*, 1989.

28   Z. Xue, Z. Hua, N. Yao and S. Chen, *Sep. Sci. Technol.*, 1992, **27**, 213.

29   C. J. Brown, 'Separation Processes in Jydrometallurgy', ed. G. A Davies, Ellis Horwood, Chichester, 1987, p.379.

30   N. J. Bridgewater, C. P. Jones and M. D. Neville, *J. Chem. Technol. Biotechnol.*, 1991, 469.

31   S.Niessen, F. Persin and M. Rumeau, 'Abstract 3.7–18, 4th World Congress in Chemical Engineering', Karlsruhe, Germany, 1991.

32   A. S. Sairafi, 'Electrocementation of Silver Using a Fluidised Bed', PhD Thesis, University of Newcastle, 1993.

CHAPTER 7

The Treatment of Industrial Process Streams and Effluents.
Part II Organic and Inorganic Species

This chapter, describes electrochemical methods for the treatment of waste waters, which contain organic and inorganic species and hazardous substances, and of flue and waste gases. The methods include direct and indirect oxidation, photoelectrochemical oxidation, electroreduction, electrosorption and electroflotation.

7.1 TREATMENT OF ORGANIC COMPOUNDS

Processes for the removal of organic species from solutions include:

*   chemical oxidation, using ozone, UV, *etc*
*   chemical reduction, *e.g.* catalytic dehalogenation
*   biological treatment
*   concentrating methods, *e.g.* membrane or adsorption
*   incineration

Incineration is a high cost method when dealing with dilute waste waters due to high transportation and fuel costs. Methods for concentrating solutions depend on further treatment steps and have limited applications. Chemical oxidation methods are useful, although they are not generally used for one major class of toxic compounds, *i.e.* halogenated organics. Chemical reduction methods are expected to be particularly suitable for concentrated waste streams. Biological methods using special microorganisms are at a relatively early stage of development.

The electrochemical treatment or destruction of organics in waste waters is potentially a powerful method of pollution control, offering a low temperature alternative to incineration and a clean process. In many cases no other chemicals are released into the waste waters. Owing to the nature of the process streams and the implications of the electrochemistry, the methods which are applied either partially or completely detoxify or decompose the organic species. Complete destruction often means the oxidation of the organic species to carbon dioxide. This oxidation can be achieved either directly or indirectly at the anode, or indirectly using appropriate anodically generated reagents.

In certain cases the complete destruction of an organic molecule by anodic oxidation is not an appropriate or easy process to perform and a partial transformation may be more desirable. The complete oxidation of large organic molecules requires a large number of electrons and energy consumption can be relatively high, as will the cost of the cell hardware. Ideally, current efficiencies for the process should be high, but more importantly any products from the treatment should be either non-toxic, readily biodegradable, or in sufficiently low concentrations to satisfy environmental legislative requirements.

7.1.1   Direct Anodic Oxidation

Phenolic wastes occur in many process industries and as a result they have featured prominently in attempts to demonstrate the usefulness of anodic oxidation. Early attempts to utilise anodic oxidation for phenolic waste waters used a lead dioxide packed-bed anode.[1,2] Complete oxidation of phenol was demonstrated, but complete total organic carbon (TOC) removal was not achieved. The sequence of reactions is believed to be:

$$C_6H_6O \rightarrow C_6H_4O_2 \rightarrow C_4H_4O_4 + 2CO_2 \rightarrow 4CO_2 \qquad (1)$$
$$\text{phenol} \quad \text{benzoquinone} \quad \text{maleic acid}$$

Overall almost complete conversion of phenol and *p*-benzoquinone can be achieved with optimum conditions of pH, temperature and current density. It is believed that adsorption of phenol occurs as part of the reaction mechanism. The rate of phenol oxidation is higher on lead dioxide than on either graphite, carbons, platinum or nickel.

Comninellis[3] has studied the oxidation of approximately two dozen organic aromatic compounds (see Table 1), and proposed two measures for estimating the electrochemical oxidation of organics: the electrochemical oxidation index (EOI) and the electrochemical oxygen demand (EOD).

The EOI relates to the evolution of oxygen during electrolysis and is defined as:

$$EOI = ICE \, dt/t \qquad (2)$$

where t is the time of electrolysis and ICE is the instantaneous current efficiency of phenol calculated from the relationship:

$$ICE = (v - v_{org})/v \qquad (3)$$

Table 1   *The electrochemical oxidation index for various organic compounds*

Compound	Pt anode	SnO$_2$ ABB-^{a}
Ethanol	0.02	0.49
Acetone	0.02	0.21
Acetic acid	0.00	0.09
Formic acid	0.01	0.05
Tartaric acid	0.27	0.34
Oxalic acid	0.01	0.05
Malonic acid	0.01	0.21
Maleic acid	0.00	0.15
Benzoic acid	0.10	0.79
Naphthalene-2 sulfonic acid	0.04	0.51
Naphthalene-1-sulfonic acid	—	0.41
Phenol	0.15	0.60
Aniline	0.56	0.43
Benzenesulfonic acid	<0.05	0.28
5-Methyl-3-aminonisoxazole	—	0.25
Organe II		0.58
Anthraquinone sulfonic acid		0.18
Nitrobenzene		0.80
Nitrobenzenesulfonic acid		0.46
Triamiontriazin		0.02
EDTA	0.30	0.30
p-NDMA	0.30	0.37
4-Chlorophenol	—	0.35

^{a} ABB = Aseo Brown Boveri.

where v is the volumetric rate of oxygen evolved in the absence of the organic species and v_{org} is the oxidation rate in the presence of the organic.

The EOD is calculated from the relationship:

$$\text{EOD} = 8(\text{EOI I } t)\,/\,\text{F [PhOH]} \quad (\text{g O}_2/\text{g organic}) \qquad (4)$$

where [PhOH] is the initial concentration of the organic species.

Using platinum anodes, it was found that where one of the substituents is electron donating (*e.g.* $-NH_2$), only benzene derivatives are efficiently oxidised giving maleic acid as the principal product. Benzene derivatives with powerful electron withdrawing groups ($-COOH$, $-NO_2$, $-SO_3H$) have a low EOI value. Platinum anodes have a limited range of oxidation potentials and thus attention has focused on SnO$_2$-coated titanium materials. The tin oxide material, particularly when doped with Sb (*ca.* 5%) to impart the appropriate electrical conductivity, has oxygen over-potentials some 600 mV greater than those of platinum.

Tin oxide is (Figure 1) a superior anode material to platinum in removing TOC during the oxidation of phenol. With SnO$_2$, only residual

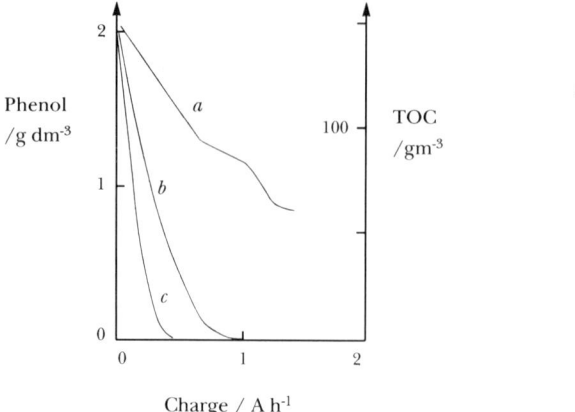

Figure 1   Comparison of the oxidation of phenol on platinum and tin oxide anodes; a:
TOC, Pt; b: TOC, SnO₂; c: Phenol conc. Pt and SnO₂

amounts of intermediates (hydroquinone, catechol and benzoquinone) remain after oxidation. Aliphatic acids (oxalic, fumaric and maleic) are rapidly oxidised at a SnO_2 anode, but are almost inactive at platinum.

To explain this behaviour the reaction mechanism of Figure 2 has been proposed[4] for the electrochemical oxidation. For both of the electrodes the hydroxylation is relatively fast, whilst the ring-opening step on platinum is much slower than that on tin oxide. The final electrochemical combustion of the aliphatic acid is very slow on platinum. The mechanism of the oxidation of species such as phenol on lead dioxide and tin oxide is not a pure charge transfer reaction,[5] but rather it is suggested that the electrodes produce the very reactive hydroxyl radicals, which perform a 'homogeneous' oxidation reaction of the organic. The standard reduction potential of the OH radical is 2.8 (vs. NHE), which makes it a more powerful oxidising agent than ozone (2.07 V) and atomic oxygen (2.42 V).

Tin oxide has also been proposed as a material for electrochemical oxidation of biorefractory organics in waste water.[6] This electrochemical oxidation is seen as an alternative to chemical treatment, using powerful oxidants such as ozone or hydrogen peroxide, prior to discharge to a biological/mechanical sewage plant. Stucki et al.[7] have investigated the influence of the cathode reaction using undivided cells for the oxidation of benzoic acid. The high over-voltage of the tin oxide anode leads primarily to irreversible oxidations and thus the reaction products are not reduced to any significant degree. In the case of phenol oxidation, electrode materials such as platinum may lead to the formation of the quinone/hydroquinone reversible couple and thus a potential loss in efficiency. From the point of a low reactor cost and low cell voltage the use of an undivided cell is the best option, although this will not be the case for all organic species.

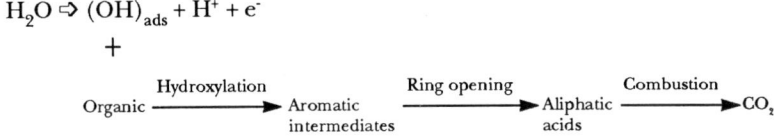

Figure 2   *Mechanism of the anodic oxidation of phenol*

The reactor used bipolar coated titanium anodes, the anode side was coated with tin oxide and the cathode side coated with platinum. Plastic end extension were fitted to the exposed edges of the electrodes to reduce the current bypass. The quoted performance of this unit is a space time yield of 6.4 kg COD $h^{-1}m^{-3}$, at an assumed EOI of 40% and at a current density of 300 A m^{-2}. It is claimed that the process competes with wet oxidation and combustion processes at relatively high chemical oxygen demand (COD) concentrations and with adsorption at the lower values of COD. Typically COD levels in the range 500–1500 ppm are accessible by the method with specific energy consumptions of the order of 50 kWh kg^{-1}.

7.1.1.1 Other Applications. Anodic oxidation has been researched as a pollution control device for several applications (as is indicated in Table 1) and for several waste waters.

(i)   the spent wash from distillery effluent containing high biological oxygen demand (BOD) and dark-brown colour;
(ii)   wastewaters containing alcohols and sugars;
(iii)   from the production of amino ether;
(iv)   nitration waste sulfuric acid emanating from the manufacture of trinitrotoluene.

A significant area for the application of anodic oxidation is the treatment of dyes and surfactants. The consumption of water in dyeing operations is large, for example in the proceessing of 1.0 tonne of cloth, *ca.* 100–400 m^3 of water can be used. There are several methods for the treatment of waste waters, which contain dyes and surfactants, that include:

(i)   Coagulation, by lime, or by aluminium or iron salts. This treatment involves increasing the degree of mineralisation of the effluent, a large consumption of reagents and the production of a wet solid.
(ii)   Extraction, with solvents such as diethyl ether and chloroform, which requires a large consumption of the solvent.
(iii)   Biological methods.
(iv)   Adsorption on activated carbon, although the adsorbent can rapidly become de-activated.

(v)   Oxidation using chlorine or ozone, which require large amounts of
        reagents, *e.g.* up to 9 kg of ozone to oxidise 1 kg of dye.

7.1.1.2 Treatment of Dyes. The following classes of dyes have been considered in the electrochemical treatment of waste waters;[8] anthraquinone,
thiazines, oxazines, phenazines, triphenylmethanes, xanthenes, vat and
azo dyes. Purification through electrolysis is usually obtained by electrooxidation with non-dissolving anodes or electrocoagulation, using dissolving anodes. Direct reduction has been proposed for the decolourisation
of the dyes although the energy costs are significantly greater than anodic
oxidation and the process leads to the formation of the corresponding
amine. The degradation products from the oxidation of azo dyes are typically CO_2, N_2 and Na_2SO_4, with the possible formation of aromatic esters,
phenols, aliphatic carboxylic acids, cyclic and aliphatic hydrocarbons and
aromatic amines *etc.* The oxidation of the azo groups takes place first followed by the further oxidation of the decomposition products. The oxidation rate is dependent upon the anode material. A range of coated titanium metal anodes formed from ruthenium oxide, cobalt oxide, manganese dioxide and lead dioxide and graphite have been successfully
employed. In alkaline solution nickel or an alloy of nickel with titanium is
suitable.

   In the presence of chloride ions the oxidation of azo dyes is largely
through the generation of 'active' chlorine, although direct anodic oxidation does play a part. The energy consumption of the process, which is a
major economic factor, using electrochemically generated active chlorine
is in the range 25–35 kWh kg^{-1}. Several reactor designs have been used in
the treatment of dyes; packed bed anodes of lead spheres, carbon felts
and carbon fibres.

   Soluble metal anodes are applied in an electrocoagulation process in
which the resultant metal hydroxides adsorb the dye. Steel has been used
mainly as the anode material. The rate of anode consumption is in the
region of 5–200 g m^{-3} of effluent. There is some evidence for the partial
direct oxidation of the dye at the anode. The electricity costs for complete removal of azo dyes at a concentration of 0.1 g dm^{-3} are between
30–100 kWh kg^{-1}. The cell design for coagulation must incorporate a
froth collector and a flotation chamber.

   Electrochemical reduction is attractive in the case of metal chelate
complexes where the toxic metal ion may be eliminated as solid metal or
the valence state is changed to that more suitable to ion exchange.

7.1.1.3 Removal of Surfactants. Surfactants can, as in the case of dyes, be
removed by direct anodic oxidation or by oxidation with electrolytically
generated 'active' chlorine. In addition the surfactants can readily be
removed by flotation, which is discussed later in section 7.5. Anode materials are typically graphite or ruthenium oxide coated titanium. Electrocoagulation, using steel or aluminium anodes, is also used for the treat-

ment of waste waters containing surfactants. There is evidence that, in the case of anionic surfactants, steel anodes are more effective, giving a greater degree of removal of surfactant and reduced volumes of sediment and flotation concentrate. Typically the consumption of metal is in the range of 100–700 g m^{-3} of a solution containing surfactant of concentrations 10–100 g m^{-3}. The energy consumption is dependent upon the concentration of the surfactant and, in the range of concentrations of 0.1–0.02 g dm^{-3}, varies from 30–600 kWh kg^{-1}. The anodic oxidation of nonionic surface active substances has been demonstrated on platinum and ruthenium oxide coated titanium electrodes. This group of surfactant materials is toxic in the aqueous environment and biodegrades very slowly. The products of this oxidation have a much lower toxicity to aquatic life than the surfactant.

7.1.2   Chloride and Chlorinated Compounds

An important issue in the electrochemical treatment of waste waters and related streams is the presence of halogens, and in particular chloride. Electrochemical oxidation of chloride ions will readily produce chlorine which may react, either as free chlorine or as monoatomic chlorine, with the dissolved organics species or any oxidation intermediates, to produce halogenated compounds. These halogenated compounds are generally more toxic than their unchlorinated counterparts and thus would make anodic oxidation unsuitable. Materials, which are poor electrocatalysts for chlorine evolution, are needed, *e.g.* tin oxide, although at sufficiently high concentrations of chloride, chlorine may be generated. Thus the potential formation of halogenated organics cannot be excluded in the use of anodic oxidation. Tin oxide has been used in the oxidation of a waste water from the chlorine bleaching step of a pulp plant which contained both organic and halogenated organic species.[7] Both the COD and TOC were reduced and an adsorbable organic carbon was also removed.

The anodic dehalogenation of 1,2-dichloroethane has been demonstrated[9] at a smooth platinum electrode. The major products of this are CO_2 (60% current efficiency), $HClO_4$ (20%) and chlorine. Some trichloroethane and other chlorinated species are formed in small amounts. However, these may not be considered negligible for waste treatment applications. The smooth platinum anode was found to be superior to electrodes made of RuO_2, TiO_2, PtO_x and PbO_2-coated materials.

7.1.2.1 Electrochemical Reduction. The direct electrochemical reduction of organic species can be used to detoxify the wide range of chlorinated compounds on the EC list of priority compounds.[10]

$$R-Cl + H^+ + 2e^- \rightarrow R-H + Cl^- \tag{5}$$

Dehalogenation changes the toxicological properties of the wastes, generally decreasing the toxicity and enhancing the biodegradability. For example, pentachlorophenol has an EC-50 ppm limit of 0.1, whereas phenol has a value between 22–42 ppm. An effective electrode material is carbon fibre, which has the following attributes:

(i)   Very large specific area for dealing with low concentrations of organics (100–1000 ppm).
(ii)   Readily available and relatively cheap.
(iii)   Tolerant to clogging by small solid particles.
(iv)   Reasonably high over-potential, as required to minimise hydrogen evolution.
(v)   Non-toxic in comparison to materials with high cathodic over-potentials such as lead or cadmium.

For example, with an electrode composed of 10 μm diameter carbon fibres, dehalogenenation of a waste water containing 50 ppm pentachlorophenol reduced the pentachlorophenol concentration to below 0.5 ppm, toxicity was decreased by a factor of 20, with the final product being phenol and some monochlorophenol. Similarly, *p*-chloronitrobenzene and dichlorvos (DDVP) were dehalogenated to < 0.1 ppm and < 1 ppm, respectively.

7.1.3   Indirect Oxidation Processes

There are three general categories of indirect oxidation, heterogeneous oxidation at oxide anodes, considered in section 7.1, generation of short life species, *e.g.* $O^\bullet$, $OH^\bullet$, O_3 and regenerable solution redox couples.

Indirect oxidation offers the advantage of removing the mass transfer limitations associated with electrochemical surface reaction at low concentrations. The active reagent can be generated at high concentrations electrochemically and the reaction carried out in homogeneous solution (see Figure 3). However, owing to the solubility of some redox reagents, in the waste water, the method may be more complex to operate.

The potential oxidation power of certain reagents has resulted in a number of processes based on 'chemical oxidation'. A major force in water purification applications is the use of hypochlorite or chlorine, discussed in section 7.3.4.1. Among other oxidants, ozone is an attractive alternative for water treatment as it does not release any additional reagents into the waters. Ozone can be generated electrochemically from relatively pure water according to the reaction:

$$3\ H_2O - 6\ e^- \rightarrow O_3 + 6\ H^+ \tag{6}$$

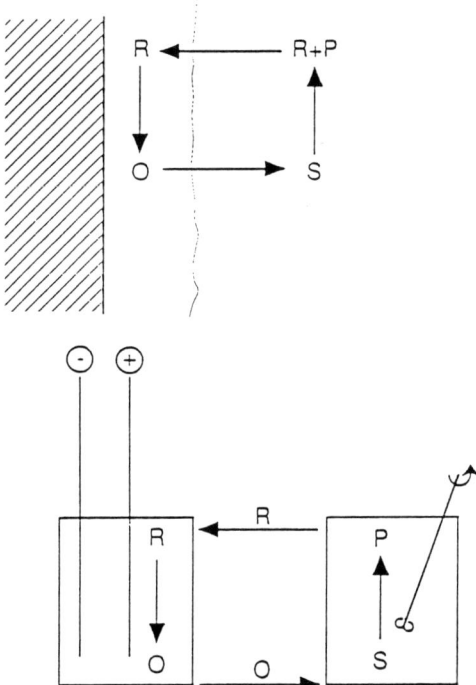

Figure 3   *Indirect oxidation process for effluent treatment;* (a) *generation of short-lived intermediates;* (b) *homogeneous solution oxidation*

There are two commercial methods of generating ozone electrochemically, discussed further in chapter 8, using either a solid-polymer electrolyte cell or a cell fitted with a gas diffusion cathode. An advantage of electrochemical generation is the on-site, on-demand production of ozone at a concentration much higher than can be achieved by other methods. The principal applications of ozone-generation cells are for water sterilisation and oxidation of process liquors containing organics (phenols, dyes and pesticides) and cyanides.

Attempts to couple electrochemical ozonisers with UV light have been investigated, for example at the ERDC Capenhurst (now EA Technology), Cheshire.[11] UV-light can be used separately for water purification. However, when dissolved ozone is irradiated with 254 nm wavelength UV light, a photoenhanced oxidation process occurs due to the increased reactivity caused by the hydroxyl radical (which is formed by photolysis of the ozone). This method totally oxidises a wide range of persistent organics (pesticides and polychlorinated biphenyls) and pyrogens. Other methods of generating hydroxide radicals include the coupling of hydrogen peroxide with a UV-light, H_2O_2–O_3 mixture and semiconductor photocatalysis.

7.1.3.1 *Metal Ion Oxidants.*

The AEA technology division at Dounreay, Scotland have developed a process (shown in Figure 4) for destroying organic hazardous wastes.[12] The process uses a silver salt/nitric acid electrolyte, selected as a very efficient electroactive system because of:

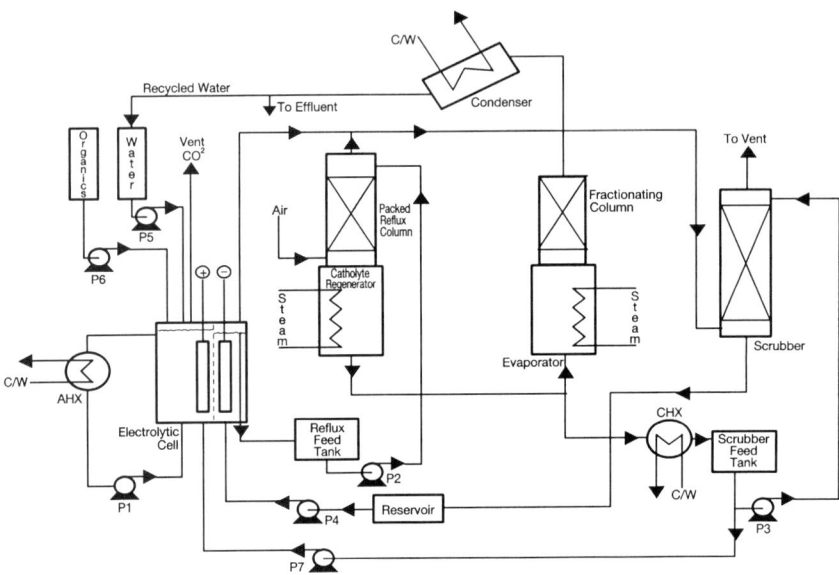

Figure 4 *Schematic diagram of the Ag ion oxidation process*
              (With permission AEA Technology)

1   The very high E^0 value of 1.98 V of the reaction $Ag^{II} \rightarrow Ag^{I}$
2   Fast electrode kinetics
3   Fast solution oxidation kinetics
4   Both ionic species are very soluble

At the anode of the cell a very highly reactive chemical species of Ag^{II} ions is formed.

$$Ag^{I} \rightarrow Ag^{II} + e^{-} \tag{7}$$

This reagent, or the free radical(s) generated from the reaction of Ag^{II} ions with water, attack organic species or other species dissolved in the waste water. The formation of radicals from the reaction of Ag^{II} is through the initial formation of a silver nitrate complex $AgNO_3^{+}$ (brown complex) which subsequently reacts with the water in the anolyte.

$$AgNO_3^+ + H_2O \rightarrow OH^\bullet + H^+ + Ag^+ + NO_3^- \tag{8}$$

The organic species then reacts with the hydroxide radical

$$Organic + OH^\bullet \rightarrow CO_2 + CO + H_2O \tag{9}$$

The overall reaction of Ag^{II} can be written as:

$$Organic + Ag^{II} \rightarrow CO_2 + H_2O + Ag^I \tag{10}$$

There also may be a contribution to the overall oxidation from direct oxidation at the anode. Any inorganic species associated with the organics, are oxidised *e.g.* sulfur to sulfate. In the cathode compartment of the cell, nitric acid is reduced to nitrous acid.

$$NO_3^- + 3\,H^+ + 2\,e^- \rightarrow HNO_2 + H_2O \tag{11}$$

The positive E° value of the cathodic reaction of nitrate to nitrite is a benefit to the operating cell voltage. The nitric acid is recovered by the thermal decomposition of the nitrous acid with air (or oxygen) in a packed column. With suitable recovery processes there are negligible losses of the silver and nitric acid reagents. Thus the overall process can be represented as:

$$Organics + O_2 \rightarrow CO_2 + CO + H_2O \tag{12}$$

At the centre of the process development are the ICI FM01 and FM21 series of cells for small scale and pilot scale operation. The anolyte and catholyte streams are separated by a Nafion cation-exchange membrane. Current is passed through this membrane by hydrated protons and to a lesser extent by silver ions.

The operation of the plant requires attention to the water management. Water is either produced from the oxidation of the organic species or is originally present in the waste feed. Water formed from the oxidation is transferred by electroosmosis through the membrane of the cell into the catholyte. This water must be distilled from the catholyte and either returned to the anolyte, in the case of concentrated feeds, or discharged, in the case of dilute feeds. There is a lower limit to the organic species concentration for optimum operation of the plant, which is approximately 5% in the case of phenol. Below this limit the feed would need to be either concentrated or blended with a more concentrated solution. A large range of compounds have been successfully destroyed using this process (see Table 2). The performance of the plant is inevitably affected by the type of compounds to be oxidised. Table 3 gives an indication of the relative charge requirements and throughputs for a plant. The equivalent power consumption is in the range 2.3–3.5 kWh kg^{-1}.

Table 2 *Species oxidised by the silver ion oxidation process*

	Product
Dodecane	Epoxy resin (Tufnol, Bakelite and Epophen)
Kerosene	Styrene divinylbenzene ion-exchange resins
Benzene	Reillex HPQ anion-exchange resin
Toluene	Phenol/formaldehyde cation-exchange resin
Octanoic acid	Polyurethane
Phenol	Cellulose (as tissues)
	Tri-*n*-butyl phosphate
Detex	Lithium based grease
Oils (cutting, lubricating, pulp, hydraulic fluids)	
Rubbers and plastics (polyethene, PVC)	
Xylene	Chlorobenzene
Acetic acid	1,2,4-Trichlorobenzene
Dinitro stilbene	1,1,2,2,-Tetrachloroethane
m-Nitro *p*-toluidine	Various polychlorinated biphenyls
p-Aminophenol	(2-Chloroethyl)ethyl sulfide
p-Nitrosophenol	Various mixed organic amines
Dimethylformamide	3-Chloropropan-1-ol
n-Butanol	
2-Methoxyethanol	
Triethanolamine	2-Chloro-4-fluorobenzoic acid

Table 3 *Relative performance of oxidation by Ag^{II} ions*

Substrate	$Charge^a$/F kg^{-1}	$Throughput^b$/kg h^{-1}
Dodecane	435.3	200
Aniline	387.1	225
Cresol	314.8	277
Nitrobenzene	243.9	257
Parathion	206.2	422
Pentachlorobiphenyl	162.3	536
Pentachlorophenol	86.3	1009
Chloroform	41.8	2085
Carbon tetrachloride	26	3350

^{a} Required for oxidation to carbon dioxide *etc.* ^{b} Of a plant sized to destroy 200 kg h^{-1} of dodecane.

In the treatment of chlorinated compounds the release of chlorine results in the formation of AgCl and thus its possible precipitation. The process can be operated without any significant precipitation of AgCl, although this is at the expense of efficiency. The treatment of bromine and iodine species would be more problematic resulting in greater losses of Ag as a precipitate. The presence of sulfur and phosphorous compounds results in the formation of sulfate and phosphate, respectively, which would build up in the system and thus require purging. Similarly metal ions, which would be oxidised to the highest oxidation state in the electrolyte, would have to be purged from the solution and recovered in other reprocessing steps.

7.1.3.2 Other Oxidants. Other metal ion oxidants, which can be generated electrochemically, include cobalt and iron. A study[13] of the electrochemical oxidation of ethylene glycol has compared the use of Co^{II} and Fe^{III} with Ag^{II} in a nitric acid electrolyte. The effectiveness of these species in the oxidation followed the reversible potentials of the associated couples Ag^{II}/Ag^{I},1.987 V; Co^{III}/Co^{II}, 1.842 V; Fe^{III}/Fe^{II}, 0.77 V, which are a measure of their oxidising power. The use of both the Ag and Co ions resulted in the complete destruction of the ethylene glycol and intermediate products (formaldehyde and formic acid). The coulombic efficiency in the case of Co^{III} was 55–64% in comparison to 99% for Ag^{II}. The use of Fe^{III} resulted in very little oxidation of the glycol and performed little better than the use of nitric acid alone.

Although Fe^{III} ions have relatively low oxidation capabilities, advantages of this species are that it is generally less toxic and requires a lower oxidation potential in regeneration. The oxidation capabilities, however, can be enhanced by higher operating temperatures of 120–180 °C. At these temperatures it is very efficient in the oxidation of polycyclic aromatics, tarry and oily wastes, chlorinated aromatics *etc.*[14] The technology is used by Chevron Chemical Co. (California).

An alternative use of the Fe^{III}/Fe^{II} couple is in the electrogeneration of Fenton's reagent. Fenton's reagent is produced by the simultaneous cathodic reduction of Fe^{III} and O_2 and subsequent reaction to the hydroxide radical:

$$O_2 + 2 H^+ + 2 e^- = H_2O_2 \tag{13}$$

$$Fe^{3+} + e^- = Fe^{2+} \tag{14}$$

$$Fe^{2+} + H_2O_2 = Fe^{3+} + OH^- + OH^\bullet \tag{15}$$

Furthermore the combined use of anodic oxidation with the generation of hydrogen peroxide and Fenton's reagent through the electroreduction of oxygen, is also proposed for the oxidative degradation of organic compounds. The use of a cocatalyst/catalyst combination has also been studied,[14] such as the use of Fe^{III}/Co^{II} in the oxidation of waste organic species, *e.g.* wood chips.

An alternative process is based on the electrooxidation of barium peroxide in aqueous surfactant suspensions which produces the reactive intermediate barium superoxide.[15] The system reaction has been applied

to the oxidation of several halogenated organics, *e.g.* 1,2,4-trichloroben-
zene, hexafluorobenzene *etc.* Destruction is initiated by nucleophilic susti-
tution of the halide by the superoxide ion, the resulting product is either
chemically or electrochemically oxidised. The superoxide ion is stabilised
by the barium ion and the surfactant. The reagents do not react with aromatic compounds containing nitro groups or aliphatic compounds containing amine or nitrile groups.

7.1.3.3   The Electrochemical Treatment of Polychlorobiphenyl (PCB). Chlorinated
organics have over the years been released into the environment as a result
of several manufacturing practices. They contaminate groundwater in
regions near to toxic dump sites, are formed by the chlorination of humic
acid in waters and are used (PCBs) as dielectric insulating fluids for trans-
formers and capacitors. The use of these materials has stopped in many
countries but still large quantities exist which require disposal. For concentrated solutions of PCBs disposal can be by incineration or involve dechlorination using sodium metal. For dilute solutions, of the order of 1000 ppm,
these technologies are expensive and alternative techniques are desirable.

The direct unmediated electrochemical reduction of chlorinated
biphenyls requires quite negative electrode potentials in organic solvents,
e.g. dechlorination of 4-chlorobiphenyl in dimethylformamide on Hg at
ca. -2.3 V *vs.* SCE. The treatment of these species will often have to occur
in the presence of water which, because of the proximity of the water
decomposition reaction, limits the use of direct reduction. A technique,
which overcomes this limitation, is to mediate the reactions in micellar
solutions. To increase the solubility of the PCBs and catalyst, surfactant
dispersions and microemulsions of surfactant, oil and water can be used.
The procedure has been demonstrated using zinc phthalocyanine as
mediator in dispersions and microemulsions of didodecyl dimethylammo-
nium bromide. Although Hg was the most effective material, carbon felt
was preferred for environmental reasons. The method is also applicable
to other pollutants such as chloroacetic acid and organohalide pesticides.
The use of insoluble surfactants and surface clay composite films coated
onto cathodes which incorporate metal phthalocyanine mediators has
also been used for similar reactions.[16]

Transformer washings can contain PCBs at concentrations up to 1000
ppm in mineral oils which are inappropriate for direct electrolysis. A
process has been developed in which the PCBs are solvent extracted into
an immiscible solvent and is suitable for electroreduction.[17] Propylene
carbonate containing tetraethylammonium chloride has been used as one
possible extractant. In operation an emulsion containing the oil is
pumped through an electrochemical cell which contains a high surface
area carbon felt electrode, to efficiently reduce the low concentrations of
PCBs. The cell operates in a flow-through mode using a perforated
graphite backplate, which also acts as the electrical feeder. In the reduction of the PCBs, the concentration can be reduced from 700– < 1 ppm.

7.2   PHOTOELECTROCHEMICAL OXIDATION

The heterogeneous photocatalytic oxidation of aqueous based organic compounds is currently of some considerable interest. Photocatalysis with semiconductor particles can overcome the energy barrier of thermodynamically feasible reactions due to the excess energy stored in the electron-hole pairs which are generated in the semiconductor by the absorption of light. The valence band (VB) holes may react with potentially oxidisable solution species, solvent or oxidisable lattice sites. The photogenerated conduction band (CB) electrons may reduce solution species and thus oxygen reduction, for example, could be used to scavenge the photogenerated electrons and thus reduce the tendency for electron-hole recombination. The holes are then left free to oxidise the dissolved species *e.g.*

$$4(\text{Semiconductor} + h\upsilon \ \longrightarrow \ h_{VB}^+ + e_{CB}^-) \tag{16}$$

$$O_2 + 4\ H^+ + 4\ e_{CB}^- \ \longrightarrow \ 2\ H_2O \tag{17}$$

$$2\ (CN^- + H_2O + 2\ h_{VB}^+ \ \longrightarrow \ 2\ H^+ + \ CNO^-) \tag{18}$$

To be able to function continuously with such a catalytic action the semiconductor has to be stable against photodecomposition. This limits the materials which are suitable to only a few oxides such as TiO_2 and SnO_2. These semiconductors all have wide band gaps (*e.g.* TiO_2, 3 eV) and absorb light only in the UV wavelength and the most commonly used semiconductor is TiO_2.

The photooxidation of several organic species, *e.g.* 4-chlorophenol, trichloroacetate, hydroquinone, *p*-aminophenol, analine, ethanol and several inorganic species, (CN^-, S^{2-}, I^-, Br^-, Fe^{II}, Ce^{III} and Cl^- ions) has been achieved.[18] In many cases the complete mineralisation of the organic has been demonstrated. In terms of the engineering of the semiconductor system two main methods have been employed; particulate materials in the form of suspensions or slurries and immobilisation of the catalyst on a suitable substrate.

7.3   TREATMENT OF WASTE WATERS CONTAINING INORGANIC COMPOUNDS

7.3.1   Cyanides and Thiocyanates

Several methods for cyanide destruction have been adopted and proposed based on the oxidation of the cyanide to cyanate, which then decomposes further to innocuous products such as carbonate and nitrogen. Electrochemical methods of cyanide treatment use direct oxidation and indirect oxidation. The electrochemical method offers the advantages of reduced handling of reagents and in many applications the possibility of simultaneous recovery of dissolved metal ions.

7.3.1.1 Direct Oxidation. The electrochemical oxidation of cyanide in an alkaline environment is believed to proceed according to the following reactions:

$$\text{discharge:} \qquad 2\ CN^- \rightarrow 2\ CN^\bullet + 2\ e^- \tag{19}$$

$$\text{dimerisation:} \quad 2\ CN^\bullet \rightarrow (CN)_2 \tag{20}$$

$$\text{overall:} \qquad 2\ CN^- \rightarrow (CN)_2 + 2\ e \quad (E^o = -0.18\ V) \tag{21}$$

The cyanogen formed can then undergo alkaline hydrolysis to cyanate

$$(CN)_2 + 2\ OH^- \rightarrow CN^- + CNO^- + H_2O \tag{22}$$

At high pH, cyanide oxidation could result in cyanate directly

$$CN^- + 2\ OH^- = CNO^- + H_2O + 2\ e^- \quad (E^o = -0.97\ V) \tag{23}$$

The cyanate reacts rapidly with alkali to give NH_4OH, NH_4HCO_3 and Na_2CO_3. Electrode materials which are effective for cyanide oxidation are platinised titanium, copper, stainless steel, magnetite and graphite, although the loss of platinum from the former is high. Kinetic studies of cyanide oxidation suggest that the oxidation of cyanide is zero order in the concentration of CN^- ion at relatively high concentrations of ≥ 40 mg dm^{-3}. The direct oxidation of cyanide is very slow on platinum and graphite anodes. The reaction is apparently hindered by adsorbed species and is thus kinetically limited.

The early industrial practice for cyanide oxidation used tank (mild steel) electrolysers with copper electrodes operating in a batch mode with air agitation. Relatively high temperatures, up to 100 °C, and high current densities of typically 400 A m^{-2} were used. The cyanide concentrations were reduced from values of 20 000 to 100 000 ppm to below 1 ppm. Several anode materials are known to oxidise/corrode during the treatment of cyanide, including graphite, PbO_2 and RuO_2. Lead dioxide-coated titanium electrodes exhibit high current efficiencies in the oxidation of sodium cyanide solutions of concentration of 0.23–1.0 mol dm^{-3}, at current density of 500 A m^{-2}, although corrosion rates of 0.4 mg A h^{-1} occurred. Nickel has recently been considered as an anode material in cyanide oxidation,[19] in alkaline electrolysis, as it is known for its high corrosion resistance in alkaline solutions. High current efficiencies were achieved although the current densities were low.

7.3.1.2 Indirect Oxidation. The indirect oxidation of cyanide is based primarily on the oxidation of chloride ions to produce hypochlorite. In practice the cyanide feed solution can be dosed with sodium chloride as a saturated solution and passed continuously through the cell. The indirect process is said to have several potential advantages over direct oxidation, which include:

(i)   a lower cell voltage, through the increased conductivity;
(ii)   chloride ion oxidation is an inherently fast reaction;
(iii)   use of platinum or DSA type coated electrodes gives lower over-voltages and reduced wear.

However, the indirect process must avoid the formation of chlorate and minimise the formation of slime in the cell, and thus any requirements for sludge handling. The economics of the process are quoted to be superior to the use of direct dosing with chlorine, presumably due to the latter requiring chlorine storage and handling. Energy consumptions are quoted at around 4–10 kWh kg^{-1} CN$^-$.

The treatment of cyanates and thiocyanates can generally use technology identical with that for cyanides. In the treatment of thiocyanate at low pH the SCN$^-$ species is oxidised to cyanide which can be recovered as HCN or may be further oxidised to give relatively harmless products.

7.3.1.3 Treatment of Metal Cyanides. The treatment of the metal cyanides has commonly relied on the sludging of the metal, as hydroxide, after the removal of the cyanide, by alkaline chlorination. This, however, must not be used for concentrations above *ca.* 7.5 g dm^{-3}, when the toxic cyanogen chloride can be formed or when the oxidation is slow as the metal cyanide may precipitate in the sludge.

Metal cyanides can also be decomposed by electrochemical oxidation: at low concentrations of cyanide (<500 ppm) by *in-situ* hypochlorite oxidation, and at higher concentrations of cyanide (>1000 ppm) by direct oxidation. This, however, can be unsatisfactory if the uncomplexed metal is insoluble and may precipitate, especially if this occurs on the anode.

e.g.

$$[M(CN)_n]^{(n-z)-} + 2(n+z)\ OH^- \rightarrow M\ (OH)_z + n\ CNO^- + nH_2O + 2\ ne^- \quad (24)$$

For example the oxidation of cuprocyanide complexes [Cu(CN)$_n$$^{(n-1)^-}$, *n*=2,3 or 4] results in the formation of copper oxides (red and black) at values of pH >12. At lower pH (7–11) the formation of azulmin (HCN polymer) occurs along with cyanate. At pH between 5–7 the oxalate ammonium ions and white oximide can be produced. The current efficiencies for direct oxidation of several cyanides Cu^I, Cd^{II} and Zn^{II} are low and ferricyanide complex oxidation is not possible on most materials *e.g.* PbO$_2$ on Ti.

In the treatment of metal cyanide complexes it can be preferable to first liberate the cyanide by metal deposition

$$[M(CN)_n]^{(n-z)^-} + z\ e^- \rightarrow M + n\ CN^- \tag{25}$$

and then oxidise the cyanide. This is because the rate of the metal deposition reaction is greater than the rate of oxidation of the cyanide, which is kinetically controlled on most anodes. The oxidation and deposition may be carried out simultaneously in a divided cell where the metal depleted catholyte is used as the anolyte although it is desirable to use undivided tank cells.

In certain applications ozone is considered as an alternative oxidant because of the potential for undesirable formation of chlorine compounds with the latter. It has been successful in the treatment of simpler cyanides of Na, K, Cu, Cd and Zn but not for tighter complexes. The rate of oxidation with ozone is generally improved by using catalysts, such as copper, in bubble contactors. There is, however, a higher direct cost associated with ozone in comparison to chlorine.

7.3.2   Treatment of Chromium Liquors

Liquors containing dissolved chromium are used in a number of sectors of the process industries in applications such as plating and coating, metal finishing and as oxidising and reducing agents. The most common form is hexavalent chromium, which is highly toxic to man and to aquatic life. The concentration limit in public water supplies is of the order of 50 ppb. Electrodeposition of chromium is not efficient and therefore other electrochemical methods are used[20] such as, anodic oxidation of Cr^{III} to Cr^{VI}, combined Cr^{III} oxidation and cation removal using electrodialysis and precipitation of chromium using anodically generated Fe^{2+} ions.

7.3.2.1 Precipitation of Chromium. Precipitation is used to remove the chromium from metal-finishing streams and cooling waters. The process, (Andco Chemical Corp), uses an undivided cell with cold-rolled steel-plate electrodes connected in a bipolar configuration, the anode side of which dissolves to form the Fe^{II} ions. The cathode side of the electrodes generates hydroxide ions through the formation of hydrogen gas. The Fe^{II} and OH^- ions combine in the overall process which involves the reduction of Cr^{VI} to Cr^{III} by Fe^{II} with the subsequent precipitation of the insoluble hydroxides of chromium and Fe^{II} and Fe^{III}.

$$3\ Fe^{2+} + CrO_4^{2-} + 4\ H_2O = 3\ Fe^{3+} + Cr^{3+} + 8\ OH^- \tag{26}$$

The power consumption of the commercial unit is approximately 11 kWh kg^{-1} of heavy metal removed, and the concentration of heavy metal

Table 4   *Comparison of methods for hexavalent chrome ion treatment*

Process	Annual cost without credit [a]	Annual cost with credit
Sulfide precipitation	$148 300	$148 300
Hydroxide precipitation	$132 200	$132 200
Electrochemical	$107 300	$107 300
Liquid-ion exchange	$116 700	$96 600
Evaporation	$86 400	$72 500
Ion exchange	$86 600	$59 300

[a] Credit — market value of recovered chrome.

at discharge is less than 1 ppm. An economic comparison of alternative methods of chrome removal from metal plating waste water has been reported[21] in which six processes were considered; hydroxide ppt, sulfide ppt, electrochemical ppt (Andco Process), evaporation, ion exchange and liquid-ion exchange. The costs of the processes for a 13.5 dm^3 min^{-1} operation with 30 ppm hexavalent chrome are compared in Table 4. The cheapest process was ion exchange which is capable of direct recycling of the chrome as chromic acid to the process, as well as recycling of the process water. The fact that this operation eliminates the sludge formation is an addition environmental incentive to the process.

The precipitate, which is formed in the Ando process, tends to collect on the electrodes and this can be removed by intermittent acid washing. The formation of this precipitate on the electrode has been used in a process in which the Cr^{VI} is reduced directly at a high surface area cathode (*e.g.* carbon felt).

$$CrO_4^- + 4\ H_2O + 3\ e^- = Cr(OH)_3 + 5\ OH^- \tag{27}$$

The resulting precipitate remains in the cell chamber adhered to the surface as a charged colloid. The chrome is recovered from the cell by either chemical dissolution or by chemical oxidation with hypochlorite (or possibly by electrochemical oxidation).

$$2\ Cr_3^+ + 4\ H_2O + 3\ NaOCl \rightarrow Cr_2O_7^{2-} + 8\ H^+ + 3\ NaCl \tag{28}$$

The application of this recovery process is in the industrial production of sodium chlorate where Cr^{VI} is added to the electrolyte to improve the current efficiency (see chapter 8).

7.3.2.2 Anodic Oxidation of Cr^{III}. When Cr^{VI} is used as an etchant, Cr^{III} ions accumulate in the process liquors. This Cr^{III} can be anodically oxidised to Cr^{VI} to regenerate the etchant:

$$2\ Cr^{3+} + 7\ H_2O - 6\ e^- \rightarrow Cr_2O_7^- + 14\ H^+ \tag{29}$$

The electrolysis takes place in a divided cell to prevent the reduction of the Cr^{VI} and thus a loss in the current efficiency. The oxidation of Cr^{III} can also be applied to the regeneration of precipitates of chromic oxide, or hydroxide, by first dissolving them in chromic acid prior to oxidation.

7.3.2.3 Combined Oxidation and Electrolysis. There are cell designs used for the recycling of chromium ions which combine oxidation with ion-exchange membranes. These cells are used for the removal of ionic impurities which have been picked up in the chrome solutions by, for example, dissolution of the workpiece. The method shown schematically in Figure 5, uses a cell with a cation-exchange membrane to separate the chromic acid solution from the catholyte contaminant reservoir.

The contaminant cations are transported through the membrane whilst the chromium anions remain in the anolyte and are concentrated by the anodic oxidation of Cr^{III}. There are two types of cell designs used commercially; plate-in-tank or filter press designs (Chromium Oxidation and Purification Systems, Scientific Control Labs Inc.) and concentric cell

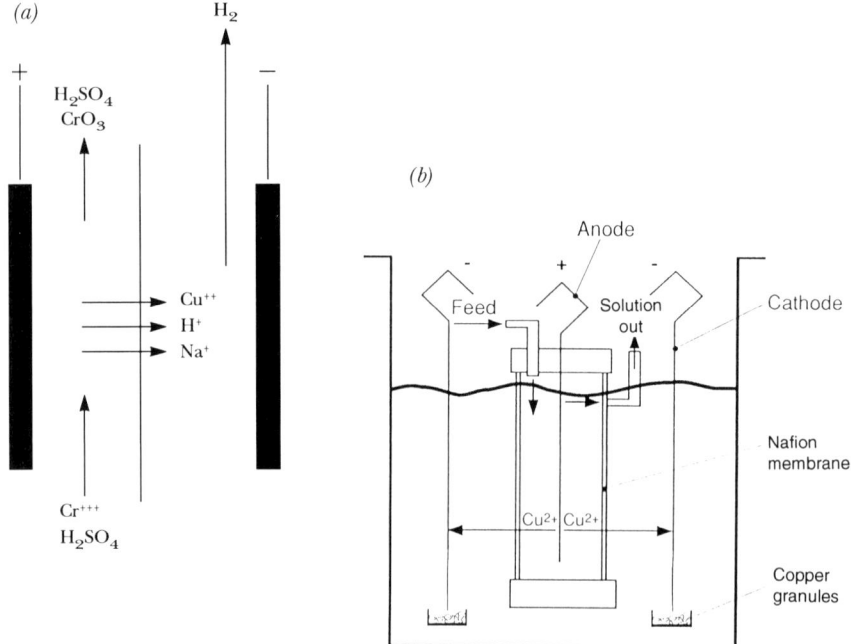

Figure 5   *Cell processes for the recycling of chromium ions*

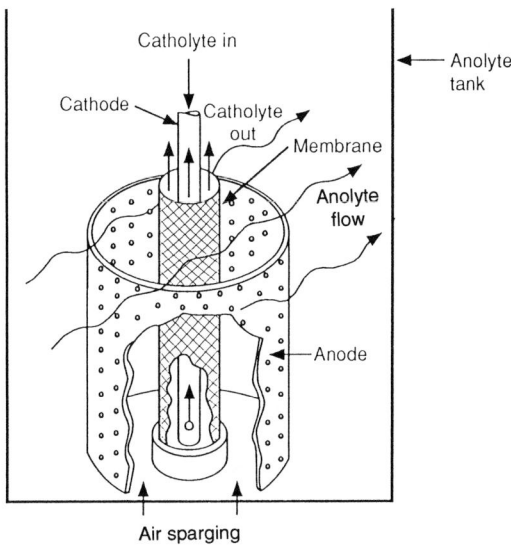

Figure 6   *Concentric cylinder cell for chromium recycling*

designs (The Pure Tec System, Ultrasonics Inc.). The surface area of the anodes (Pb or Pb alloys) is generally much greater then that of the cathodes to ensure low current densities which minimise the anodic generation of oxygen. In the concentric cylinder arrangement (see Figure 6) the cathode can be a hollow cathode rod and the anode a perforated tube and a series of these unit cells can be placed in a common tank with built in heat exchange coils. Cross flow of anolyte through the cell and also possible air sparging can be used to improve the mass transfer and mixing characteristics of the unit.

   With appropriate contaminant cations it is possible to duel up the function of the cell and, for example, in the case of copper remove the contaminant continuously by electrodeposition of the base metal as a powder.

7.3.2.4 Galvanic Reduction. A method for removing hexavalent chromium from waste waters using galvanic reduction with iron scrap is described by Abdo and Sedahmed.[22] The method claims to generate electrical energy from the reactions:

Steel anode

$$3\ Fe \rightarrow 3\ Fe^{2+} + 6\ e^- \qquad E_o = 0.44V \qquad (30)$$

Cathode

$$2\ CrO_4^{2-} + 16\ H^+ + 6\ e^- \rightarrow 2\ Cr^{3+} + 8\ H_2O \quad E_o = 1.33V \qquad (31)$$

The standard equilibrium constant for this reaction is high and in principle the cell can almost completely remove the chromate ions. The energy output of the cell varies from 0.14–1 kWh kg^{-1} chromate, using chromate ion concentrations in the range 0.1–1.2 mol dm^{-3}. Current densities for this system are low, of the order of 35 A m^{-2} and lower.

7.3.3   Other Inorganic Compounds

The valence states of most inorganic compounds can be changed by either cathodic reduction or anodic oxidation enabling, in principle, appropriate species to be recycled electrochemically. Sulfur and nitrogen compounds are thus frequently encountered *e.g.* the reduction of nitrates and nitrites. The oxidation of sulfur compounds such as sulfite, thiosulfate and dithionite can be carried out quite effectively by anodic oxidation. The oxidation of sodium dithionite in an undivided cell with graphite anodes and stainless steel cathodes has been performed at a pilot plant level.[23] This application was for a waste water from a dithionite production plant and contained 40 g dm^{-3} Na$_2$S$_2$O$_4$ and 249 g dm^{-3} NaCl. The process was capable of reducing the concentration of dithionite to 1 g dm^{-3} at a current density of 5000 A m^2 with very high current efficiencies and with energy consumptions of *ca.* 3.5 kWh kg^{-1}. The scheme is a staged cell with combined air oxidation. The cell was also used for the oxidation of sulfite and thiosulfate and high current efficiencies were obtained.

7.3.3.1 Hydrometallurgical Treatment of Sulfide Ores. The hydrometallurgical treatment of sulfide ores (e.g. Hg, Ag, Mo, Cu and Sb) using electrooxidation is a means of eliminating SO$_2$ emissions from smelting operations. The ore, after grinding and pulping with brine solution is electrolysed to generate hypochlorite which oxidises the sulfide mineral to sulfate

$$MS + 4\ OCl^- \rightarrow MSO_4 + 4\ Cl^- \qquad\qquad (32)$$

Subsequent recovery of the metal is achieved by, for example, precipitation with an active metal such as zinc. The cell is undivided using either lead dioxide coated titanium anodes or graphite anodes. The slurry is agitated mechanically to force it through the electrodes (interelectrode gap — 1 cm) to achieve effective generation of hypochlorite.

7.3.3.2 Bromine Generation. A system for the regeneration of bromine from spent NaBr streams has been developed by Great Lakes Chemicals. The system, based on the Chloropac^R cell (Electrocatalytic), has applications in the gold extraction industry as an alternative to the conventional cyanide process. The use of bromine is kinder to the environment than cyanide and the NaBr effluent can be re-used to regenerate bromine.

Other technologies can reprocess the liquors from organic brominations to recover the Br_2 and NaOH. Furthermore a system for the disinfection of swimming pools based on hypobromite is available. This is a modification of a hypochlorite generator system marketed by Olin, called the Lectranator[R].

7.3.4   Sterilisation of Water and Waste

There are several waste water treatment applications which involve electrochemically generated oxidants. Suitable oxidants includes ozone, hydrogen peroxide and notably hypochlorite, which is widely used.

7.3.4.1 Hypochlorite. Hypochlorite can be produced by the reaction of chlorine gas with sodium hydroxide solution. Although storage of the active chlorine as hypochlorite is safer than as chlorine gas, the capacity for storage per unit of chlorine is reduced. On-site generation of hypochlorite avoids storage and transport difficulties associated with chlorine gas, and is convenient for many applications, including: sewage treatment, sterilisation of water, disinfection and biological growth prevention. The basic reaction involved in the electrogeneration of hypochlorite can be summarised as:

Anode: $2 \ Cl^- - 2 \ e^- \rightarrow Cl_2$ (33)

Cathode: $2 \ H_2O + 2e^- \rightarrow H_2 + 2 \ OH^-$ (34)

Chemical: $Cl_2 + 2 \ OH^- \rightarrow H_2O + OCl^- + Cl^-$ (35)

The primary cell reactions are the kinetically controlled formation of chlorine at the anode and hydrogen at the cathode. The chlorine evolved is rapidly hydrolysed, the dissolved concentration of chlorine in commercial cells is negligible. The performance of the electrolysers can be affected by several factors which include so called loss reactions (see chapter 8) and the anodic evoluton of oxygen, which may result from poor electrolyte agitation or too low a concentration of chloride ion. The important reactor performance criteria in electrochlorinators are:

(i)   current efficiency: influenced by current density, anode material, temperature, chloride concentration and conversion;
(ii)   electrical energy consumption;
(iii)   anode life;
(iv)   frequency of system cleaning.

Good commercial electrocatalysts for chlorine generation are typically

DSA, RuO_2 based coated titanium material or a Pt/Ir coated Ti material. A range of commercially available coated metal anodes is:

- Optima PTA — platinum plated, clad and thermally deposited
- Optima PTA — platinum iridium
- Optima RUA — ruthenium oxide
- Optima IOA — iridium oxide
- Optima COA — combined metal and oxide
- Optima NOA — nickel oxide
- Optima PDA — palladium
  (With permission of Electrode Products Ltd)

These coatings demonstrate some energy benefits over other materials such as graphite and lead dioxide. The lifetime of these electrodes is now established as several years. The cathodes in hypochlorite cells must be good electrocatalysts for hydrogen evolution, and be stable under extended periods of service. Ti, Hastelloy and nickel alloy cathodes are commonly used although the former can suffer from hydrogen enbrittlement by hydrogen diffusing interstitially from the cathode surface. If the coated titanium electrode is operated in a bipolar manner this enbrittlement could lead to early anode failure. This problem can be resolved by sandwiching a conducting material, impervious to hydrogen, between the anode and the cathode surface. In some applications the ratio of the cathode area to anode area is low to minimise the back reduction of OCl^- ions.

The majority of cell designs use parallel plate or mesh electrode geometries connected either in a monopolar or a bipolar configuration. These cells can be mounted in rectangular tanks or in cylindrical vessels. A number of commercial cell designs are available (*e.g.* Krebs, Cumberland Engineering) for hypochlorite generation based on an undivided configuration. Figure 7 shows a typical cell process, used in sea water electrolysis, which operates on a once through basis. A single pass operation is used because the feedstock is essentially free. If there is significant cost associated with the feedstock, then recycle may be necessary to increase the conversion of the chloride. In most applications the cell product contains between 0.01–1.0 wt% of hypochlorite ions. In the use of sea water the formation of insoluble precipitates (hydroxides) on the electrodes has to be minimised by good hydrodynamic cell design. Even with this there is still a need for periodic electrode and system cleaning, which can be done by acid washing or in some cases reversal of the cell polarity. In the latter case due care is required to prevent corrosion of the electrodes. The energy consumption of sea water hypochlorite generators is in the range of 3.5 –4.0 kWh kg^{-1} at current densities between 1000–5000 A m^{-2}.

Small scale units for on-site generation of hypochlorite (or chlorine) for disinfection and other applications are available. For example Electrocell have marketed a skid mounted unit, shown in Figure 8, for the electrochemical generation of chlorine by the electrolysis of industrial

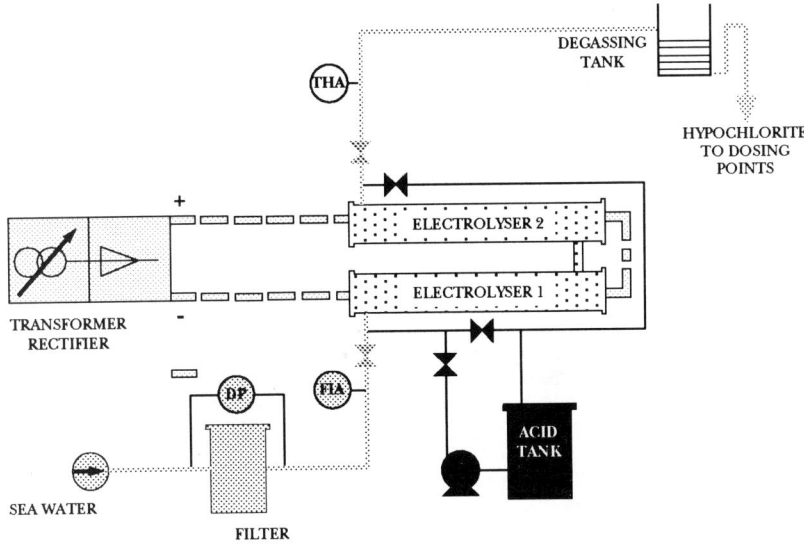

Figure 7   *Industrial electrolyser for the generation of hypochlorite*
(Courtesy of Cumberland Engineering Ltd)

grade HCl. The cells operate with internal gas circulation and separation of the Cl_2 and H_2 gas is achieved by a proprietary separation unit supported by multiple sensors to ensure safe operation. The unit can produce up to 0.6 kg h^{-1} of chlorine depending on the number of cells and current density used.

7.3.4.2 Sterilisation Applications. An important use of electrochlorination is as a disinfectant in water supplies and water storage facilities. The scale of the operation can be viewed by the applicaton at the Thames Water Hambleden site where it is used, following the treament of bore hole water by UV radiation, to destroy bacteria. The provision of water here is of the order of 12–16 million gallons per day. The electrochlorination system (Electrocatalytic) supplies a dosing of 0.2 mg dm^{-3} of chlorine from three electrogenerators fed with brine at the rate of 45 dm^3 h^{-1}.

The *in-situ* generation of sodium hypochlorite was applied for direct oxidation of sewage on the island of Guernsey, after extensive development work by Constructors John Brown (CJB) Ltd. The electrolysis of the sea water also resulted in the generation of magnesium hydroxides, which formed useful flocculating agents. Problems encountered with this type of application include excessive wear of coated anodes at low operating temperatures and fouling of the electrodes by alkaline earth salts. The application of electrolysis in the treatment of mixtures of secondary treatment plant effluent and sea water has been described[24] with an aim of reducing the algal growth in the Adriatic sea.

Figure 8   *Small scale hypochlorite generator*
(Courtesy of Electrocell AB)

7.3.4.3 Water Disinfection. The disinfection of water by 'direct' electrolysis
is in effect a modification of indirect oxidation, probably chlorination.
Disinfection is effective at residual chloride ion concentrations of > 0.7 g
m^{-3} for surface water and 0.3 g m^{-3} for ground water. The effectiveness of
disinfection is controlled by the coli index and the chlorine content of
the water.

An electrolytic cell, for the disinfection and purification of water and
fluids, using an array of staggered bipolar bars as the electrodes (see
Figure 9) has recently been developed.[25] Because of the low conductivity
of the water to be electrolysed the interelectrode spacing is small to minimise energy consumption. The electrolyte flow through the cell and the
high rate of electrochemical gas (hydrogen and oxygen) generation lead
to turbulent conditions, which assists the bubble removal from the electrodes. Bacteria are effectively removed completely. A major problem
with this application is the formation of insoluble precipitates (hydrated
oxides or hydroxides of Mg and Ca) which foul the electrode and the cell.
The problem can be resolved using regular current reversal in which the
polarity of the bipolar cells is reversed at regular intervals. To withstand
this aggressive electrolysis operation a stable conducting ceramic material, Ebonex (magneli phase Ti_4O_7), is used.

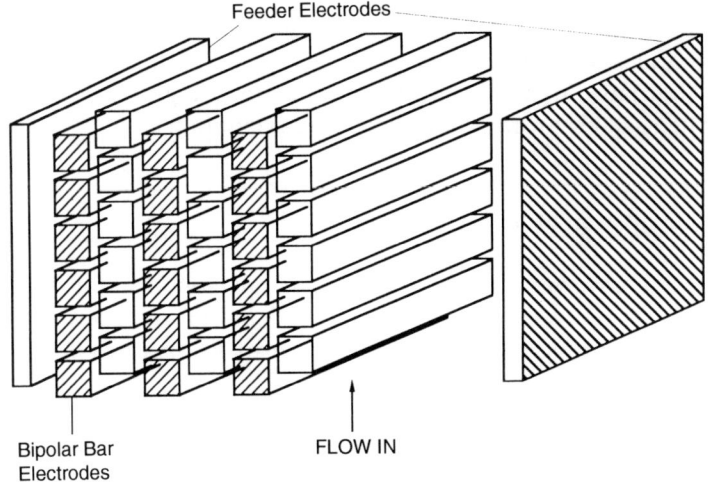

Figure 9 *Bipolar rod cell for water purification*

An alternative method, which potentially could eliminate the problem of low electrolyte conductivitiy of the waters, is to use solid polymer electrolytes. This can be applied in the post-treatment of reclaimed wastes, final polishing, prior to re-use and constitutes the removal of organic impurities at low levels of 100 ppm to 500 ppb total organic carbon (TOC).[26]

7.4   ELECTROCHEMICAL TREATMENT OF GASES

The emission of a number of acid gases, *e.g.* SO_2, NO_x, H_2S and HCl, is of environmental concern throughout the world. These potential gaseous pollutants can be converted electrochemically to species which are more environmentally acceptable and that may have some commercial value. Sulfur dioxide can be oxidised to sulfuric acid or reduced to sulfur, chlorine can be reduced to chloride ions, hydrogen sulfide can be oxidised to sulfur and nitrous oxides can be oxidised to nitric acid.

$$Cl_2 + 2e^- \rightarrow 2Cl^- \qquad E^o = + 1.36 \text{ V} \tag{36}$$

$$SO_4^{2-} + 4 H^+ + 2e^- \rightarrow SO_2 + 2 H_2O \quad E^o = + 0.169 \text{ V} \tag{37}$$

$$S + 2 H^+ + 2e^- \rightarrow H_2S \qquad E^o = +0.142 \text{ V} \tag{38}$$

$$NO_3^{2-} + 4H^+ + 3e^- \rightarrow NO + 2 H_2O \quad E^o = +0.958 \text{ V} \tag{39}$$

The standard electrode potentials for the above reactions are in the range of the water decomposition reaction and thus the electrochemical conversions are feasible in the aqueous phase. Consequently there have been several studies and proposals for electrochemical treatment of these gases. Alternative treatment processes operate at high temperatures and use molten salt electrolytes immobilised in membranes.

7.4.1   Oxidation of Sulfur Dioxide

A worldwide environmental problem is the emission of sulfur dioxide gas from power stations, and from the chemical and metallurgical industries. Available technologies for desulfurisation of these gases can be summarised as:

(i)   Conversion to calcium sulfate (gypsum) using wet gas scrubbing with lime and oxidation with air.

(ii) Concentration by aqueous absorption and desorption (Wellman–Lord process)

$$Na_2SO_4 + H_2O + SO_2 \longrightarrow 2\ NaHSO_3 \tag{40}$$

which may be followed by the catalytic oxidation of the concentrated SO_2 to sulfuric acid.

(iii) High temperature reduction with natural gas to produce elemental sulfur.

(iv) Oxidation of SO_2 to sulfuric acid using high-temperature catalysis, or other methods utilising redox couples, *e.g.* ammonia, hydrogen peroxide or manganese chelate compounds.

Electrochemical analogues of the above processes exist although the only technically advanced one is based on indirect oxidation *i.e.* the ISPRA Mark13A process, resulting from the ECs hydrogen programme (1977–1980).

7.4.1.1 Electrochemical Oxidation. Interest in the anodic oxidation of SO_2 has arisen from two quite different areas; removal from flue and waste gases and in the 'electrolytic' generation of hydrogen in the sulfur–hydrogen hybrid cycle. The latter arose due to research on hybrid thermochemical cycles in connection with the 'hydrogen energy concept' in which hydrogen is proposed as the major energy vector for the future.[27] Much of the research in this area is relevent to the treatment of waste gas streams containing SO_2 and is thus briefly discussed here.

7.4.1.2 Sulphur–Hydrogen Cycle. The sulphur–hydrogen cycle production process, is essentially composed of a thermochemical and an electrochemical reaction. In the electrolysis stage the sulfur dioxide, dissolved in aqueous solution, is anodically oxidised while hydrogen is produced at the cathode.

$$SO_2 + 2\ H_2O \rightarrow H_2SO_4 + H_2 \tag{41}$$

The sulfuric acid produced in the electroyser is then concentrated using thermal energy from a high temperature heat source. The sulfuric acid is then catalytically decomposed to sulfur dioxide and the SO_2 returned to the electrolyser.

$$H_2SO_4 \rightarrow H_2O + SO_2 + \tfrac{1}{2}\ O_2 \tag{42}$$

The standard thermodynamic potential of the electrochemical reaction is only 0.17 V and rises with temperature. In a 50 wt% solution of sulfuric acid, at 1 bar SO_2 pressure and 80 °C, the value is 0.3 V. This is attractive when compared to the 1.23 V for the electrolysis of water, reducing the thermodynamic energy demand by *ca.* 75%.

7.4.1.3 Anode Materials. Platinum and activated charcoal are typical anode materials chosen for SO_2 oxidation. A comparitive kinetic study of SO_2 oxidation[28] using several materials showed that Ir, Re and Rh anodes were relatively inactive during oxidation. Au and Ru anodes showed approximately the same activity as Pt, and Pd was the best of the anode materials tested. The DSA type of materials such as $RuO_x–TiO_2/Ti$ and $IrO_x–TiO_2/Ti$ were inactive.

More recent studies of the direct oxidation of SO_2 include platinum supported (1.5 mg cm^{-2}) flow through porous graphite sheet anodes[29] and oxygen reduction cathodes. The rate of oxidation was significantly enhanced by a reductive pre-treatment, which produced a sulfur modified surface. The oxidation of SO_2 in neutral and alkaline electrolytes gives sulfate and dithionate as the oxidation products. The reaction can be carried out in a divided cell as sulfite ions are not reduced to sulfur at these values of pH; the cathodic product is hydrogen. The mechanism of the anodic oxidation of SO_2 in 100 mol m^{-3} H_2SO_4 solution has been investigated on platinum and lead dioxide.[30] Rather than a simple model, based on a single Tafel equation for the overall reaction of sulfite to sulfate, two mechanisms are proposed for the oxidation, based on a sulfite intermediate and an oxide mechanism involving platinum oxide:

$$SO_3^{2-} \rightarrow SO_3^{2-*} \tag{43}$$

$$SO_3^{2-*} + PtO \rightarrow SO_4^{2-} + Pt \tag{44}$$

$$Pt + H_2O \rightarrow PtO + 2\ H^+ + 2\ e^- \tag{45}$$

These processes are proposed to occur in parallel and the experimental results showed that on platinum the reaction occurred mainly by the direct mechanism, whereas on lead dioxide the reaction occurred mainly by the oxide mechanism.

7.4.1.4 Electrolyser Design. The laboratory cell used in the Westinghouse sulfur-hydrogen cycle process[27] is shown schematically in Figure 10. The cell used platinum activated carbon electrodes separated by a porous diaphragm, used in preference to a cation-exchange membrane to prevent the diffusion of SO_2 towards the cathode. This was achieved by having a flow of catholyte through the separator under the influence of a slightly higher pressure on the catholyte side of the diaphragm. This procedure is essential to stop the cathodic reduction of SO_2 to sulfur which would precipitate onto the cathode and cause de-activation. This will occur even using electrolytic cells fitted with an ion-exchange membrane materials due to the transport of sulfur dioxide in solution through the material. Under practical operating conditions the Westinghouse cell operated with a voltage of 0.68 V at a current density of 2000 A m^{-2}.

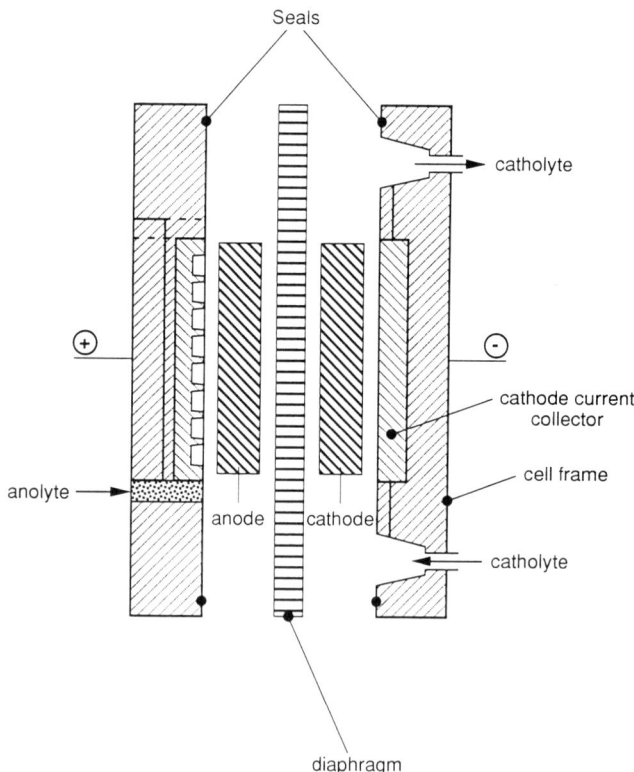

Figure 10 *Laboratory scale HBr electroysis cell*

Overall in the electrolyser design a determining factor is the prevention of de-activation of the cathode by sulfur and so other methods were also considered:

(i)   the use of gas diffusion electrodes, where the SO_2 supplied on the gas side is mainly depleted at the anode and so the scope for diffusion to the catholyte is small;

(ii)   the use of cathode materials less sensitive to sulfur (or H_2S) formation such as tungsten carbide;

(iii)   the use of a third electrolyte compartment between two-cation exchange membranes in which the SO_2 is transported out of the cell by the flow of this third electrolyte. The cell voltage is not severely penalised if the concentration of sulfuric acid used is at its maximum conductivity *i.e.* at 35 wt% and 80 °C.

7.4.1.5 Sulfur–Bromine–Hydrogen Cycle. An alternative to the Westinghouse sulfur–hydrogen cycle is based on the sulfur–bromine cycle (ISPRA Mark 13) which comprises of three reactions:

$$2\ HBr \rightarrow H_2 + Br_2 \tag{46}$$

$$Br_2 + SO_2 + H_2O \rightarrow 2\ HBr + H_2SO_4 \tag{47}$$

$$H_2SO_4 \rightarrow H_2O + SO_2 + \tfrac{1}{2}\ O_2 \tag{48}$$

Although this reaction cycle contains an additional reaction step in comparison to the sulphur–hydrogen cycle, the electrochemical step for the production of bromine is simpler than the electrochemical oxidation of SO_2. In the above cycle reaction (47) produces gaseous HBr as one product and highly concentrated sulfuric acid (70–80%) as the other. This can almost eliminate, the need to concentrate sulfuric acid in comparison to the Westinghouse process. The reaction between SO_2 and Br_2 must produce HBr as a vapour in high purity and produce sulfuric acid at a concentration as close to 85% as possible. The chemical equilibria of the reaction does not allow this to be achieved in one stage, without considerable amounts of sulfur dioxide and bromine present in the vapour. In practice this can be carried out in two stages as shown in Figure 11.

In the electrochemical step the stability of the electrode material in the corrosive, highly concentrated HBr solutions (47.5%) with Br_2 is a major factor. Only Ta, Pd and Pt metals were considered feasible. Ta was considered to be suitable but however it exhibits both high hydrogen and bromine over-voltages and requires the addition of expensive electrocatalyst coatings. Palladium metal interstitially absorbs hydrogen which causes a shift to more negative potentials. Platinum coated with Pt black was considered the best of the metal electrodes because of the low hydrogen over-voltage. However, the severe problems associated with the anodic

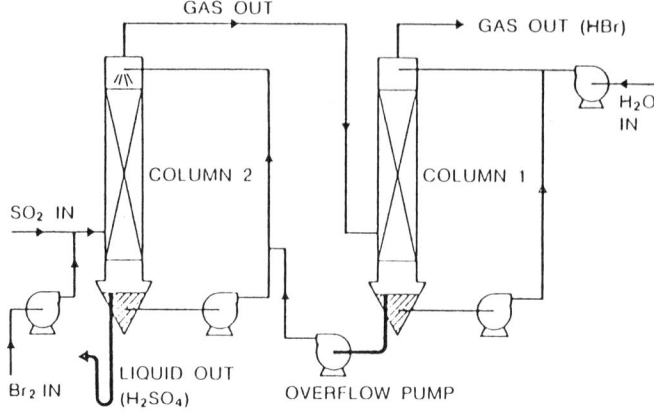

Figure 11 *Two-stage arrangement for the reaction of SO₂ and Br₂*

generation of bromine in concentrated HBr solutions and those associated with the cathode, which even in a divided cell would be exposed to bromine, resulted in the selection of graphite for both the anode and cathode. Corrosion tests carried out at a temperature of 100 °C, at a current density of 10 000 A m^{-2} in aqueous electrolyte containing 40% HBr and 10% Br$_2$ identified several types of graphite as suitable anode materials. Anodic over-voltages of Br$_2$ on good electrographites are less than 0.05 V at current densities up to 10 000 A m^{-2}. The corresponding hydrogen over-voltage for graphite is too high if acceptable energy consumption is to be achieved. The deposition of a thin layer of suitable electrocatalyst onto the graphite cathode was therefore applied.

7.4.1.6 Electrolyser Performance. The research and development of the cell design for HBr electrolysis considered both an undivided monopolar design and a divided bipolar design. The divided cell used a thermostabilised 0.5 mm thick PVC tissue separator as the diaphragm, which has the required stability in HBr/Br$_2$ solutions at 80 °C. This material and the cell design is similar to the industrial cells used for the electrolysis of HCl by the F.Uhde company in Dortmund. The cell interelectrode gap was 3 mm and resulted in a typical voltage performance of 0.85–0.9 V at 3000 A m^{-2} with a 45 wt% solution of HBr at a temperature of 100 °C.

The undivided cell for HBr electroysis, shown schematically in Figure 12, used graphite electrodes. The design requirement was for the separate production of hydrogen gas and the Br$_2$. To achieve efficient hydrogen gas release the cell is mounted horizontally and the electrolyte flow is laminar to provide controlled convective diffusion. The interelectrode gap was at least 4–5 mm and to prevent H$_2$ disturbing the electrolyte flow and causing mixing and thus recombination of hydrogen and bromine. With this larger interelectrode gap, in comparison to that of the diaphragm cell, the cell voltages of both units were virtually the same.

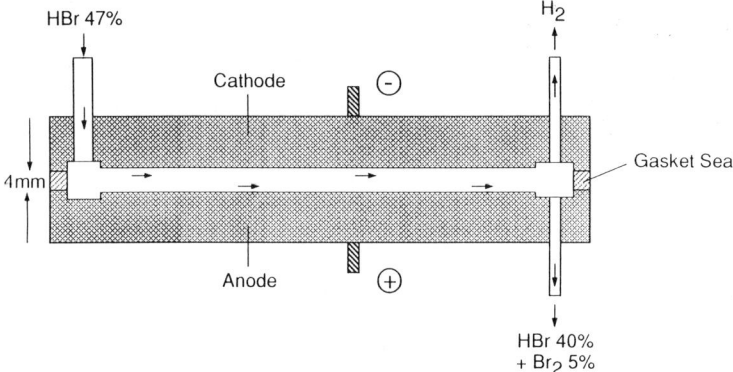

Figure 12 *Rectangular undivided cell for HBr electrolysis*

The current efficiencies for Br$_2$ generation were similar — 90–100% for the undivided cell producing 1–5% bromine and near 100% for the diaphragm cell producing 2–5% bromine. The energy efficiency of the diaphragm cell electrolysis could be improved by operating at temperatures above 100 °C and with HBr electrolyte concentrations up to 60% but material durability under these conditions are a problem.

7.4.2 Flue Gas Desulfurisation (FGD) Processes

7.4.2.1 The ISPRA Mark III Process. The ISPRA process[30] is based on reactions (46) and (47) of the thermochemical cycle. The reactive agent produced by electrolysis is a dilute solution of bromine (<0.56 wt%) in an aqueous solution of 1 wt% H$_2$SO$_4$ and 15 wt% HBr. The bromine is used as the oxidising agent and converts the sulfur dioxide to sulfuric acid. There are three basic stages to the process, (shown in Figure 13) external to the electroysis section:

1 The flue gas bearing SO$_2$ is contacted with product solution from the electrolytic reactor, in which the SO$_2$ is absorbed and reacts with the bromine, which is 100% converted.
2 The reactor product solution is then concentrated by evaporation using the sensible heat contained in the entering flue gas. All the HBr and the majority of the water are vaporised and an 80–85% sulfuric acid solution is produced. This solution may be furthur concentrated if desired.
3 The desulfurised gas leaving the reactor is scrubbed with water to remove the HBr and the acid droplets from the vapour.

A demonstration plant has been in operation at the SARAS refinery in Sarroch near Cagliari (Sardinia, Italy) since 1989 and has an operating

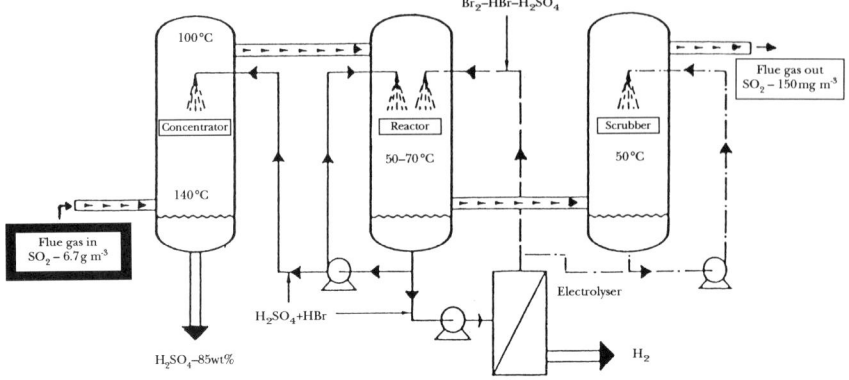

Figure 13 *Block diagram of the ISPRA mark 13A process for FGD*

capacity of 32 000 Nm³ h⁻¹. The bromine production rate required by this plant is 320 kg h⁻¹ using an electrolysis current of 110 kA. The process development of this FGD includes the parallel testing of two types of electrolytic reactors:

(a)   The DCAG cell(Deutsche Carbone Aktion Gesellschaft) based on vertically mounted bipolar parallel plate graphite electrodes in a plastic container with a PVDF lining (see Figure 14). The electrodes are 4 cm thick and the interelectrode gap is 1 cm. Electrolyte is supplied to the cells through a common distributor plate at the base of the cell and hydrogen and the bromine solution exit from the top. The demonstration plant uses several banks, of 41 cells each, having the overall dimensions; width 0.54 m, height 0.9 m .

(b)   The DEM cell (dished electrode membrane) shown in chapter 4 is operated as an undivided configuration with monopolar metallic electrodes. The cathode material is Hastelloy C and the anode is a DSA material. The demonstration plant consists of 32 bipolar cells each with a width and height of 1.0 m.

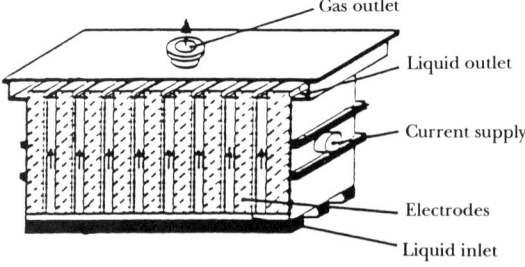

Figure 14 *The DCAG cell for bromine generation*

In the ISPRA plant the current density of operation is 2000 A m^{-2} and at a temperature of 50 °C the cell voltage is between 1.3–1.4 V and the current efficiency for bromine production is 90%. The major cause of the inefficiency is due to the reduction of bromine at the cathode:

$$Br_2 + 2\ e^- \longrightarrow 2\ Br^- \tag{57}$$

The back reduction of bromine is a mass transfer controlled reaction and is thus independent of cell current density but is particularly dependent on the bromine concentration in the cell. Thus operating at moderately low mass-transfer rates with low concentrations of bromine and at high current densities reduces the extent of back reduction of bromine. In practice the cells are designed to give good hydrogen gas bubble release from the cathode, moderately low bubble hold-up and efficient conditions for the anodic production of bromine.

The energy consumption of the electrolysis is a major factor in the process operation and is clearly influenced by the required current density of operation, which determines the operating cell voltage. Typical cell voltage characteristics for both of the cell designs are shown in Figure 15. At current densities above 600 A m^{-2} the voltage–current characteristics are approximately linear, indicating that cell internal resistance is the major determining factor. The larger interelectrode gap of the DCAG cell is reflected in the larger slope of the voltage–current density curve, *i.e.* 0.32 V/(kA m^{-2}) compared with a value of 0.095 V/(kA m^{-2}) for the DEM cell. The performance of the DEM cell fitted with platinum coated Hastelloy cathodes give a significant reduction in the cell voltage. A typical value of 1.5 V at a current density of 2500 A m^{-2} has been demonstrated under long term tests (1500 h). Both the Hastelloy cathode and the

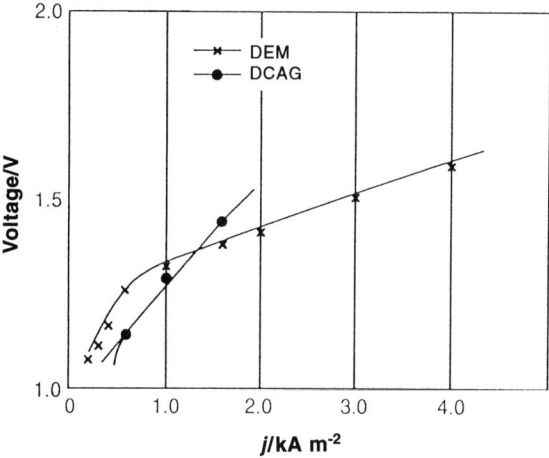

Figure 15 *Current density–voltage curves for bromine cells*

DSA anode have performed well with only small amounts of wear. Estimated lifetimes of the materials are in excess of three years.

7.4.2.2 Other FGD Processes. Table 5 summarises electrochemical methods for FGD into direct and indirect processes.

Table 5   *Electrochemical processes for FGD*

Direct Processes	Indirect Processes
Adsorption and regeneration of the adsorbant by electrochemical SO_2 oxidation	Homogeneous redox mediators: outer-cell processes inner-cell processes
Absorption with separate electro-chemical SO_2 conversion	Heterogeneous redox mediators
Absorption within the electrolysis cell	Catalytic oxidation with oxygen and electrochemical regeneration of the catalyst
Electrochemical reaction at a gas diffusion electrode	Chemical absorption with electro-chemically produced acid/alkali

(a) Direct processes — During direct electrochemical oxidation of flue gases the electrocatalysts are succeptible to poisoning from flue gas impurities and this factor must be considered in any process developed.

In the absence of dissolved catalysts, or oxidants, the low partial pressure of the sulfur dioxide in flue gas and the low solubility can cause limitations in the absorption of SO_2. The absorption of sulfur dioxide into aqueous solution is a slow process; the value of Henry's constant for this step is *ca.* 1.13 mol dm^{-3} bar^{-1} for a dilute sulfuric acid solution. In the design of electrochemical reactors for direct oxidation of SO_2 high interfacial areas and high mass-transfer coefficients are required in the absorption stage. The absorption may be carried out in the electrolytic reactor, as an inner-cell process or in an external scrubber, as an outer-cell process.

A packed bed absorption–electrochemical reactor has been applied to the direct oxidation of SO_2. The bed, of graphite spheres, gives a high interfacial electrode area, which is needed to overcome the low kinetic current densities, and promotes good mass transport between the gas and the liquid. At the steady state the reactor was able to reduce the concentration of SO_2 in the gas phase from 8000 to 200 ppm.[30]

A major cost factor in electrolytic FGD is the energy consumption of the cells. Thus attempts have been made to minimise this factor by developing processes based on gas diffusion electrodes, in which oxygen reduction takes place at the cathode. Processes that combine electrochemical and catalytic oxidation have also been explored. The electrogenerative oxidation of dissolved sulfur dioxide has been demonstrated with packed-bed anodes.[31] Packed bed, composite Teflon–Pt gas diffusion electrodes, platinum deposited on graphite particles and graphite packed beds were studied in conjunction with oxygen reduction cathodes. Current densities up to 1000 A m^{-2} were achieved and the only product of the process was sulfuric acid. The thermodynamic potentials of this process are given by the following Nernst equation:

$$E = E^o + RT/2F \ln[a_{H_2O}a_{SO_2}(a_{O_2})^{0.5}/a_{H_2SO_4}] \qquad (49)$$

where $E^o = 1.06$ V.

The activity of water in sulfuric acid decreases with an increase in acid strength and is small in an 80 wt% solution concentration. Thus there is a corresponding reduction in the cell potential that restricts the concentration of sulfuric acid, which can be obtained if the electrogenerative process is to be attractive. It has thus been proposed that the process could be used for the production of low concentration sulfuric acid (35%) as used in the battery industry.

A method for electrocatalytic desulfurisation of flue gases and waste gases (ELCOX process) is based on the electrochemical reaction between sulfur dioxide and oxygen (from the gas and water). The reaction can take place in an electrochemical fuel cell, and oxygen reduction proceeds at the cathode. The reaction product is 40% sulfuric acid.

(b) Indirect electrochemical processes — Electrocatalyst poisoning in direct electrochemical oxidation can be overcome by using indirect electro-chemical methods. There have been several studies and patents relating to the oxidation of SO_2 in aqueous solutions using metal ions in the absence or presence of oxygen. These include systems based on Cr, Mn, Fe, Cu and Pb as oxides or reducible metal ions.

One method uses Cr^{VI} ions and will simultaneously treat the NO$_x$ components of the flue gas. Both the SO_2 and the NO$_x$ are oxidised to the corresponding acids:

$$2 H_2SO_4 + 2 NO + H_2Cr_2O_7 \rightarrow Cr_2(SO_4)_2(NO_3)_2 + 3 H_2O \quad (50)$$

$$3 SO_2 + H_2Cr_2O_7 \rightarrow Cr_2(SO_4)_3 + 3 H_2O \qquad (51)$$

Electrolysis of the Cr^{III}-bearing solution regenerates the chromic acid solution in a separate stage of the process:

$$Cr_2(SO_4)_3 + 7 H_2O \rightarrow 3 H_2SO_4 + H_2Cr_2O_7 + 3 H_2 \qquad (52)$$

$$Cr_2(SO_4)_2(NO_3)_2 + 7 H_2O \rightarrow 2 HNO_3 + H_2Cr_2O_7 + 3 H_2 + 2 H_2SO_4 \quad (53)$$

In this process both of the product acids are recovered by distillation.

A similar process is based on the use of the Ce^{III}/Ce^{IV} redox couple. In this system the nitric acid, formed from the oxidation of NO with the electrogenerated Ce^{IV}, is cathodically reduced to NO in an electrolytic cell. In the anolyte chamber the Ce^{IV} is anodically generated from the Ce^{III}. Overall this system effectively produces concentrated NO.

Lurgi have developed a process (Peracidox) in which the SO_2 is oxidised by a redox mediator peroxydisulfate, formed by anodic oxidation of sulfuric acid (see Table 6). The lead–lead dioxide system can also be used as a heterogeneous redox mediator system.[30] Current efficiencies of 97% can be achieved at electrode potentials of 1.61 V. The electrochemically formed lead dioxide can also remove the sulfur dioxide from the gas stream in a separate chemical step.

Table 6   *Reactions in the peracidox process*

	Reaction
Anode:	$2\,HSO_4^- \rightarrow H_2S_2O_8 + 2e^-$
Cathode:	$2\,H^+ + 2e^- \rightarrow H_2$
Overall cell:	$2\,H_2SO_4 \rightarrow H_2S_2O_8 + H_2$
Hydrolysis:	$H_2S_2O_8 + H_2O \rightarrow H_2SO_5 + H_2SO_4$
	$H_2SO_5 + H_2O \rightarrow H_2O_2 + H_2SO_4$
SO_2 Oxidation:	$H_2S_2O_8 + SO_2 + 2\,H_2O \rightarrow 3\,H_2SO_4$
	$H_2SO_5 + SO_2 + H_2O \rightarrow 2\,H_2SO_4$
	$H_2O_2 + SO_2 \rightarrow H_2SO_4$

A system based on a copper redox system has been devised[30] which is a combined electrolytic and catalytic process carried out in a three-compartment cell. The cell, shown schematically in Figure 16, consists of a central packed bed anode of graphite separated, by two ion-exchange membranes, from a packed cathode bed of copper and a packed bed of Cu which acts as an absorber. In the absorption chamber the copper, after first being oxidised to copper(I) oxide, is responsible for the oxidation of SO_2 to sulfuric acid.

$$4\,Cu + O_2 \rightarrow 2\,Cu_2O \tag{54}$$

$$Cu_2O + SO_2 + H_2O \rightarrow 2\,Cu + H_2SO_4 \tag{55}$$

$$Cu_2O + H_2SO_4 \rightarrow Cu + CuSO_4 + H_2O \tag{56}$$

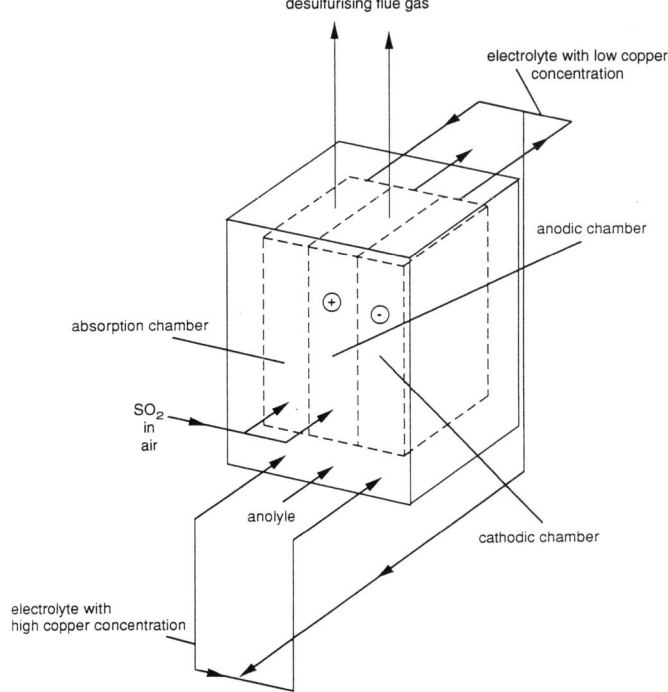

Figure 16 *Schematic of three-compartment cell for SO₂ oxidation*

As a side reaction the copper(I) oxide is oxidised to copper sulfate. The copper is recovered from the sulfate by electrodeposition in the packed-bed cathode chamber. At the anode sulfur dioxide is oxidised to sulfuric acid. In the practical operation, at a certain time the function of the two chambers on either side of the anode chamber is reversed so that freshly deposited copper can react with SO_2 again.

7.4.3 Cathodic Reduction of Sulfur Dioxide

The cathodic reduction of SO_2 in alkaline solutions can produce sodium dithionite[32]

$$2\,SO_2 + 2\,NaOH \longrightarrow Na_2S_2O_4 + H_2O + \tfrac{1}{2}\,O_2 \tag{57}$$

However, the instabilities and secondary electroreductions of product species cause problems in the development of a practical process. Thermodynamically the electroreduction of dithionite is favoured over its formation, but this reaction is relatively slow and is increased by increasing dithionite concentration and current density. The thermochemical decomposition of dithionite decreases with increasing pH, whilst the electrochemical formation decreases, and thus an optimum pH exists for the accumulation of the dithionite. To achieve operation at practical current densities a packed bed electrode with trickle flow of electrolyte can be used. With this type of electrode, dithionite solutions with concentrations of up to 780 mol m^{-3} have been obtained at current densities of approximately 1600 A m^{-2}, but yields were less than 70%. The significant quantities of thiosulfate and sulfide formed detract from the use of this product solution in certain commercial applications such as brightening wood pulp. The presence of oxygen in the SO_2 gas adversely affects the process and thus applications in the FGD treatment of SO_2 are impractical.

However, the absorption of NO into dithionite solution produces N_2 and N_2O, and sulfite ions. The sulfite ions can be electrochemically reduced back to dithionite and thus effectively recycled. This therefore opens a way for the integration of both flue gas treatments into one combined overall electrochemical process. Clearly there are many technological challenges to the implementation of this concept.

7.4.3.1 Indirect Reductions. A process (shown in Figure 17) based on indirect electrochemical reduction has been reported[33] in which SO_2 solutions are reduced to elemental sulfur by electrochemically generated reductants such as Ti^{III}, V^{II} and Cr^{II} ions in the reaction:

e.g.

$$SO_2 + 4\ Cr^{II} + 4\ H^+ \rightarrow S + 4\ Cr^{III} + 2\ H_2O \qquad (58)$$

$$Cr^{III} + e^- \rightarrow Cr^{II} \qquad (59)$$

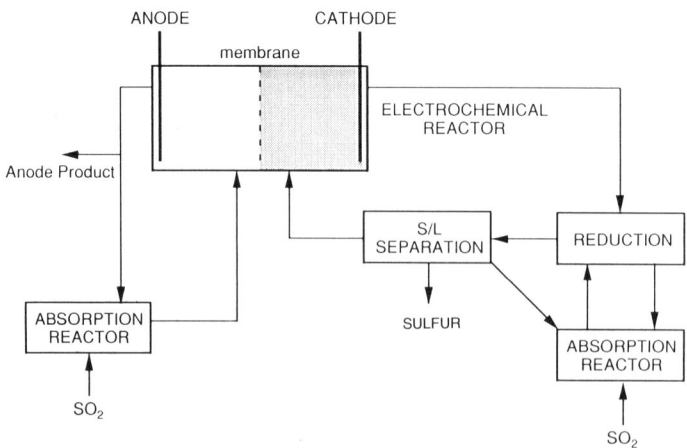

Figure 17 *Schematic flowsheet of a process for the indirect electrochemical reduction of SO_2 to sulfur*

The electrochemical reactor uses a Nafion cation-exchange membrane to separate a cathode (either lead, mercury, tin or carbon) from the anode. The chemical reduction step takes place in an external reactor and can produce yields of sulfur greater than 90%. The colloidal sulfur product must be separated from the aqueous solution electrolyte prior to introduction into the electrochemical stage to avoid contamination of the cathode.

7.4.4   Treatment of Nitrogen Oxides

The process described above in which both SO_2 and NO_x are simultaneously treated, indicates that electrochemistry is at least capable of treating flue and waste gases containing NO_x. In the electroreduction of NO, high ammonia selectivities at platinum are possible. The electroregenerative reduction of NO^{34} to ammonia with a selectivity of 70% is possible. The addition of carbon monoxide results in the selective formation of hydroxylamine. An electrochemical flue gas scrubbing procedure has been suggested[35] which incorporates electrochemical oxidation of sulfur dioxide with the electrogenerative reduction of NO_x as shown schematically in Figure 18. This utilises the reduction of NO at a Pt-black gas-diffusion electrode, in the presence of adsorbed sulfur (to partially poison the surface).

Overall the treatment of NO_x using electrochemical means is, at present, not well developed and highly selective and energy efficient processes are needed. If electrochemistry has a future, then perhaps either the simultaneous treatment by wet oxidation or high-temperature electrochemical-membrane separation discussed later offer the greatest promise. There may however be applications where the scale of operation and the gas composition generated is less demanding than in flue gas.

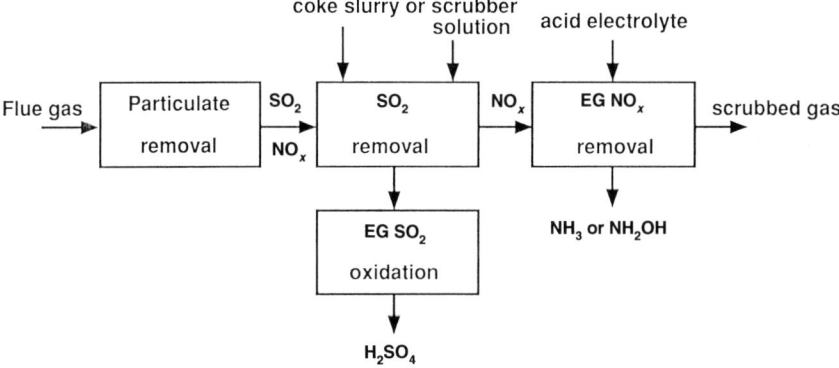

Figure 18 *Proposed flowsheet for electroregenerative treatment of flue gas*

7.4.5   Treatment of Hydrogen Sulfide

Hydrogen sulfide is present in a number of industrial process streams *e.g.* coal gas and oil reservoir sour gas. British Gas developed the Stretford process for treating coal gas which contained relatively low amounts of hydrogen sulfide. In the process the H_2S is absorbed into aqueous carbonate solutions, at pH 8–9, and in the presence of atmospheric oxygen, bubbled through the solution, elemental sulfur is produced by catalytic oxidation.

absorption

$$H_2S + OH^- \rightarrow HS^- + H_2O \qquad\qquad (60)$$

reaction (catalysed by V^V and anthraquinone)

$$8\ HS^- + 4\ O_2 \rightarrow S_8 + 8\ OH^- \qquad\qquad (61)$$

The reaction mechanism is in fact electrochemical[36] — oxygen reduction is catalysed by vanadium species and anthraquinone, producing hydrogen peroxide. The peroxide oxidises the polysulfide ion intermediate to elemental sulfur and thereby regenerates the oxidised form of the catalyst.

There have been several electrochemical methods proposed for the recovery of sulfur and hydrogen from hydrogen sulfide, based on either direct or indirect routes.

7.4.5.1 Direct Oxidation. The direct oxidation of the sulfide ion, formed from H_2S absorption, generally results in a sulfur deposit which blocks the anode and requires the use of a solvent at 80 °C. However, the direct oxidation of sulfide ions in a reduced pH solution (resulting from the dissolution of H_2S in alkali solution) has been demonstrated[37] at 85 °C using a carbon anode. This produced polysulfides in solution, with no passivation of the anode, and no subsequent precipitation of sulfur.

A proposed process has thus been suggested for 'the low temperature removal of hydrogen sulfide from sour gas and its utilisation for hydrogen and sulfur production', which has several claimed advantages:

(i)   unlimited sources of H_2S;
(ii)   low thermodynamic potential for H_2S decomposition, $E^0 = 0.17$ V;
(iii) environmental clean-up;
(iv)   valuable products S and H_2, a valuable fuel for a future 'hydrogen economy'.

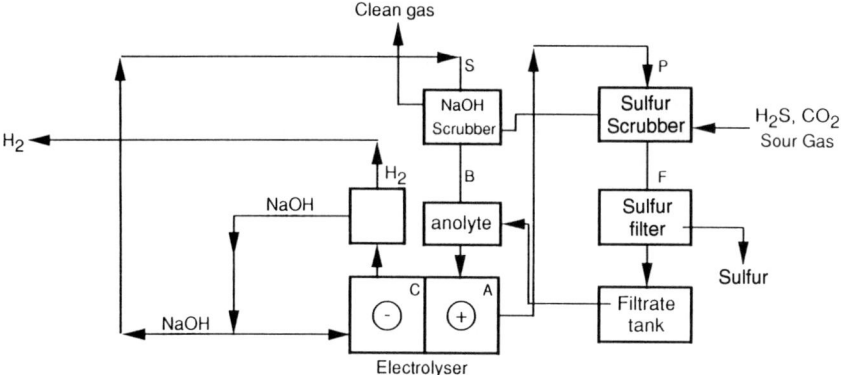

Figure 19 *Process flowsheet for hydrogen sulfide removal from sour gas*

A process diagram of the low temperature removal of H_2S from sour gas is shown in Figure 19. The key steps in the process are:

(a) H_2S is scrubbed from the gas in concentrated caustic solution to form an electrolyte containing NaHS and NaOH. The caustic is produced by the cathodic cell reaction: hydrogen evolution or O_2 reduction.

(b)   At the anode of the divided electrolytic cell polysulfide is formed without the precipitation of sulfur, as the operating temperature is 80 °C. The cell voltage in this system is typically 1.0 V at a current density of 3000 A m^{-2}.

(c)   The anolyte product solution is then contacted, in a separate vessel, with waste gas containing H_2S and CO_2 at room temperature. With the high concentration of polysulfide in the electrolyte, the low temperature and the low pH, the precipitation of sulfur is induced.

(d)   The sulfur is then filtered off from the caustic solution, which is returned to the electrochemical cell as anolyte.

7.4.5.2 Indirect Treatment. Indirect electrochemical oxidation of H_2S has been suggested[38] based on reaction with:-

(a) I_2 (as I_3^- ion). The I^- ion was oxidised to regenerate the I_3^- ion during the process:

$$I_3^- + H_2S -> 3\ I^- + S + 2\ H^+ \tag{62}$$

$$3\ I^- -> I_3^- + 2\ e^- \tag{63}$$

(b)   Iron(III) complex. Regeneration of Fe^{III} from the resulting Fe^{II} by oxidation is in a compartment separate to that used for absorption and oxidation.

(c)   Solution of $FeCl_3$ which absorbs hydrogen sulfide and oxidises it to
       sulfur. After sulfur removal, the $FeCl_3$ is regenerated by the oxida-
       tion of the $FeCl_2$ formed in the solution (in HCl) in a bipolar elec-
       trolysis cell (operating at 50 °C and a current density of 1020 A m^{-2}).
       Sulfur purity of 99.99% is obtained, and it is claimed that the H_2S
       absorption rate is so high that tail-gas treatment, necessary for the
       conventional Claus method, can be omitted.

An interesting method for the treatment of H_2S is the application of
photoelectrolysis which, in addition to the formation of sulfur and hydrogen, gives the possibility of converting sunlight to electricity. The concept
has been demonstrated[39] using *n*-CdSe thin film photoelectrode. With the
utilisation of the hydrogen generated in a hydrogen–air fuel cell a practical energy efficiency of 10.6% is claimed.

7.4.5.3 Other Species. An air purification system has been introduced by
Electrosynthesis Co., called the Electrocinerator™.[40] The system is a general technology for the separation and destruction of airborne chemical
and biological pollutants at ambient temperatures. It consists of an electrochemical reactor, which produces appropriate redox agents, such as
Ag^{II}, $S_2O_8^{2-}$ and Co^{III} ions for oxidations, and a high efficiency gas scrubbing system. The electrochemically generated redox reagent is fed to the
scrubber, as an aqueous electrolyte solution, and reacts with the absorbed
airborne pollutants and thus achieves the desired transformation. The
product species in the case of carbon dioxide and nitrogen are discharged
in the air and other species are retained in the electrolyte solution.

The electrocinerator can destroy a wide range of organic compounds,
odours and biological substances. Airborne metals can be oxidised and
solubilised in the scrubbing solution. Inorganic acid gases such as ammonia, hydrogen cyanide, phophine *etc.* can also be removed and many virus-
es and bacteria are also rapidly destroyed.

The economics of the electrocinerator are dependent upon the species
and its concentration. At lower pollutant concentrations the operating
costs can be much lower than natural gas incinerators. For example the
cost (1992) to destroy 45 kg per day of benzene is approximately $0.1 mil-
lion per year for the electrocinerator compared with $0.5 million per year
for incineration with natural gas. In the case of hydrogen sulfide the costs
to treat 0.9 kg per day at a concentration in the air of 15 ppm is estimated
at $1000 per year, which is less expensive than most available techniques.

7.4.6   Electrochemical Membrane Processes

An electrochemical membrane process for gas separation can utilise the
difference in electrochemical potentials set up across an appropriate membrane by the application of a potential gradient. The minimum voltage

required to effect separation is given from the Nernst equation and the definition of chemical potential as:

$$E = E^o + (R\,T/n\mathrm{F})\ln\{a_{i1}/a_{i2}\} \tag{64}$$

where a_{i1} and a_{i2} are activities of species i at either side of the membrane.

For example, if the species exists in the same chemical form on either side of the membrane, then $E^o = 0$, and at high temperatures of approximately 1000 K, a concentration of gas of the order of 10^{10} can be achieved by the application of *ca.* 1.0 V. In practice, the system has to be driven at potentials much greater than the theoretical thermodynamic values to overcome the effects of polarisation and internal electrical resistance.

The method has been applied in the purification of oxygen and hydrogen at the laboratory scale and in the removal of carbon dioxide in manned spacecraft life support using as a membrane caesium carbonate immobilised in a thin microporous polymer. The molten carbonate fuel cell has been adapted to concentrate CO_2, by operating in a reverse mode with the application of electrical energy.

The purification of contaminated chlorine gas has been achieved using electrochemical membrane separation. The chlorine gas is cathodically reduced to chloride ion:

$$Cl_2 + 2\ e^- \rightarrow 2\ Cl^- \tag{65}$$

The chloride ions are then transported across an aqueous HCl electrolyte, immobilised in an asbestos matrix, and then anodically oxidised to chlorine. The electrodes in this system are graphite.

The electrochemical membrane processes for effluent treatment which are attracting interest are for the removal of SO_2 and H_2S.[41] These are high temperatre processes using eutectic molten salt mixtures immobilised in a ceramic membrane.

7.4.6.1 Hydrogen Sulfide. The separation of H_2S from a mixture of carbon dioxide, water vapour, hydrogen, nitrogen and carbon monoxide in an electrochemical membrane cell (see Figure 20) uses the electronation of H_2S at a suitable cathode:

$$H_2S + 2\ e^- \rightarrow H_2 + S^{2-} \tag{66}$$

The sulfide ion is then transported across a membrane to an anode where it is preferentially oxidised.

$$S^{2-} \rightarrow \tfrac{1}{2}\ S_2 + 2\ e^- \tag{67}$$

To remove sulfur from the anode effectively, it is necessary to operate above its boiling point.

Process gas
containing   →
contaminant

Cleaned
→ process gas

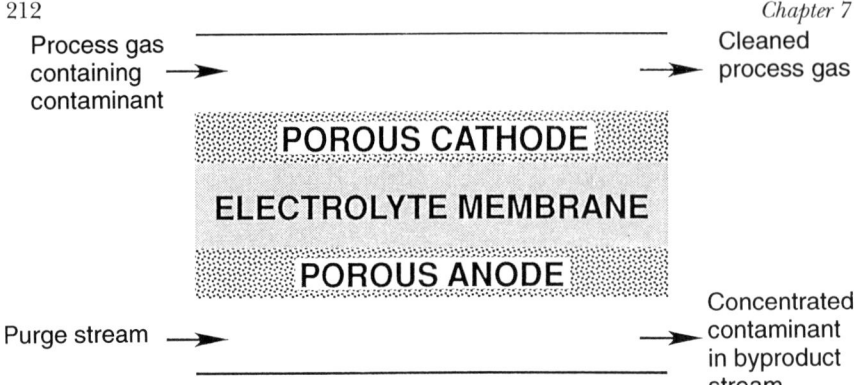

Purge stream   →

Concentrated
→ contaminant
in byproduct
stream

Figure 20 *High temperature hydrogen sulfide membrane concentration cell*

The process uses an electrode–membrane–electrode composite sand-which, the electrodes are either, carbon, nickel sulfide, cobalt sulfide or $La_{0.8}Sr_{0.2}CrO_3$, all with porosities of approximately 60% and the membrane is a ceramic molten carbonate 'fuel cell tile' composed of 55% by wt. $LiAlO_2$ matrix and a Li_2CO_3–K_2CO_3 eutectic mixture. This high-temperature process has applications for the treatment of a coal gasification product stream containing H_2S and must operate at temperatures above 700 °C to satisfy melt conditions, and to give reasonable CO_2 rejection from the melt. In operation the H_2S flows past the cathode where it is reduced to sulfide ions or hydrogen sulfide (if hydrogen is present in the anode). In this latter case the cell acts as a concentrator of H_2S, and 98.8% removal of H_2S, from a gas stream containing H_2, N_2 and H_2S, is possible. Components such as CO_2 and H_2O in the gas stream complicate the process because competitive reactions can occur, such as:

$$CO_2 + H_2O + 2\ e^- \rightarrow CO_3^{2-} + H_2 \tag{68}$$

$$CO_3^{2-} + H_2S \rightarrow H_2O + CO_2 + S^{2-} \tag{69}$$

The extent to which the chemical equilibrium reaction (69) occurs determines the amount of co-transport of CO_2 with H_2S.

At the heart of this high-temperature process is a bank of electrochemical cells, with a proposed design similar to NASAs battery stacking structure for the molten-carbonate fuel cells (see Figure 21).

7.4.6.2 Treatment of SO_2. The high-temperature membrane separator for FGD operates at temperatures much lower than that for the treatment of hydrogen sulfide, *i.e.* 350 °C. A different combination of molten salts of Li, Na and K sulfates are also used in this application (see Figure 22).

Flue gas containing SO_2 is fed to the cathode, and freely diffuses into the electrolyte pores. An applied potential drives the following set of reactions or rate steps:

(i)   selective removal of SO_x at the cathode/membrane interface;

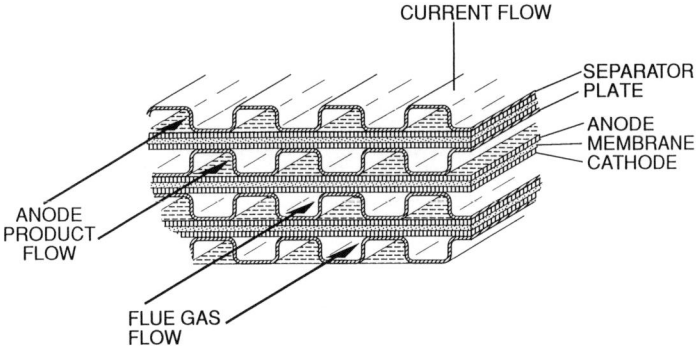

Figure 21 *Cell design concept for high-temperature flue gas treatment cell*

(*ii*)   ionic transport across a membrane interface;

(*iii*) ionic transport across a membrane (immobilised molten salt electrolyte);

(*iv*)   evolution of SO_3 at the anode/membrane interface with diffusion of SO_3 out of the anode pores.

On the treatment of a simulated flue gas (3% SO_2), 98% conversion of SO_2 can be achieved at a cell voltage of 0.5 V and a current density of 500 A m^{-2}. A product gas stream containing up to 50% SO_2 can be attained with a transport rate of gas of *ca.* 1.0 kg m^{-3}. The projected economics of the process at this scale of operation are encouraging, with estimates of capital and operating costs significantly lower than other methods such as scrubbing. With a conservative estimate of current efficiency, operating at a current density of 500 A m^{-2}, an applied current of 16 MA is needed for a 500 MW power plant. The power demand is therefore 2% of the power output of the plant, *i.e.* 8 MW. The installed cost of the FGD facility,

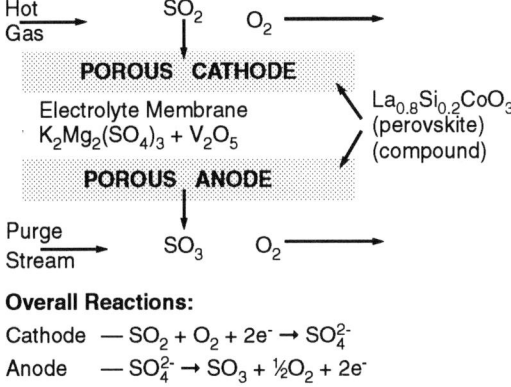

Overall Reactions:

Cathode — $SO_2 + O_2 + 2e^- \rightarrow SO_4^{2-}$

Anode — $SO_4^{2-} \rightarrow SO_3 + \frac{1}{2}O_2 + 2e^-$

Figure 22 *High-temperature process for recovering SO$_2$*

including ancillary equipment, is estimated at $10 kW^{-1} installed.

The high-temperture membrane process is also suggested as a possible method for the simultaneous treatment of NO_x and SO_2. Once proven commercially on a large scale other applications may also be realised. These include the removal of HCl from incineration flue gas and the removal of HF from process air in phosphoric acid manufacture.

7.5   ELECTROFLOTATION

Separation of solid suspensions, oils, emulsions and colloidal particles, and other organic matter in water, is essential to lower the BOD of an effluent before discharge. The addition of flocculating agents (*e.g.* Fe^{3+}) or the use of air flotation are often used to achieve faster separation than would be achieved by settling. Flotation is a technique which depends on the existence or the formation of an hydrophobic surface and thus is often based on the use of surfactants. The formation of a finite contact angle when the particle is in contact with the bubble facilitates collection of the particulate. The change in the effective density, which results, enables the particulates to be floated to the surface of the dispersion and retained in a froth layer. This layer is then recovered by mechanical means. A problem area in flotation is the treatment of particulates with a size below 20 μm.[42] A solution to this problem is the use of finer bubbles, produced by dissolved air or by electrolytic flotation. Flotation methods and techniques can be divided into several categories:

(*i*)   Ion flotation. For the removal of surface inactive ions by adding an appropriate surfactant.
(*ii*)   Precipitate flotation. This is an evolution of ion flotation involving the formation of some form of precipitate.
(*iii*) Adsorbing colloid flotation. This involves solute removal by adsoption onto a carrier floc.
(*iv*)   Dissolved air flotation.
(*v*)   Electrolytic flotation.

The advantages of using electrolytic bubbles in flotation are:

1   The formation of extremely fine dispersed gas bubbles, *e.g.* 8–15 μm, allows the flotation and removal of very fine particles.
2   The creation of a very wide range of bubble concentration is possible by varying the current density. Thus a gas medium with an overall large surface area is formed during flotation, increasing the probability of bubble/particle collision and adhesion.
3   The creation of bubbles of varying sizes to match appropriate applications. This is achieved by varying the electrode material, current density, pH and temperature.

In electroflotation, electrolytic gases are generated in a cell with two closely positioned mesh (or gauze) electrodes situated at the base of the treatment tank. The flow of waste water, or untreated water, should be slow so that the capture mechanism of the bubbles is not disturbed as they rise to the top of the tank. Low ionic strength solutions, which are required to achieve fine dispersion of particles by electrostatic means, generally conflict with requirements to minimise the electrical energy consumption. Current densities used in electroflotation cells are generally modest (0.1–10 mA cm^{-2}) and cell potentials are up to 10 V giving energy consumptions in the range 0.2–0.4 kWh m^{-3} of treated solution. The capacity of electroflotaion cells is generally not large, the maximum rate of treatment is of the order of 150 m^3 h^{-1}.

The simultaneous generation of hydrogen and oxygen does not add to the separation ability but brings a potential safety problem. Thus consumeable anodes of Al or Fe are often used and the process of electroflocculation is used in conjunction with electroflotation. The formation of the dissolved Al(OH)$_3$ or Fe(OH)$_3$ assists in coagulation of the dispersed species. The latter method has recently been introduced by Sintra GmbH, Munich as an alternative to ultrafiltration for dealing with a waste water containing heavy metals, surfactants, oils and other hydrocarbons. The electrochemical cell uses an aluminium tube anode and a steel cathode to generate hydrogen, which then flocculates the aluminium hydroxide complex. Standard Microsystems Corp (USA) have introduced electroflotation for the removal of fluoride ions from waste water. Aluminium alloy anodes are used, which in the presence of flouride cause precipitation of AlF$_3$.

7.6   ELECTROSORPTION

There is an immense amount of literature concerned with adsorption phenomena at electrodes. However, few practical applications of electrodes as sorption surfaces, are reported for the recovery of chemicals, as occurs in the classical adsorption process. Electrosorption, however, could be an alternative means of separation of small quantities of organics and other species from effluent streams. High surface-area adsorbent electrodes are clearly required. The technique has been used in the adsorption of β-naphthol onto a packed bed of glassy carbon spheres,[43] and cyclic electrosorption experimentally demonstrated.

7.7   APPLICATIONS IN THE NUCLEAR INDUSTRY

The nuclear industry can pose its own special requirements in the disposal of waste materials and in recycling. The high profile of this industry can make the development of appropriate techniques more exacting than may be the case in other industries. Many of these methods involve the electrochemical dissolution of fuels and fuel products.[44]

7.7.1   Indirect Electrochemical Dissolution

Generally the dissolution of plutonium dioxide has proven to be a difficult problem requiring the use of highly concentrated nitric acid (12 mol dm^{-3}) with hydrofluoric acid (0.18 mol dm^{-3}) at boiling temperatures. An electrochemical process for the dissolution of Pu, from high fired plutonium dioxide, and leaching of plutonium from contaminated scrap and waste is available. The proposed electrochemical process uses the strong oxidising power of either Ce^{IV}, Co^{III} or Ag^{II} ions at room temperatures. The final product of the oxidation is the PuO_2^{2+} ion formed from the reaction in nitric acid:

$$3 \; PuO_2 + 2 \; NO_3^- + 8 \; H^+ = 3 \; PuO_2^{2+} + 4 \; H_2O + 2 \; NO \qquad (70)$$

or a corresponding reaction leading to NO_2.

Oxidation by the silver catalyst (which has the highest standard redox potential), is very fast and is 5–15 times faster than the chemical dissolution process. The advantages of the electrochemical dissolution process over the chemical dissolution process, is the less aggressive operating conditions and improved efficiency of operation.

7.7.2   Uranium Recovery by Direct Oxidation

There are essentially two main steps to the recovery of uranium from irradiated nuclear fuel elements in aqueous solutions:

1   dissolution of the fuel elements in a suitable inorganic acid;
2   separation of the fissionable material by solvent extraction.

The preferred inorganic acid is nitric as it is compatible with the second solvent extraction stage in the recovery process. Although chemical dissolution of the uranium clad stainless steel, or aluminium, is possible this usually requires the addition of other reagents, such as chloride in the case of stainless steel. These additives cause later difficulties in operation such as the corrosion of stainless steel equipment. Thus electrodissolution, in the absence of such additives, has been developed using direct oxidation in a nitric acid electrolyte. The equipment consists of a V-shaped dissolving chamber in which the element to be dissolved is held in a niobium basket. This basket sits between the platinum/iridium coated niobium anode and the titanium holding vessel, which acts as the cathode, and is used to prevent electrical shorting between the anode and cathode due to the metal piece to be dissolved. Thus in operation the piece becomes a bipolar element and the side facing the cathode dissolves anodically. Current densities used in this application are high, 25 kA m^{-2}, to promote high rates of dissolution.

7.7.3   The Purification of Discharged Fuels by Electrorefining

The electrowinning and electrorefining of uranium in a plutonium molten electrolyte was demonstrated several decades ago with the latter able to produce high purity uranium >99.99%. The electrorefining of plutonium to recover and purify scrap or aged weapons grade plutonium has been practiced for many years. A process has been developed, which enables the electrorefining of both U and Pu, using a liquid cadmium anode. This anode acts as solvent for the fast reactor metal fuel, which after dissolution is transported in ionic form through a molten chloride electrolyte, of mixed stable chlorides (at temperatures less than 500 °C), to a cathode (molybdenum) where electrodeposition can take place. The uranium and plutonium can be successively deposited (uranium first) although the deposition of plutonium is susceptible to oxidation by uranium trichloride:

$$Pu + UCl_3 = PuCl_3 + U \qquad\qquad (71)$$

An alternate cathode material is molten cadmium, which would require vaporisation of the cadmium to release the Pu and U metals. The process can be used to recover both of the metals simultaneously.

7.7.4   Reductive Separation of Plutonium and Uranium (The Purex Process[45])

An important step in the work-up procedures of spent nuclear fuels is the separation of uranium and plutonium. These components are present together in a tributyl–kerosene solvent in a complex form. The Pu^{IV} can be reduced to Pu^{III} using a reducing agent, in this case U^{VI}, and the Pu^{III} extracted into an aqueous solution, in which it is preferentially soluble. The U^{IV} is formed by the cathodic reduction of U^{VI} in an aqueous electrolyte phase using a pulsed perforated plate column. In this column the organic phase and aqueous phase flows are countercurrent with the plates serving as the cathodes and the anode a central shaft supporting the cathode plates. A recent improvement (see ref. 46) in the design of the electrochemical reactor–extractor is the use of a concentric cylinder electro-pulse column. This design improves the current distribution in the cell which otherwise can lead to formation of titanium hydride and other Ti species, during the operation of the titanium cathodes.

7.7.5   Electrochemical Processing of Alkaline Nitrate and Nitrite Solutions

The processing of high level radioactive wastes will produce, as a byproduct, an alkaline salt solution containing sodium nitrite and sodium nitrate (17%). Two electrochemical methods have been proposed to reduce the amount of nitrate and nitrite in solution and thus to reduce the possible leaching of these components from the saltstone into which they are to incorporated prior to landfill:

(*i*)   Electrolysis. The electrochemical reduction of nitrate and nitrite to ammonia and nitrogen *e.g.*

$$3\ NaNO_3 + 3\ H_2O = N_2 + NH_3 + 3\ NaOH + 9/2\ O_2 \qquad (72)$$

occurs with the corresponding formation of oxygen and caustic at the anode. The process is able to convert over 99% of the nitrate and nitrite, although the efficiency is adversly affected by the presence of chromium ions. In addition the electrolytic reduction process can also provide for the decontamination of technetium-99 and ruthenium-106 radionuclides by their electrochemical deposition.

(*ii*)   Electrochemical membrane separation. The electrochemical separation of the sodium nitrate solution into nitric acid and sodium hydroxide using a three-compartment membrane cell has been demonstrated. However the actual process wastes contain the aluminate ion which results in the formation of aluminium hydroxide in the interstitial volume of the membrane. This problem of membrane fouling means practical operation is only feasible with pretreatment of the feed.

REFERENCES

1   V. Smith De Sucre and A.P. Watkinson, *Can. J. Chem. Eng.*, 1981, **59**, 52.
2   H. Sharifian and D.W. Kirk, *J. Electrochem. Soc.*, 1986, **133**, 921.
3   C. Comninellis and E. Plattner, *Chimia*, 1988, **42**, 250.
4   C. Comninellis, 'Electrochemical Engineering and the Environment', *Int. Chem.Eng. Symp. Ser.*, 1992, **127**, 189.
5   D.Wabner and C. Grambow, *J. Electroanal. Chem.*, 1984, **195**, 95.
6   R. Kotz, S. Stucki, and B. Carcer, *J. Appl. Electrochem.*, 1991, **21**, 14.
7   S. Stucki, R. Kotz, W. Suter, and B. Carter, B, *J. Appl. Electrochem.*, 1991, **21**, 99.
8   T.A. Kharlamova and N.I. Mitashova, *Sov. Chem. Ind.*, 1986, **18**, 18.

9   F. Beck, H. Shulz and B. Wermeckes, *Chem. Eng. Technol.*, 1990, **13**, 371.

10   D.Schmal, J. van Erkel and P. J. van Duin 'Electrochemical Engineering', *Int. Chem. Eng. Symp. Ser.*, 1986, **98**, 259.

11   P.D. Francis, Water treatment via UV/Ozone Techniques, 'Electrochemical Technique for a Cleaner Environment', SCI Meeting, London, 19 April, 1991.

12   D.F. Steele, *Chem. Br.*, 1991, **27**, 915.

13   J.C. Farmer, F.T. Wang, P.R. Lewis and L.J. Summers, *Int. Chem. Eng. Symp. Ser.*, 1986, **98**.

14   R.L. Clarke, 'Electrochemistry for a Cleaner Environment', ed. J.D. Genders and N.L. Weinberg, The Electroysnthesis Co, Buffalo, 1992.

15   P.C. Franklin, J. Darlington, T. Soluki and N. Tran, J. *Electrochem. Soc.*, 1991, **138**, 2285.

16   J.F. Rusling, N. Hu, D.J. Howe, C.L. Miaw and E.C. Couture, in 'Electrochemistry in Microheterogeneous Fluids' ed. J. Texter and R. A. Mackay, VCH Publishers, 1992.

17   D.Z. Mazur and N.L. Weinberg, US Pat., 4 702 804, 1987.

18   S.N. Frank and A.J. Bard, *J. Am. Chem. Soc.*, (14), 1977, **99**, 4667.

19   G. H. Kelsall, S. Saage and D. Brandt, *J. Electrochem. Soc.*, 1991, **138**, 117.

20   D. Pletcher and F.C. Walsh, 'Industrial Eletcrochemistry', Chapman and Hall, London, 1990.

21   D. Golub, and O. Yoram, *J. Appl. Electrochem.*, 1989, **19**, 311.

22   M.S.E. Abdo and G.H. Sedahmed, Abstract 3.7, 4th World Congress in Chemical Engineering, Karlsruhe, 1991, Germany.

23   P. Meszaros, I. Orszang, B. Kovacs, Gy. Malovecczky and Z. Kovacs, *Hung. J. Ind. Chem.*, 1984, **12**, 163.

24   M. Dell Monica, A. Agostiano and A. Ceglie, *J. Appl. Electrochem.*, 1980, **10**, 527.

25   Department of the Environment Contract, PECD 7/7/138, January 1991.

26   L. Kaba, C. E. Verostoko and M. G. Duncan, 1st Int. Conf. on Environmental Systems, July 15–18, 1991, published by SAE USA, pp. 131–140

27   B.D. Struck, G.H. Schutz and D. Van Velzen, 'Electrochemical Hydrogen Technologies', ed. H. Wendt, Elsevier, Amsterdam, 1990.

28   P.W.T.Lu and R.L.Ammon, *J. Electrochem Soc.*, 1980, **127**, 2610.

29   S.E. Lyke and S.H. Langer, *J. Electrochem. Soc.*, 1991, **138**, 1682.

30   G. Kreysa and A. Storck, 'Electrochemical cell design and optimisation procedures', Dechema Monograph, 1990, vol.123, p. 225.

31   J.C. Card, M.J. Foral and S.H. Langer, *Environ. Sci. Technol.*, 1988, **22**, 1499.

32   C.Oloman, B. Lee and W. Leyton, *Can. J. Chem. Eng.*, 1990, **68**, 1004.

33   G.H.Kelsall and D.H. Robbins, *Trans. Int. Chem. Eng.*, 1991, **69**, 43.

34   S.H. Langer, and K.T. Pate, *Ind. Eng. Chem. Prod. Res. Dev.*, 1983, **22**, 264.

35   S. H. Langer, M. J. Foral, J. A. Coucci and K.T.Pate, *Environ. Progress*, 1986, **5**, 277.

36   G.H. Kelsall and I. Thompson, *J. Appl. Electrochem.*, 1993, **23**, 296.

37   B. Dandapani, B.R. Scharifker, and J. O'M Bockris, 'Proceedings of the Symposium on Diaphragms, Separators and Ion-Exchange Membranes', ed. J.W. van Zee, R.E. White, K. Kinoshita and H.S. Burney, 1986, pp. 228–237.

38   K. Scott, *Dev. Chem. Eng. Min. Processes*, 1993, **1**, 185.

39   R.C. Kainthla and J. O'M Bockris, I*nt. J. Hydrogen Energy*, 1987, **12**, 23.

40   N.L. Weinberg, 'Electrochemistry for a Cleaner Environment', ed. D. Genders and N. Weinberg, Electrosynthesis Co. Inc. NY, 1992, ch. 16, p. 323.

41   J. Winnick, 'Advances in electrochemical science and enginering', ed. H.Gerischer and C.W.Tobias, VCH, 1990, vol. 1.

42   A.I. Zouboulis, K.A. Matis and G.A. Stalidis, 'Innovations in Flotation Technology', ed. P. Mavros and K. A. Matis, Nato ASI Series E: Appl. Sci., Kluwer Academic, 1992, vol. 208, p. 475.

43   R.C. Alkire, and R.S. Eisinger, *J. Electrochem. Soc.*, 1983, **130**, 85.

44   R.E. White, R.F. Savinell, and A. Schneider, 'Electrochemical Engineering Applications', *A.Int.Chem.Eng. Symp. Ser.*, 1987, **254**, 83.

45   F. Baumgartner, *Chem. Eng. Technol.*, 1977, **49**, 756.

46   F. Goodridge and K. Scott, ' Electrochemical Process Engineering', Plenum Press, New York, 1995.

CHAPTER 8

Electrochemical Synthesis

Electrosynthesis has long been established as a method of inorganic chemical manufacture. Such processes are well known, involve relatively straightforward reactions and the selectivity is often high, In organic synthesis a range of products can often result from one compound. This can make product work-up more difficult, due to both the removal of the organic species from aqueous electrolyte and further processing required to separate the organic species. One of the advantages of performing organic synthesis electrochemically can be an improved selectivity over catalytic or chemical routes. In addition organic electrochemistry has several other important attributes.

*   Mild conditions of operation, *e.g.* low temperature and pressure
*   Clean synthesis avoiding other chemical reagents
*   Availability of novel chemical transformations
*   Reduction in the number of synthesis steps
*   Improved management of potential pollutants
*   Avoidance of aggressive and hazardous reagents
*   Use of alternative feedstocks

An important area where inorganic electrosynthesis and organic synthesis overlap, for processes which utilise inorganic oxidising agents, reducing agents, nitrating agents, *etc.* is described. Considerable interest is now being shown in the electrochemical regeneration and recycling of a range of such mediating agents for organic synthesis.[1–4]

8.1 ORGANIC ELECTROCHEMICAL PROCESSES

8.1.1 Types of Organic Electrosynthesis

A wide range of electroorganic synthesis reactions have been demonstrated in laboratories around the world. These include hydrogenation, oxidation, substitution, reduction and oxidative coupling, cleavage, cyclization and polymerisation. New electrochemical organic chemical reactions continue to be developed, the vast majority of which use aprotic electrolyte media. Amongst other factors, this offers a wider electrode potential range for transformation than aqueous-based media. However, aqueous-based electrosynthesis often results in less complex electrochemistry, better pH control (through the splitting of water), and less problems due to solvent loss and has taken a predominant role in industrial electroorganic synthesis. The types of electrochemical reactions used extensively in industrial electroorganic synthesis are:

- Reduction of carbon–carbon double bonds
- Reduction of carbonyl groups
- Reduction of nitrile or nitro groups
- Reductive coupling
- Reductive cleavage
- Oxidation of hydrocarbons
- Oxidation of functional groups
- Oxidative coupling
- Oxidative substitution
- Electrochemical fluorination
- Indirect oxidation or reduction

   The reactions, which are of interest from an industrial viewpoint, are those that cannot be carried out by conventional chemistry or those that offer certain advantages over chemical routes such as reduced cost through higher selectivity, improved product work-up, a reduction in the number of overall steps and a reduction in the amount of waste products. In the case of anodic oxidations, unless the conventional chemical route requires relatively expensive oxidants such as chromate, or peroxide as opposed to air or nitric acid, then it will be hard to compete. This is accentuated if the oxidation requires many Faradays of charge to convert one mole of the species. A similar argument follows for cathodic hydrogenation reactions, where molecular hydrogen is obtained more cheaply from steam reforming of hydrocarbons. Thus catalytic hydrogenations are favoured unless high selectivity is not achieved. Electrochemical synthesis will be adopted where it offers unique, or novel, methods, such as partial hydrogenation of aromatics, electroreductive couplings or anodic coupling reactions and electrofluorinations.[5]

8.1.2   Limitations in Solubility

The commercial development of electrochemical organic synthesis in aqueous electrolyte is frequently hindered by the low concentration of the organic reagent (depolariser) in the electrolyte and thus by the low value of diffusion limiting current density. Also performance is limited by the low total amount of organic species in the electrolyte and the ability of the reactor to rapidly replenish the material consumed by the reaction in the aqueous phase. Both these factors introduce mass transfer limitations to the electrosynthesis. In terms of the mass transfer requirements at the electrode surface order of magnitude calculations put a lower limit of organic depolariser concentration at *ca.* 100 mol m^{-3} if a practical current density of *ca.* 100 A m^{-2} is to be realised. Thus for many organic chemicals special techniques are adopted, in attempts to remove this limitation and include:[6]

(i)   The addition of organic solvents to the aqueous electrolyte.

(ii)   The addition of acids or bases to increase solubility.

(iii) The use of micellar solubilization of the organic species using for example McKee salts.

(iv)   The use of an organic solvent together with lipophilic electrolyte.

(v)   The use of high surface area porous or particulate electrodes.

(vi)   The addition of mediating redox systems to the aqueous electrolyte. This may be assisted by the use of phase-transfer catalysts which move the reaction site from the aqueous phase to the organic phase, benefiting the reaction by the increase in concentration.

However, the use of any of the above techniques can sometimes introduce other limitations in terms of the subsequent chemistry or engineering of the process. Clearly the addition of further components to the electrolyte, to increase the solubility of the organic depolariser, adds to the cost of the process and may result in more complex product recovery and separation as the resultant product will itself usually have an increased solubility. Furthermore this can make this product more susceptible to additional reactions.

In some cases the organic is added as a dispersion and thus acts as a product sink for the synthesis, thereby protecting it from further reaction. The phase separation of the product is then an attractive feature in terms of product recovery. The use of organic solvents may not be feasible when water is needed as a reactant. The use of micelle-solubilised substances such as long carbon chain alkylammonium salts (cationic surfactants) or long carbon chain sulfonic acids (anionic surfactants) or polyoxyolefins (non-ionic surfactants) can greatly enhance the solubility of the organic reactants. Usually the micelles of the surfactant incorporates only a few organic molecules and thus relatively large concentrations are needed. In practice the solubilising effect has to be balanced against the mass transfer characteristics of the micelles, which are large species and thus have relatively low diffusion coefficients, *ca.* 50–100 times smaller than those of the unsolubilised organic molecules. Thus to gain any benefit in terms of mass transport the increase in solubility will need to be of the order of 100 times greater or approximately 1 kmol m^{-3}.

The use of mediating redox systems in indirect electrosynthesis is an attractive way of removing mass transfer limitations of a low concentration of depolariser.

8.1.3 Indirect Electrosynthesis

A common procedure for oxidation and reduction processes in organic synthesis is by homogeneous reaction with inorganic redox reagents, many of which can be prepared electrochemically. The appropriate regeneration of these agents by electrochemical oxidation or reduction can lead to simplifications in the process operation, reduction in any problems of effluent treatment and eliminate, or reduce, the need for bulk storage of hazardous or toxic chemicals on site. The electrochemical mediator must fulfill several requirements for it to be effective:

(i)   The oxidised and reduced forms of the mediator must be chemically stable and not undergo side reactions which are irreversible.
(ii) The electron transfer with the electrode should ideally be fast and reversible. This minimises cell voltages and the possibility of undesirable side reactions.
(iii) Redox reactions with species other than the target molecules should be minimal.
(iv) The cycle time for regeneration of the mediator should be high.
(v)   A high solubility in the aqueous supporting electrolyte is desirable to maximise the reaction rate with the organic species.

Many processes have been evaluated and developed to the pilot level or to full scale manufacture especially in the fine chemical and pharmaceutical industries. Table 1 lists a selection of the processes.

Indirect electrochemical reactions can be classified as either in-cell or ex-cell processes. In the ex-cell process reaction between mediator and substrate occurs in a chemical reactor separate from the cell. Following the reaction, the electrolyte containing used mediator and the second product phase are separated and the electrolyte phase is returned to the cell for redox regeneration. This is often preferred in order to minimise contact of (say) the organic phase with the electrode, which can result in electrode fouling and de-activation. For example in the oxidation of anthracene to anthraquinone with Cr^{IV}, the spent Cr^{III} electrolyte is treated in a charcoal column to remove organics, which would otherwise poison the lead dioxide anode used to regenerate the chromium oxidant.

'In-cell' processes also occur when the mediator and reactant phases are soluble in the electrolyte phase. Generally, in both the in- and ex-cell cases, if reaction is slow then the cell must include a separator between the anode and cathode to prevent the counter-electrode causing an electrochemical back-reaction of the redox agent. A strategy, which has been used to try and eliminate the need for a cell separator (and thus reduce the cost), is to introduce a phase-transfer agent into the emulsion phase, this selectively removes the active redox agent after regeneration. This regeneration is carried out on a continuous basis for as long as the reaction proceeds.

Table 1 *Selected mediated electroorganic syntheses*

Reagents	Mediator	Product
Naphthalene	Ce^{IV}	Menadione
		Naphthaquinone
p-Methoxytoluene	Ce^{IV}	Anisaldehyde
		(*p*-Methoxybenzaldehyde)
D-Gluconic acid	Ce^{4+}	D-Arabinose
p-Nitrotoluene	$Cr_2O_7^{2-}$	*p*-Nitrobenzoic acid
Substituted toluenes	$Cr_2O_7^{2-}$	Dicarboxydiphenyl sulfone
	$Cr_2O_7^{2-}$	Dicarboxyphenyl ether
	Mn^{III}	Tolualdehyde
Glucose	Br_2	Gluconic acid
Butane-2, 3- diol		Acetoin
Alkenes		
Propene	Br_2 and	Epoxides
Ethene	Cl_2	
Methoxylations		2, 5-Dimethoxydihydrofurfurylethan-1-ol
Furfurylethan-1-ol	Br_2	
o-Toluene sulfonamide	Cr^{6+}	Saccharin
Oleic acid	Cr^{6+}	Azelaic acid
		Pelargonic acid
Anthracene	$Cr_2O_7^{2-}$	Anthraquinone

8.1.3.1 Phase-transfer Catalysts. Phase-transfer catalysts are used in organic synthesis involving liquid–liquid dispersion where the two phases each contain reactant which is preferentially soluble. The phase-transfer catalyst combines with the redox agent and imparts solubility of the redox agent in the organic phase (or organic–solvent phase) thus enabling oxidation or reduction to take place at the relatively high concentration of the organic phase. The phase-transfer catalysts are ionic in nature and lipophilic and form ion pairs with the inorganic counter-ion mediators. These ion pairs are then transferred into the organic phase (see Figure 1) where reaction takes place. Typically tetraalkyl or alkylaryl ammonium salts are used when lipophilic cations are required and long carbon chain carbonic acids, alkylsulfates and sulfonates are used when anionic phase-transfer catalysts are required. The use of phase-transfer catalysts in electroorganic synthesis has been reported in for example, the oxidation of benzyl alcohol (with hypobromite) and alcohols and aromatic hydrocarbons (with Cr^{III} ions). The cationic, tetrabutylammonium ion was used and resulted in an order of magnitude increase in the rate of oxidation in comparison to that in the absence of the catalyst and gave better yields

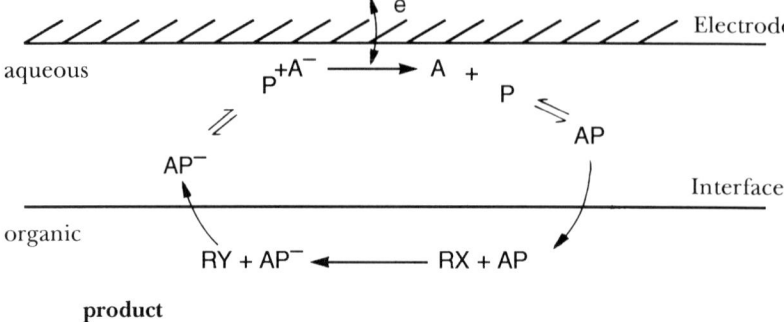

Figure 1   *Phase-tramsfer catalysis*

and selectivity. In certain cases a high solubility of the ion pair is not achieved and preferential adsorption at the liquid/liquid interface may occur.

8.1.3.2 Redox Catalysts. The reaction rate and selectivity, of a particular indirect synthesis will depend largely on the oxidation or reduction power of the agent, measured by the standard potential of the redox couple. The factor of chemical stability is generally more readily satisfied by inorganic ions and metal salts (and some metal complexes) rather than organic mediators. Organic mediators can often react with the the electrolyte, its solvent or intermediates formed during the reaction. Inorganic mediators can often involve slow electron transfer in regeneration, although they can offer the advantage of moderating the reactivity of intermediates and enable selective reaction.

Redox catalysts have been classified into two groups in terms of the mechanism of the homogeneous reaction step.

1   Redox catalysis. The reaction consists of a pure electron transfer between the mediator, $M^{\bullet+}$, and the substrate, RH, *e.g.*,

$$M^{\bullet+} + RH <=K=> M + RH^{\bullet+} \tag{1}$$

$$RH^{\bullet+} \xrightarrow{k} R^{\bullet} \longrightarrow products \tag{2}$$

With this type of reaction the electron transfer can take place at potentials several hundreds of millivolts lower than the electrode potentials of the substrates and thus overcome this slow electron-transfer step. The overall reaction rate is determined by the equilibrium reaction rate constant K and the rate constant of the follow-up reaction k.

2   Chemical catalysis. In this case the homogeneous redox reaction is combined with a chemical reaction, *e.g.*,

chemical reaction $\quad M^{\bullet+} + RH \longrightarrow MH^+ + R^{\bullet} \qquad (3)$

redox reaction $\quad MH^+ + B \longrightarrow M + HB^+ \qquad (4)$

$R^{\bullet} \longrightarrow$ products $\qquad\qquad\qquad\qquad\qquad (5)$

The chemical step determines the selectivity of the reaction. A typical example of redox catalyst of this type is halogen ions.

Table 2 *Redox catalyst couples and standard potentials*

Reduction			E/V (vs. *NHE*)
$Ag^{2+} + e^-$	$\rightarrow$	Ag^+	1.98
$Co^{3+} + e^-$	$\rightarrow$	Co^{2+}	1.83
$Pb^{4+} + 2e^-$	$\rightarrow$	Pb^{2+}	1.69
$Ce^{4+} + e^-$	$\rightarrow$	Ce^{3+}	1.61
$Mn^{3+} + e^-$	$\rightarrow$	Mn^{2+}	1.54
$MnO^+ + 8H^+ + 5e^-$	$\rightarrow$	$Mn^{2+} + 4H_2O$	1.51
$Ce^{4+} + e^-$	$\rightarrow$	Ce^{3+}	1.44
$RuO_4 + 4H^+ + 4e^-$	$\rightarrow$	$RuO_2 + 2H_2O$	1.39
$Cr_2O_7^{2-} + 12H^+ + 6e^-$	$\rightarrow$	$2Cr(OH)^{2+} + 5H_2O$	1.26
$MnO^{2-} + 4H^+ + 2e^-$	$\rightarrow$	$Mn^{2+} + 2H_2O$	1.22
$Tl(OH)_3 + 3H^+ + 2e^-$	$\rightarrow$	$Tl^+ + 3H_2O$	1.19
$Fe^{3+} + e^-$	$\rightarrow$	Fe^{2+}	0.77
$Cu^+ + e^-$	$\rightarrow$	Cu	0.52
$OsO_4 + 2e^-$	$\rightarrow$	OsO_4^{2-}	0.43
$Fe(CN)_6^{3-} + e^-$	$\rightarrow$	$Fe(CN)_6^{4-}$	0.36
$Sn^{4+} + e^-$	$\rightarrow$	Sn^{2+}	0.15
$Co(CN)_6^{3-} + e^-$	$\rightarrow$	$Co(CN)_6^{4-}$	-0.83
$Ti^{3+} + e^-$	$\rightarrow$	Ti^{2+}	-0.37
$V^{3+} + e^-$	$\rightarrow$	V^{2+}	-0.26
$Cr^{3+} + e^-$	$\rightarrow$	Cr^{2+}	-0.41
$Cl_2(sol.) + 2e^-$	$\rightarrow$	$2Cl^-$	1.40
$ClO^- + 2H^+ + 2e^-$	$\rightarrow$	$Cl^- + H_2O$	1.70
$ClO_2 + 4H^+ + 5e^-$	$\rightarrow$	$Cl^- + 2H_2O$	1.51
$HClO + H^+ + 2e^-$	$\rightarrow$	$Cl^- + H_2O$	1.48
$Br_2(sol.) + 2e^-$	$\rightarrow$	$2Br^-$	1.09
$BrO^- + 2H^+ + 2e^-$	$\rightarrow$	$Br^- + H_2O$	1.59
$HBrO + H^+ + 2e^-$	$\rightarrow$	$Br^- + H_2O$	1.33
$I_2 + 2e^-$	$\rightarrow$	$2I^-$	0.62
$I^+ + 2e^-$	$\rightarrow$	I^-	0.95
$IO^- + 2H^+ + 2e^-$	$\rightarrow$	$I^- + H_2O$	1.31
$3HIO + 3H^+ + 4e^-$	$\rightarrow$	$I_3^- + 3H_2O$	1.21

Typical inorganic species used to carry out oxidations include several metal ions *e.g.* Fe[III] and Co[II] (see Table 2). The use of mediated electrochemical reactions for the reduction of organic compounds is not as popular in industry as that of oxidations. The use of alkali metal amalgams in the reduction of organics, notably nitro compounds is practiced. Homogeneous reducing agents such as Eu[II], Cr[II], Ti[II], Sn[II] and V[II] could also be considered in reductions of nitro groups to amines and hydroxylamino compounds, of epoxides to olefins and cleavage of carbon–halogen bonds[7] (see Figure 2).

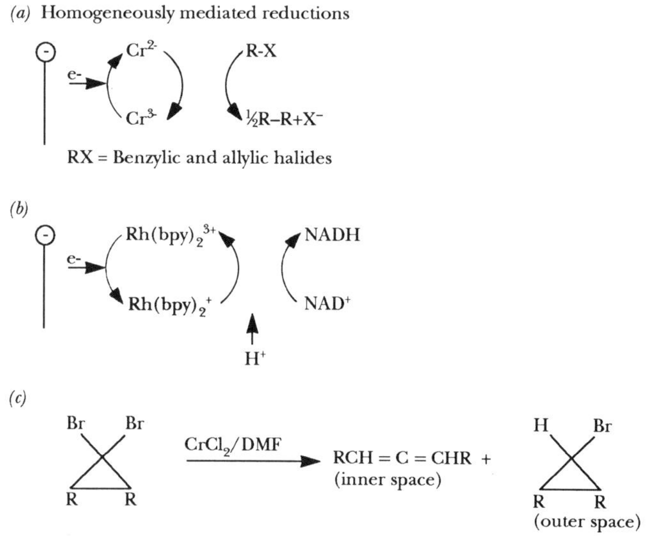

Figure 2   *Examples of indirect electrochemical reduction* (a) *alkylhalide reduction mediated by Cr[II];* (b) *NAD+ to NADH conversion mediated by Rh(bpy)$_2^{2+}$;* (c) *inner- and outer-sphere debromination mediated by Cr[II]* (Adapted from ref. 7)

The wide variety of indirect mediated electroorganic reactions has been reviewed by Steckhan.[8] These include oxidation of aromatics at the side chain and at the aromatic nucleus with metal ions, olefin oxidations and reductions with metal ions, base metals and transition metal complexes (see Table 3).

There has been significant research on the oxidation of toluenes to benzaldehydes using electrochemically generated mediators, *e.g.* Mn[III].

$$C_6H_5 + 2\ Mn^{3+} + H_2O \rightarrow C_6H_5CH_2OH + 2\ Mn^{2+} + 2\ H^+ \qquad (6)$$

$$C_6H_5CH_2OH + 2Mn^{3+} \rightarrow C_6H_5CHO + 2\ Mn^{2+} + 2\ H^+ \qquad (7)$$

$$C_6H_5CHO + 2\ Mn^{3+} + H_2O \rightarrow C_6H_5COOH + 2\ Mn^{2+} + 2\ H^+ \qquad (8)$$

Table 3 *Range of indirect electrochemical synthesis reactions*

Oxidation with metal salts	Oxidation of aromatics with side-chain *e.g.* synthesis of substituted benzyaldehydes with Mn^{3+} and Ce^{4+} Oxidation at the aromatic nucleus *e.g.* benzene to phenol with Fe^{2+} / H_2O_2. Olefin oxidations with Tl^{3+}, Pd^{2+} and Os^{8+}
Oxidation with inorganic anions	*e.g.* Oxidation of alkenes to epoxides, secondary alcohols to ketones with halogens dialdehyde starch production from starch by periodate
Reductions with metal salts and base metals	*e.g.* Ti^{3+}, V^{3+}, Sn^{2+}, Cr^{2+}, Zn, Cu, Fe, Sn powders reduction of nitro compounds an halogenated hydrocarbons
Reductions with metal complexes and metal salts	Ni^0, Ni^A, Co^+, Sn^0, Pd^0, Ph^I complexes *e.g.* reaction with alkylating agents *via* oxidative addition. Electrochemical reduction of alkyl complex cleaves metal–carbon bond
Synthesis with organic redox catalysts	Reductions with radical anions and dianions of mainly aromatics for *e.g.* cleavage of alkyl *e.g.* perylene, biphenyl, anthracene and aryl halides, sulfonates, sulfides, epoxides. Dehalogenation of polychlorinated biphenyls with 9,1- diphenylanthracene
Reduction by viologen radical cations	*e.g.* NAD to NADH and NADPH with methyl viologen
Oxidation with triarylamine radical cations	Selective oxidations of arylalkanes to benzaldehyde dimethyl acetals
Oxidation with electrochemically regenerable hydride or hydrogen abstracting reagents	*e.g.* *N*-Hydroxy phthalamide for oxidation of arylalkanes, benzlylethers and olefins

The first oxidation is the fastest followed by reaction (2) to benzalde-hyde and then by reaction (3), the oxidation of benzaldelyde. The reactions take place in two reactors, the second is a counter current extractor, with a large resistance, in which the reaction of dissolved benzyl alcohol to benzaldehyde in the aqueous phase takes place and the latter is extracted in an inert organic solvent (kerosene).

In the oxidation of toluenes the formation of the various aldehydes is dependent upon the redox agent used. For toluene, *p*-chlorotoluene and xylene, Mn^{3+} mediated oxidation is attractive at temperatures above $60\,^\circ$C. The conversion of *p*-nitrotoluene to *p*-nitrobenzaldehyde is only effective with Co^{3+}. The oxidations of methylanisole to anisaldehyde is effectively carried out with Ce^{4+}. The relative reactivities of the oxidants is in the order Co^{3+}, Mn^{3+}, Ce^{4+}.

The production of benzaldehyde by a paired oxidation of toluene using Mn^{3+} and OH free radicals generated in the anodic and cathodic compartments, respectively of the same cell, has been tested.[9] The OH free radicals are generated as Fenton's reagent by the simultaneous cathodic reduction of Fe^{3+} ions and oxygen. Maximum total current efficiencies of 171% for benzaldehyde production are reported.

Table 4   *Indirect oxidations using Ce^{4+}*

Oxidation	Conversion (%)	Selectivity (%)	Yield (A) (%)
Anthracene → 9,10-anthraquinone	98	95	93
Naphthalene → 1,4-naphthoquinone	100	98	98
2-Methylnapthalene →			
2-methyl-1, 4-naphthoquinone	94	67	63
6-methyl-1, 4-naphthoquinone		17	46
		84	79
1-Nitronaphthalene →			3
5-nitro-1, 4-naphthoquinone	92	90	
2-*tert*-Naphthalene →			
6-*tert*-Butyl-1, 4-naphthoquinone	100	65	65
2-*tert*-Butyl-1, 4-naphthoquinone	26	26	26
		91	91
Toluene → benzaldehyde	98	93	91
Toluene → benzaldehyde	10	57	6
benzyl alcohol		37	4
		94	10
p-Chlorotoluene → *p*-chlorobenzaldehyde	99	87	86
o-Chlorotoluene → *o*-chlorobenzaldehyde	92	73	67
p-Xylene → *p*-tolualdehyde	98	80	78
o-Xylene → *o*-tolualdehyde	98	81	79
m-Xylene → *m*-tolualdehyde	91	77	70
1,2,3,5-Tetramethylbenzene →			
2,4,6-trimethylbenzaldehyde	26	63	16
p-*tert*-Butyltoluene → *p*-*tert*-butylbenzaldehyde	31	88	27
p-Ethyltoluene → *p*-methylacetophenone	32	70	22
p-ethylbenzaldehyde		10	3
		80	25
1,2,3,4-Tetrahydronaphthalene →			
1-tetralone	16	62	10
p-Methylanisole › *p*-anisaldehyde	93	84	78
Styrene → benzaldehyde	98	89	87

An important example, adopted by the W. R. Grace and Co. of mediated synthesis is the oxidation of aromatic precursers to aldehydes, ketones and quinones. The products of these oxidations are used as intermediates, flavours, fragrances and dyes. Traditional oxidants used to carry out these syntheses are sodium dichromate, MnO_2 and oxygen in air. The non-electrochemical processes have limitations in terms of disposal of spent reagents, used metal oxidants or poor selectivity in the case of air oxidation. The Grace process[10] uses the Ce^{IV} species for the oxidation, because of its high selectivity to many of the desired products (see Table 4).

The novel feature of this process, in comparison to other processes based on the use of cerium, is the choice of anion and electrolyte. Rather than using sulfuric acid, the electrolyte is a solution of methane sulfonic acid which offers several advantages such as:

- Low cost, the total material cost is quoted at $8.60 kg^{-1} (cerium carbonate and methane sulfonic acid)
- Methane sulfate is unreactive with reactants and products
- Methane sulfate is stable to oxidation
- Methane sulfate has a high solubility for Ce^{III} and Ce^{IV}
- Fast organic reactions

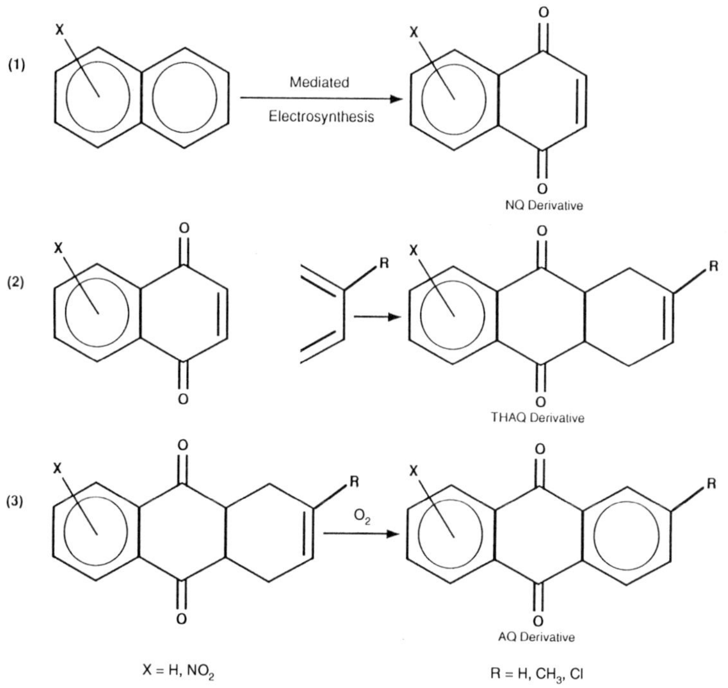

X = H, NO$_2$ R = H, CH$_3$, Cl

Figure 3 *Mediated electrosynthesis of anthraquinone derivatives*

- The high solubility of cerium in methane sulfonic acid, of the order of 2 kmol m^{-3}, enables much higher current densities to be used in the anodic regeneration of the Ce^{IV} and increases the speed of the chemical reaction, both factors reducing the capital cost of the process. The regeneration of Ce^{IV} can take place at current densities of up to 4000 A m^{-2} with current efficiencies >90%.

The process (GEMS) has been used for many syntheses (see Table 4). Good yields and selectivities are obtained in the production of many quinones, aldehydes and ketones. The process offers alternative routes to the production of anthraquinone chemicals by the initial oxidation of a naphthalene precurser (see Figure 3).

This synthesis route, carried out by a combination of mediated electrochemical oxidation, condensation with a diene and air oxidation offers several advantages over competing technologies listed in Table 5. The alternative processes can lead to large quantities of waste acid, contaminated with organics, or to large quantities of aluminium chloride waste.

Table 5   *Applications and competing technology for products of the Ce^{4+} mediated electrosynthesis*

Product	Application	Competing technology
1,4-Naphthoquinone	Intermediate to dyes, agricultural chemicals pulping catalysts and catalysts for H_2O_2 production	Catalytic air oxidation of naphthalene; ceric sulfate mediated electrochemical oxidation of naphthalene
5-Nitro-1,4-naphthoquinone	Dye intermediate	Nitration of 1,4-naphthoquinone
p-tert-Butylbenzaldehyde	Intermediate to fungicide (fenpropiomorph) and fragrance chemical (Lilial)	MnO_2 and/or electrochemical oxidation of *p-tert*-butyltoluene
p-Tolualdehyde	Flavour and fragrance chemical, polymer additive	Friedel–Crafts formylation of toluene (CO, HCl + AlCl$_3$)
p-Chlorobenzaldehyde	Photoresists, pesticide, plant growth regulator, chlorphenteramine (appetite suppressant) intermediate	Chlorination of *p*-chlorotoluene followed by hydrolysis
o-Chlorobenzaldehyde	Metal plating brightener, tear gas intermediate, (appetite suppressant) intermediate	Chlorination of *o*-chlorotoluene followed by hydrolysis
p-Anisaldehyde	Aroma chemicals, metal plating brightener, intermediate to sunscreen	MnO_2 and electrochemical oxidation of *p*-methylanisole
2-Methyl-1,4-naphthoquinone	Synthetic vitamin K$_3$ (animal feed)	Dichromate oxidation of 2-methylnaphthalene

8.1.4 Heterogeneous Redox Catalysis Processes

Several redox systems can be immobilised at the surface of electrodes, in the form of composites with metals or graphite. There is essentially no loss of the product and stereoselectivity of the redox system caused by the immobilisation, whilst the disadvantages of the alternative 'homogeneous' chemical reaction method are effectively avoided. The model of heterogeneous redox catalysis has been confirmed at various classes of solid electrodes with oxides, graphite, conducting polymers, redox polymers, inorganic semiconductors, metal chelates and ion crystals as active layers. Oxide materials, which have been used in this mode, include Cr_2O_3, PbO_2, VO_x, AgO_x and NiO_2H. Two examples[11] of the use of redox electrocatalysts are now discussed.

8.1.4.1 Lead Dioxide. Lead dioxide is known to be a strong chemical oxidant in acid solution. As an electrode it is a complicated material, which is influenced by non-stoichiometry, porosity, texture and non-conducting reduction products such as $Pb(SO)_4$, PbO and $Pb(OH)_2$ in a poorly defined way. There is, however, evidence that in many cases it functions as a redox electrocatalyst, where at the electrode surface the substrate is oxidised to an intermediate and the bivalent lead is anodically oxidised to PbO_2.

$$PbO + H_2O \rightarrow PbO_2 + 2\,H^+ + 2\,e^- \qquad (9)$$

An example of the use of lead dioxide electrodes is in the oxidation of cyclohexanone oxime to nitroso compounds in H_2SO_4 and $HClO_4$ electrolytes involving a two-electron transfer:

$$\langle\rangle{=}NOH + Nu^- \rightarrow \langle\rangle/NO + H^+ + 2\,e^- \qquad (10)$$
$$\backslash Nu$$

where Nu^- are nucleophiles such as OH^-, NO^{2-} and Cl^-.

In these reactions the electrode material reacts chemically with the oxime and it is oxidised electrochemically at electrode potentials close to the redox potential of the Pb^{II}/Pb^{IV} couple. On the other hand lead dioxide electrodes are inert regarding the oxidation of lower aliphatic alcohols. Other examples of the use of lead dioxide as heterogeneous redox catalysts are in the oxidation of benzene (and related aromatics).

$$C_6H_6 - 6\,e^- + 2\,H_2O \rightarrow C_6H_4O_2 + 6\,H^+ \qquad (11)$$

This process was developed to the pilot scale in the 1970s but did not progress to full plant operation. The reaction is carried out by the chemical oxidation of benzene at the lead dioxide surface and not the anodic electron transfer at the anode surface. The anodic process serves to regenerate the lead dioxide from the lead sulfate formed by the chemical reaction. The process used an emulsion electrolyte with turbulent flow to induce high rates of mass transport. To intensify this mass transport the anode design incorporated turbulence promoting fins. In practice the benzoquinone is susceptible to further reduction and its formation, in the dispersed benzene, is limited to approximately 3%. The benzoquinone is re-extracted into the catholyte where it is cathodically reduced to the hydroquinone product.

8.1.4.2 Nickel Hydroxide/Oxide — The Manufacture of Ascorbic Acid. Hoffman la Roche have reported the use of the anodic oxidation of a sorbose derivative as part of the manufacture of ascorbic acid (vitamin C). The reaction takes place in an aqueous sodium hydroxide solution at a nickel electrode covered by a redox catalyst oxide/hydroxide layer:

$$RCH_2OH \ + \ 4\,NiOOH \ \rightarrow \ RCOO^- \ + \ 4\,Ni(OH)_2 \qquad (12)$$

$$anode \ 4\,Ni(OH)_2 + OH^- \rightarrow 4\,NiOOH + 4\,e^- + NiOOH + H_2O \quad (13)$$

The reaction is performed in swiss-roll cells using a high surface area roughened nickel gauze anode and a steel gauze cathode in an undivided configuration. The current density (over-potential) of the oxidation is very low and required the selection of a high surface area material. On the plant a cascade of cells, operating with decreasing current density is used to achieve a conversion of 99% with a selectivity of over 90% and a current efficiency of approximately 75%. The major anodic byproduct is oxygen gas. The scale of the process is quoted at 200 tonnes per day from cells of total anode area 200 m^2. The actual electrochemical reaction is one of six overall reactions required to produce ascorbic acid from glucose in the process.

8.1.5   Electrosorbed Hydrogen

Electrochemical reductions are known to occur *via* anion radical, carbanion and other intermediates in solution. In addition, on electrode surfaces, which are able to stabilise hydrogen atoms by adsorption, a reduction process analogous to catalytic hydrogenation can occur. The adsorbed hydrogen is produced by the electron reduction of proton or solvent (water, methanol, acetic acid *etc.*)

$$H^+ \ (or \ H_2O) \ + \ e^- \ + \ M \ \longrightarrow \ M–H \quad (+ OH^-) \qquad (14)$$

on electrode surfaces such as Ni, Pt, Pd, Rh *etc.*
    The reduction of the organic molecule occurs by the chemical reaction

between the adsorbed hydrogen and the organic molecule, which itself may be adsorbed onto the surface as part of the overall reaction process. It is usual for these types of reactions to take place at low current densities and the use of high surface area electrodes are thus required to achieve reasonable space time yields. The reactions are generally very selective, the only major byproduct formed is hydrogen gas by the competitive reduction of protons at the cathode. Couper *et al.*[12] have identified several electrocatalytic hydrogenations of possible interest in electrosynthesis. (see Table 6). Cathode materials, which are frequently used, are Raney nickel based or metal blacks.

Table 6 *Electrocatalytic hydrogenations*

Cathode	Reaction
Ni, Pd, Pt	Ketones → alcohols
Ni	Aldehydes → alcohols
Ni, Pd, Pt	Acetylenes → *cis*-alkenes
Ni	Olefins → alkanes
Ni, Pd	Unsaturated ketones → ketones
Ni, Pd, Co, Fe	Nitriles → amines
Ni	Schiff bases → amines
Ni	Oximes → amines
Ni	Pyridine → piperidine
Ni	Cyclohexadiene → cyclohexane
Ni	Benzene → cyclohexane
Ni	Sugars → sugar alcohols
Pd	Unsaturated steroids → steroids
Pd	Cleavage of the benzyloxycarbonyl group from peptides
Pt	Ketones → alkanes
Pt	Butadienes → alkenes
Pt	Nitro compounds → amines
Pt	$CF_3COOH \rightarrow CF_3CH_3$
Rh	Phenols → cyclohexanols

8.1.6   Direct Electroorganic Synthesis

Although there is a wide range and diversity of electroorganic synthesis processes the mechanism by which such syntheses occur often starts with the formation of radical ions, which then react further to form intermediate species which themselves may undergo further electron transfer.

$$\text{organic} + e^- \rightarrow (\text{rad. ion}) \xrightarrow{\text{fast reaction}} \text{intermediate} + e^- \rightarrow \text{product} \quad (15)$$

The fate of the unstable ion radicals and thus the type of product(s) formed depends on the surrounding homogeneous electrolyte (solvent and other species), the electrode surface and presence of any adsorbed species. The reaction chemistry of the intermediate species is essentially no different to that of other chemical synthesis methods when performed in a similar environment. The electron transfer processes add a further, or additional, control parameter to providing possible selective synthesis. The chemical reactivity of radical ions is determined by the ionic character of the species and the radical itself. Thus the species can undergo typical radical reactions such as hydrogen abstraction, radical addition, radical disproportionation and radical–radical dimerisation. The radical cations will, because of the positive charge, also be expected to behave as electrophiles, adding to available nucleophiles. Radical cations may also add to unsaturated hydrocarbon molecules or can react further (by proton dissociation from a vicinal carbon atom) to give alkyl radicals.

The reaction paths for radical anions obtained from unsaturated hydrocarbons, *e.g.* vinyl compounds, are H abstraction, radical reaction with the electrode material (forming organo-metal compounds), radical anion disproportionation and radical dimerisation. The anionic reactions of radical anions are protonation, to form the radical, and the Michael additions of the radical anion. For radical anions the radical reactions are more important than the anionic reactions, except in the case of protonation to form the respective radical. This is because proton donors cannot be completely eliminated from electrolytes and also the lack of a proton donor is undesirable because of the need to terminate the sequence of reactions, by proton addition to a carbanion, to avoid undesirable electropolymerisation.

One of the most important industrial examples of the proton addition reaction is that associated with the electrohydrodimerisation (EHD) of acrylonitrile. In this, acrylonitrile (AN) is reduced to the radical anion:

$$CH_2=CH–CN + e^- \longrightarrow CH_2=CH–CN^{\bullet-} \tag{16}$$

Protonation of the radical anion generally proceeds very fast in most practical electrolytes. Protonated radical anions are generally readily reduced to the carbanion and in the case of acrylonitrile this gives:

$$(CH_2–CH–CN)^{\bullet-} + H^+ \rightarrow (^{\bullet}CH_2–CH_2–CN) - e^- \rightarrow CH_2\text{-}CH_2\text{-}CN \tag{17}$$

The further reactions of the carbanion determine the types of product eventually formed. The carbanion may be protonated again,which terminates the reaction,or may undergo a Michael addition to a molecule which has an activated double bond. Thus in the case of acrylonitrile the following reactions are possible: Protonation of the carbanion to give propionitrile

$$(^-CH_2CH_2-CN) + HA \longrightarrow CH_3-CH_2-CN + A^- \qquad (18)$$

Michael addition with acrylonitrile to form the adiponitrile (ADN) anion, by the C–C coupling

$$NC-CH_2-CH_2^- + CH_2-CH-CN \rightarrow (NC-CH_2-CH_2-CH_2-CH-CN)^- \quad (19)$$

with termination by a further protonation.

The factor that determines the product obtained at this stage is the composition of the electrolyte layer at the surface of the electrode. In the case of the synthesis of adiponitrile the use of quaternary alkyl ammonium salts produces an electrolyte layer which is depleted in water molecules and other proton donors due to the adsorption of the surface active lypophilic cations. The extent of this depletion is such that the protonation step of reaction (18) still takes place but the protonation of reaction (19) is suppressed to the extent that the Michael addition can still take place and thus the formation of adiponitrile at high yields and efficiencies. In the synthesis of adiponitrile there are several other products formed as a result of the activity of the intermediate species.

8.1.7 Electroorganic Reactions

To illustrate the range of organic electrosynthesis the following is a short review of those reactions which have been considered at some point as potential industrial processes and often subsequently adopted.[5]

8.1.7.1 Cathodic Reduction — C=C Double Bond System. The most industrial famous example of the reduction of C=C double bonds is in the cathodic hydrodimerisation of acrylonitrile. Generally isolated olefinic double bonds are not readily electrochemically reduced although a number of aromatic compounds and activated olefins are. Activated olefins such as acrylonitrile can also undergo cathodic C–C cross coupling reactions with carbonyl compounds. They also can be cathodically carboxylated using carbon dioxide as an electrophile.

The cathodic reductions of aromatics to the corresponding 1,4-dihydro compounds has been studied extensively:

$$\text{Benzene} \xrightarrow{2e^-} \text{cyclohexa -1,4-diene} \qquad (20)$$

The reduction can take place in non-aqueous electrolyte or in water-containing electrolytes utilising quaternary ammonium salts. Phthalic acid derivatives are more easily reduced because of the less negative cathode potentials and this has been applied in the synthesis of 1, 2-dihydrophthalic acid.

The cathodic reduction of heterocycles has been considered for the electrosynthesis of bipyridyls for *N*-substituted pyridinium salts, the products being used are herbicides and dye intermediates (in the case of indole derivatives).

8.1.7.2 Organic Halogen Compounds. The cathodic reduction of aliphatic halogen compounds, in which the cleavage of the C–halogen bond enables the selective removal of one halogen atom from polyhalogenated compounds thereby introducing the C=C double bond, has been demonstrated for many reactions:

$$Cl_2CH - COOH \rightarrow Cl-CH_2-COOH \tag{21}$$

It has also been used to produce organometallic compounds.

There is interest in electrochemical C–C coupling by reductive dehalogenation

$$Br \diagdown\diagup OH \rightarrow HO - (CH_2)_4 -OH \tag{22}$$
$$\text{butane -1,4-diol}$$

The electrochemical reduction of perhalogenated aromatics and heterocyclic compounds can be used to prepare aromatic halogen derivatives which have unusual substitution patterns.

8.1.7.3 Carbonyl Compounds and Derivatives. The cathodic reduction of carbonyl compounds can lead to the reduction of the functional group or to coupling processes. The reduction of saturated aldehydes and ketones readily produces the corresponding alcohols:

$$
\begin{array}{ccc}
R & & R \\
| & & | \\
C{=}O & \xrightarrow{2e^-} & H{-}C{-}OH \\
| & & | \\
R' & & R'
\end{array}
\tag{23}
$$

Under certain conditions reduction to the corresponding hydrocarbon is possible, which is of interest in the synthesis of carotenoids and fragrances.

$$
\begin{array}{ccc}
R & & R \\
| & & | \\
C{=}O & \rightarrow & H{-}C{-}H \\
| & & | \\
R' & & R'
\end{array}
\tag{24}
$$

There is also industrial interest in the reduction of carboxylic acids to aldehydes and alcohols. Benzyl alcohol, *o*- and *m*-hydroxybenzoic acid, and *m*-phenoxybenzaldehyde have all been produced with good yields. Glyoxylic acid is produced commercially by the cathodic reduction of oxalic acid.

$$
\begin{array}{ccc}
\text{COOH} & & \text{COOH} \\
| & \xrightarrow{2e^-} & | \\
\text{COOH} & & \text{CHO}
\end{array}
\tag{25}
$$

and can be produced in the absence of other electrolyte.

Aromatic and heteroaromatic esters and carboxamides can be reduced cathodically to alcohols and amines, respectively.

e.g.

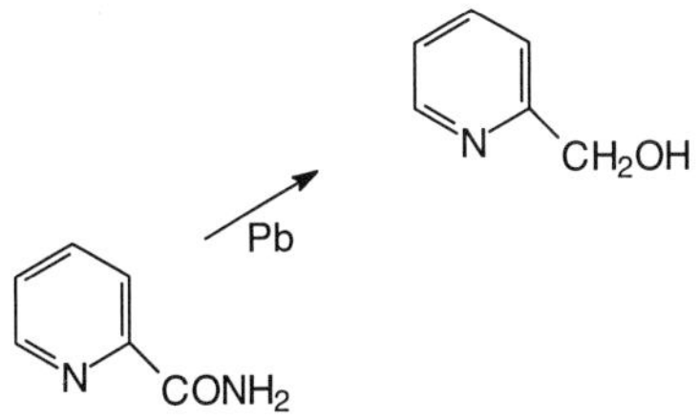

(26)

Nitriles can also be reduced cathodically to amines or alcohols.

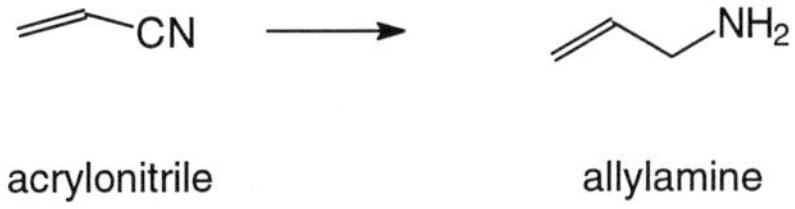

(27)

acrylonitrile   allylamine

The coupling reactions of the cathodic hydrodimerisation of aliphatic aldehydes are of interest. The production of pinacol from acetone has been widely investigated and the formation of ethylene glycol from formaldehyde is attracting significant commerical interest.

$$
\text{CH}_2\text{O} \xrightarrow{\text{graphite}}
\begin{array}{cc}
\text{CH}_2\!\!-\!\!\text{CH}_2 \\
| \quad\quad | \\
\text{OH} \quad \text{OH}
\end{array}
\tag{28}
$$

Finally CO_2 can be reduced to oxalic acid in aprotic electrolytes and to formic acid in aqueous electrolyte.

There are several examples of the reduction of compounds with C=N double bonds — oximes to amino compounds, imines to amines and hydroxylamines with N–O bond cleavage.

8.1.7.4 Nitro Compounds. Products that can be obtained from the cathodic reduction of aromatic nitro compounds are typified by those formed from nitrobenzene and similar compounds. These products are anilines, aminophenols, azoxybenzenes, azobenzene and hydrobenzene.

$$\Delta\text{-NO}_2 \xrightarrow{2e} \Delta\text{NO} \xrightarrow{2e} \Delta\text{-NHOH}$$

with branches from Δ-NHOH:

- $\xrightarrow{2e} \Delta\text{-NH}_2$
- $\xrightarrow{acid} \text{OH--}\Delta\text{-NH}_2$ (29)

From the bracket under ΔNO / Δ-NHOH:

$$\downarrow pH>8$$
$$\overset{O}{\underset{|}{\Delta\text{--N=N--}\Delta}}$$
$$\downarrow 2e$$
$$\Delta\text{--N=N-}\Delta$$
$$\downarrow$$
$$\Delta\text{--N}\overset{H}{|}\text{--N}\overset{H}{|}\text{-}\Delta$$

where Δ = R— (phenyl group)

The corresponding alkoxy-substituted anilines can be formed if the electrochemical reduction is performed in alkanols, under equivalent acid conditions. In alkaline electrolyte the corresponding azoxy, azo or hydrazo species are formed.

The reduction of aliphatic nitro compounds gives either amines or hydroxylamines depending upon conditions:

$$e.g. \quad \text{CH}_3\text{-NO}_2 \xrightarrow[\text{Cu}]{\text{H}_2\text{O} - \text{HCl}} > \text{CH}_3\,\text{NHOH} \tag{30}$$

Nitroso compounds can also be reduced to the corresponding amines or aminophenols.

$$\text{(phenyl)--NH--(phenyl)--NO} \longrightarrow \text{(phenyl)--NH--(phenyl)--NH}_2 \tag{31}$$

$$\text{HO--(phenyl)--NO} \longrightarrow \text{HO--(phenyl)--NH}_2 \tag{32}$$

Both these compounds are dye intermediates.

In the presence of ketones the reduction of nitroso compounds can be applied in the alkylation of amines. *N*-Nitroso compounds can be reduced to hydrazines and *N*-oxides can be reduced electrochemically to the corresponding amines.

8.1.7.5 Sulfur Compounds. The electrochemical reduction of organic disulfide compounds can give high yields of the mercaptans. The reduction of cystine to cysteine is of industrial significance.

$$RS - SR \xrightarrow{\dfrac{Pb}{HCl}} RSH \tag{33}$$

$$R = CH_2CH(NH_2\,HCl)COOH$$

The synthesis of substituted 4-oxo azetidines is an example of the reduction of sulfones to sulfinic acids. The reduction of sulfonates can be used for the alkylation of phenylacetic acids. The reduction of sulfonium salts has been used for C–C coupling and the reductive cleavage of sulfonium salts in the synthesis of a new herbicide.

$$\tag{34}$$

8.1.7.6 Anodic Oxidations — Anodic Functionalisation. Anodic oxidation of olefins in the presence of CH_3OH, CH_3COOH and other nucleophiles permits alkyl oxidation and C–C coupling reactions. The anodic methyoxylation of citronelli is applied as a key step in a rose oxide synthesis. Anodic methoxylation is also used in the preparation of 2,5- dimethoxy-2,5-dihydrofuran from butene-1,4-diol using carbon anodes.

$$\tag{35}$$

The anodic functionalisation of aromatics is of significant commercial interest. The oxidation of benzene to *p*-benzoquinone has already been discussed in terms of heterogeneous electrocatalysis. The similar reactions to substituted quinones (and hydroquinones) is also known, as is the use of indirect oxidation. The use of anodic methoxylation is also a possible route

to substituted quinones. The direct oxidation of naphthalenes to naphtho-
quinones is not as effective in terms of yield as indirect oxidation, which
are discussed in section 8.1.3.2 and Table 4. The anodic oxidation of substituted phenols has been demonstrated as a route to speciality chemicals.

An important group of reactions is the nuclear acyloxylation of aromatics. An example is the production of naphthyl acetate from naphthalene
in an HOAc–(CH$_3$)NOAc electrolyte. The naphthlyl acetate is readily
converted to α-naphthol. Anodic acetoxylation has been used for the synthesis of acyloxyalkoxy aromatics and diphenyl esters and applied in the
presence of trifluoroacetic acid for trifluoroacetoxylation.

Side chain substitution reactions, particularly electrosynthesis of substituted benzaldehydes, are carried out on an industrial scale. There are several examples of anodic acetoxylations:

$$(36)$$

With the use of controlled amounts of water in the electrolyte, the acetoxylation can give the aldehydes in good yields. If methanol is used, in
place of acetic acid, in the electrolyte the corresponding benzaldehyde
dimethyl acetals are obtained. The reaction can also be used for the synthesis of aromatic aldehydes. As has already been seen the use of indirect
electrosynthesis has been widely considered for aromatic aldehyde formation.

The electrochemical methoxylation of heterocyclic compounds, *e.g.*
furans to produce the corresponding dimethoxydihydrofurans, is practiced on an industrial scale *e.g.*,

$$(37)$$

which is another example of an indirect electrosynthesis. The synthesis of
biocides, cyclopentenones, prostaglandin intermediates and alkoxy-
isochromons are other examples. The oxidation of 2-methylpyridine to
picolinic acid is also practiced on an industrial scale.

$$(38)$$

8.1.7.7 Oxidation of Carbonyl Compounds. The Kolbe coupling reaction of carboxylic acid

$$2RCOO^- \xrightarrow{-2e^-} R\text{--}R + 2CO_2 \tag{39}$$

gives the $2n$-2 hydrocarbon. However, if the oxidation of the half-ester of a dibasic acid is employed, the corresponding diester of the dibasic acid is formed.

$$2\ CH_3O_2C(CH_2)_4CO_2Na \xrightarrow[\text{Pt}]{CH_3OH} CH_3O_2C(CH_2)_8CO_2CH_3 + 2\ CO_2 + 2\ Na^+ + 2e^- \tag{40}$$

The reaction has also been applied to the synthesis of sebacates of higher alcohols (diethyl hexyl sebacate) and to higher carboxylic acids (diethyl hexyl suberate). The Kolbe reaction of carboxylic acids with certain functional groups has also been carried out. Examples include the coupling of cyanocarboxylic acid to the dinitrile, and the synthesis of diols, diamines and perfluorianted alkanes.

The anodic oxidated of α-substituted carboxylic acids often results in the two-electron oxidation product, rather than a Kolbe coupling. This reaction has been used for novel isocyanate electrosynthesis, for vanillin and 2-hydroxytetrahydrofuran.

$$\text{(structure) COOH} \xrightarrow[\text{C anode}]{H_2O\text{---}THF\text{---}KOH} \text{(structure) OH} \tag{41}$$

The anodic oxidation of aldehydes to carboxylic acids has been featured in the oxidation of lactose to calcuim lactobionate and glucose to gluconates. Generally catalytic air oxidation is effective for these reactions although the formation of glyoxylic acid from glyoxal is of interest.

$$\begin{array}{c} CHO \\ | \\ CHO \end{array} \xrightarrow[\text{C anode}]{HCl} \begin{array}{c} COOH \\ | \\ CHO \end{array} \tag{42}$$

The anodic alkoxylation of carboxamides has produced some interesting syntheses. The anodic oxidation of N-ethylcarboxamides is a route to n-vinylamides

$$CH_3\text{--}CH_2\text{--}NHCHO \xrightarrow{CH_3OH\text{--}(C_2H_5)_4NBF_4} \underset{\underset{OCH_3}{|}}{CH_3\text{--}CH\text{--}NHCHO} \tag{43}$$

N-Vinyl-urethanes can be produced by similar routes. The reaction is also applied to *N*-acyl derivatives of cyclic amines and similarly to the formation of alkoxylated urethanes.

A further example of C–C coupling is the anodic dimerisaiton of CH-acidic compounds, which has been applied to malonates and the production of cyclobutane derivatives.

8.1.7.8 Oxidation of Alcohols and Ethers. Carboxylic acids can be formed from the anodic oxidation of alcohols, in good yields, with nickel oxide anodes in alkaline media.

$$(CH_3)_2 \, CH\text{–}CH_2OH \rightarrow (CH_3)CH\text{–}COOH \tag{44}$$

The reaction is practiced for the synthesis of diacetone-2-ketogulonic acid from diacetone-2-sorbose. The procedure has been applied to the preparation of diglycolic acid and phenoxyacetic acid, propynoic acid, and acetylene dicarboxylic acid.

The oxidation of secondary alcohols to ketones has been demonstrated, but is generally inferior to catalytic hydrogenation except in special cases. One example is the regioselective oxidation of an *endo*-hydroxy group in 1,4,3,6-dianhydrohexitol. There are several oxidations of ethers which can be highly selective when glassy carbon anodes are used in a $CH_3OH\text{–} (CH_3)_4NSO_4CH_3$ electrolyte, *e.g.*,

$$\tag{45}$$

8.1.7.9 Oxidation of Nitrogen, Sulfur and Phosphorous Compounds. There has been specific interest in the oxidation of sulfur compounds to tetra-alkylthiuram disulfides:

$$(CH_3)_2 \, N\text{–}CS_2Na \rightarrow (CH_3)_2 \, NCS_2S_2CN(CH_3)_2 + 2Na^+ + 2e^- \tag{46}$$

In the presence of amines, sulfenamides can be produced. Other coupling reactions, which have been investigated, include the synthesis of dibenzothiazyl disulfide and benzo thiazolysulfenamides.

The anodic cleavage of disulfides is also considered as a route to vulcanisation enhancers, and the synthesis of phenyl sulfinates and intermediates for penicillins and cephalosporins. Interest has also been shown in the electrosynthesis of sulfoxides.

$$CH_3SCH_3 \rightarrow CH_3 - SOCH_3 \tag{47}$$

The oxidation of nitrogen and phosphorous compounds has seen a number of interesting syntheses. The oxidation of amines is well known although only a few can give satisfactory yields. Examples include the oxidation of *p*-phenylenediamine adrenaline derivatives, enamines and the anodic dimerisation of substituted naphthylamine. The oxidation of hydrazones and oximes of α, β-unsaturated carbonyl compounds to isoxazoles are also known.

The elimination of phosphomethyl groups has been performed at high selectivity

$$N(CH_2-PO_3H)_3 \rightarrow HN(CH_2-PO_3H)_2 \tag{48}$$

and is a generally applicable route to metal ion sequestering agents.

The oxidation of triphenyl phosphine in the presence of heterocycles or aromatics gives the corresponding triphenylheteroaryl phosphonium and triphenylaryl phosphonium salts. The C–C coupling of triphenylphosphonium salts is a route to β-carotene.

Finally the use of phosphorous based anodes enables the synthesis of phosphates, *e.g.* triethyl phosphate.

8.1.7.10 Electrochemical Fluorination. The electrolysis of organic compounds in anhydrous hydrogen fluoride offers a means of introducing fluorine into several molecules, whilst maintaining important functional groups. There are two variations of this synthesis:

(*i*) The Simon's process, which uses Ni anodes and operates at low temperatures, usually with alkali fluoride added to the electrolyte to improve conductivity.

(*ii*) The Philip's process, which uses porous carbon anodes and a eutectic mixture of KF–HF operating at 60–105 °C.

The Philip's process[3] operates with the organic feedstock fed through the pores of the carbon anodes and is more effective when the substrate is not appreciably soluble in the molten eutectic. The process operates with near 100% current efficiencies and has been used in the production of partially or completely fluorinated alkenes, chloralkanes, cyclic alkanes, acids, ethers and esters.

Important industrial electrofluorination products are perfluorooctanoic acid $[CF_3(CF_2)_6COOH]$ and perfluorooctane sulfonic acid $[CH_3(CF_2)_7SO_3H]$, which are produced in the Simon's process.

Table 7 *Commercial electroorganic synthesis*

Reaction	Country	Estimated scale /ton per year	Conditions
Hydrodimerization			
$2\,CH_2CHCN \xrightarrow{2e^-} (CH_2CH_2CH)_2$	USA Europe Japan	3×10^5	Two phase aqueous phosphate buffer + acrylonitrile. Undivided bipolar parallel plate cell. Cd cathode, steel anode, j=2000 A m^{-2}
Hydrogenation of heterocycles, *e.g.*			
(a) pyridine $\xrightarrow{6e^-}$ piperidine	Europe	100	Pb cathode in H$_2$SO$_4$ medium; divided cell
(b) N-methylindole $\xrightarrow{2e^-}$ N-methylindoline	Europe	20	Pb cathode in H$_2$SO$_4$ medium; filterpress divided cell
Hydrogenation of nitriles			
4-CN-pyridine $\xrightarrow{4e^-}$ 4-CH$_2$NH$_2$-pyridine	USA	30	Pb cathode in acid sulfate medium; divided cell including high-area
2-CN-pyridine $\xrightarrow{4e^-}$ 2-CH$_2$NH$_2$-pyridine	USA	70	cathode j=5–20 mA cm^{-2}
Reduction of carboxylic acids			
(a) CO$_2$H/OH benzene $\xrightarrow{4e^-}$ CH$_2$OH/OH benzene	Japan	100	Pb cathode in aqueous acid
(b) CO$_2$H/OH benzene $\xrightarrow{2e^-}$ CHO/OH benzene	India	20	Pb cathode in acid media
(c) anhydride $\xrightarrow{2e^-}$ o-CO$_2$H benzene	Europe	100	Pb cathode in H$_2$SO$_4$; dioxan as co-solvent; divided filterpress cell
(d) CO$_2$H/CO$_2$H $\xrightarrow{2e^-}$ CO$_2$H/CHO	India	30	Pb cathode in H$_2$SO$_4$

Reaction	Country	Estimated scale /ton per year	Conditions
Cathodic cleavages			
(a) [structure: HO-phenyl-$CH(OH)CCl_3$] $\xrightarrow{2e^-}$ [structure: HO-phenyl-$CH=CCl_2$]	Japan	120	Acid CH_3CN–H_2O; Pb cathode; divided filterpress cell, current density $= 1500\ A\ m^{-2}$
(b) [structure: pyridine ring with $CH_2N(CH_3)_3$, F, F substituents] $\xrightarrow{4e^-}$ [structure: ring with CH_3, F]	Europe	200	H_2O — no electrolyte; product precipitates; Zn cathode; $100\ A\ m^{-2}$; divided FM21 cell
(c) [structure: CO_2H, $CHNH_2$, CH_2–S–S–CH_2] $\xrightarrow{2e^-}$ [structure: $2\ CO_2H$, $CHNH_2$, CH_2SH]	Europe Japan	30	Pb cathode in acid; divided cell
(d) [structure: Cl-substituted pyridine with CO_2H] $\xrightarrow{4e^-}$ [structure: pyridine with Cl, CO_2H]	USA		Roughened Ag cathode in aqueous acid
Nitro-group reduction			
[structure: NO_2-phenyl-CO_2H] $\xrightarrow{6e^-}$ [structure: NH_2-phenyl-CO_2H]	India	3	Pb cathode in acid medium
[structure: phenyl-NO_2] $\xrightarrow{10e^-}$ [structure: H_2N-biphenyl-NH_2]	India	30	Stronger acid
Base generation			
$(CH_3)_4NCl \xrightarrow{e^-} (CH_3)_4NOH$	Japan USA	24	Chlor-alkali membrane cell and conditions
Fluorinations perfluorination of $RCOOH$, RSO_3H	USA Japan UK	>102	Simons process. Ni anode in liquid HF
Methoxylations			
(a) [structure: furan] $\xrightarrow{-2e^-}$ [structure: CH_3O-dihydrofuran-OCH_3]	Europe	100	$NaBr$–CH_3OH in narrow gap, bipolar C disc cell. $j = 1500\ A\ m^2$
(b) [structure: furan-$CH(CH_3)OH$] $\xrightarrow{-2e^-}$ [structure: dimethoxy dihydrofuran-$CH(CH_3)OH$]	Japan	150	$NaBr$–CH_3OH, C cylinder anode in steel pipe cathode 1 mm gap; $j = 1000\ A\ m^{-2}$
(c) [structure: pyridine-CHO] $\xrightarrow{-2e^-}$ [structure: dihydropyridine-OCH_3, CHO]	Europe	25	Undivided bipolar parallel plate reactor

Reaction	Country	Estimated scale /ton per year	Conditions
Oxidation of polynuclear aromatic hydrocarbons	Europe USA	100	Indirect *via* $Cr_2O_7^{2-}$; Pb anode in divided parallel plate reactor
Oxidation of methyl aromatics			
(a) CH_3-C_6H_3(OBu^t) $\xrightarrow{-4e^-}$ CH(OCH$_3$)$_2$-C_6H_3(OBu^t)	Europe	1000	CH_3OH–1% KF; undivided C disc cell; j=400–1000 A m^{-2}
(b) CH_3-aromatic $\xrightarrow{-2e^-}$ CH_2OCH_3-aromatic	Japan	120	NaOH–CH_3OH; C tube in pipe cell
(c) pyridine-CH_3 $\xrightarrow{6e^-}$ pyridine-CO_2H	USA Europe	10	PbO$_2$ anode in H_2SO_4–Pb shot anode
(d) CH_3-C_6H_4-NO_2 $\xrightarrow{6e^-}$ CO_2H-C_6H_4-NO_2	India	30	Indirect *via* $Cr_2O_7^{2-}$
Kolbe coupling of half esters			
$2\ (CH_2)_n\!\!\begin{smallmatrix}CO_2^-\\CO_2CH_3\end{smallmatrix} \xrightarrow{-2e^-} (CH_2)_{2n}\!\!\begin{smallmatrix}CO_2CH_3\\CO_2CH_3\end{smallmatrix}$	Europe India	—	Pt anode in methanol; acid partially neutralized but no electrolyte added
Sugar chemistry $-2e$ Glucose $\rightarrow$ Gluconic Acid	Europe India	10^3	Indirect oxidation *via* Br$_2$; acid precipitated as Ca^{2+} salt
$2e$ Glucose $\rightarrow$ Sorbitol	Europe India		Pb cathode in aqueous acid

8.1.8 Examples of Electroorganic Syntheses

From the relatively large catalogue of organic electrosynthesis described in the previous section there are approximately 60 commercial processes in operation[10] (see Table 7), with about twice that number having reached the pilot stage. The majority of these are relatively small tonnage processes, with the notable exception of the production of adiponitrile from acrylonitrile. Even with this relatively small number of processes there is a diverse range of reactor and process technologies used in practice. This is largely brought about by:

(i)   the variation in the electrolyte composition in terms of pH, oxidation capability and 'inert' solvent, either protic or aprotic;
(ii) the requirement of the electrode materials in terms of activity and stability;
(iii) the requirements for anode and cathode separation;
(iv) the electrode configuration;
(v)   the type of product;
(vi) the type of reagent and/or reacting species.

BASF AG have developed several industrial electroorganic syntheses.[4] Many of these required the use of co-solvents to increase the solubility of reactant in an otherwise aqueous phase electrolyte. The production of 1,2-dihydrophthalic acid from phthalic acid is typical:

$$\text{(structure: phthalic acid with two COOH groups)} \quad \xrightarrow{2e + 2H^+} \quad \text{(structure: 1,2-dihydrophthalic acid with two COOH groups)} \tag{49}$$

Several operating problems for this synthesis had to be overcome to make it commercially viable. These included low current efficiency and cathode poisoning. The poisoning of the high over-voltage lead cathode was solved by the selection of an appropriate electrolyte which in addition would enhance the solubility of the phthalic acid to enable practical, high current densities to be employed. From co-solvent screening tests, the use of dioxane was found to give high product yield (98%) with a 70% current efficiency with almost complete elimination of coupled byproducts. The electrolyte composition (55% co-solvent, 25% H_2SO_4, and 15% phthalic acid) enabled an acceptable current density of 1000 A m^{-2} to be used.

The synthesis uses a parallel plate divided cell because both reagent and product can be anodically oxidised. Good electrical conductivity, low permeability to dioxane and water, and stability were prerequisites for a satisfactory membrane. Several membranes showed high dioxane loss to the anolyte, which was compounded by the resulting rapid destruction of the lead dioxide anode. This latter feature was the deciding factor for membrane selection (Ionics 61 AZG or Tokuyama CLE-E), which under batch conditions of operation lasted *ca.* 6–8 months before replacement. The energy consumption of the process is quoted at 4.3 kWh kg^{-1} of 1,2-dihydrophthalic acid.

As a result of the research and development efforts of BASF, which started in the 1960s they have produced a range of some 21 electrochemical intermediates (see Table 8) either as commercial scale or pilot scale products.

Table 8 *The BASF range of electrochemical intermediates*

Commercial products

4-Methoxybenzaldehyde
CAS No.: 123-11-5
EINECS
TSCA
MITI

4-tert.-Butylbenzaldehyde
CAS No.: 939-97-9
EINECS
TSCA
MITI

2,5-Dimethoxy-2,5-
dihydrofuran
CAS No.: 332-77-4
EINECS
TSCA

Pilot scale products

p-Tolylaldehyde
CAS No.: 104-87-0
EINECS
TSCA
MITI

o-Tolylaldehyde
CAS No.: 529-20-4
EINECS
MITI

Phthalaldehyde
CAS No.: 643-79-8
EINECS
TSCA
MITI

Terephthalaldehyde
CAS No.: 623-27-8
EINECS
TSCA

4-Ethoxybenzaldehyde
CAS No.: 10031-82-0
EINECS
TSCA
MITI

p-Benzoquinonetetra-
methylketal
CAS No.: 15791-03-4
EINECS
TSCA
MITI

Lab scale products

4-(1,1-Dimethylpropyl)-
benzaldehyde
CAS No.: 67468-54-6

4-Phenoxybenzaldehyde
CAS No.: 67-36-7
EINECS

2-tert.-Butyl-p-benzoqui-
none tetrametylketal
CAS No.: 134962-83-7

2,2-Dimethoxycylohexanol
CAS No.: 63703-34-4

1,1-Dimethoxy-1-phenyl-2-
butanol
CAS No.: 882-53-1

2-(Dimethoxymethyl)-2,5-
dihydro-2,5-dimethoxyfuran
CAS No.: 59906-91-1
EINECS

2,5-Dimethoxy-2,5-dihydro-
2-furanmethanol
CAS No.: 19969-71-2

2,5-Dihydro-2,5-dimethoxy-
furfurylamine
CAS No.: 14496-27-6

1,1,6,6-Tetramethoxyhexane
CAS No.: 54286-89-4
EINECS

1,1,8,8-Tetramethoxyoctane
CAS No.: 7142-84-9

5-Methoxy-2-pyrrolidinone
CAS No.: 63853-74-7

2-Methoxy-1-pyrrolidine-
carboxaldehyde
CAS No.: 61020-06-2

8.1.8.1 Electrosynthesis of Adiponitrile. The large scale electrosynthesis of adiponitrile from acrylonitrile is operated by three companies using one of two cell technologies, developed by either Monsanto Chemicals or Asahi Chem. Co. The first cell design developed by Monsanto used a divided filter press configuration with cation-exchange membranes to separate the cathode and anode reactions and thus prevent degradation of organics. It was the problems associated with the relatively poor performance of the cation-exchange membranes available at the time (1960s) that hastened the development of the undivided cell process.

The reactions in the electrosynthesis of adiponitrile (ADN) are discussed in section 8.1.6, where it was noted that the major byproducts were propionitrile (monomer) and tricyanohexane (trimeric species). Other oligomers are formed in relatively small amounts as are species such as hydroxypropionitrile and biscyanoethyl ether, formed by base promoted cyanoethylation reactions of acrylonitrile. A correct selection of pH was thus essential in this process and in addition the build up of hydroxide ions at the cathode as a result of the cathodic reaction processes had to be avoided. This was achieved due to turbulence and using high electrolyte velocities of the order of 1–2 m s^{-1}.

The selection of conditions and materials for the electrohydrodimerisation (EHD) reaction is typical of many organic syntheses in aqueous media.

1   The cathode material needed to have a high hydrogen over-potential to minimise hydrogen evolution. Of the electrode materials tested lead proved to be a good choice, although materials such as mercury and cadmium gave comparable efficiencies.

2   The catholyte needed to contain salts which were conductive and solubilised the reactant acrylonitrile and the products to a high level. Quaternary ammonium salt was selected with tetraethyl ammonium ethyl sulfate finally chosen because of its ease of synthesis (from materials available at the plant).

3   The anode material selected was a silver/lead alloy for the anodic evolution oxygen in a sulfuric acid electrolyte.

The cell design used in the Monsanto process is discussed in chapter 4. The Asahi process is based on a divided cell using the more durable membranes, which were available to Asahi. The anode and cathode both had Sb (6%) added to them to aid stability. The electrolyte was an emulsion of acrylonitrile in an aqueous electrolyte containing tetraethylammonium sulfate (7%). The electrolyte had a higher conductivity than the equivalent used by Monsanto and thus produced competitive energy consumptions of 4–6 kWh kg^{-1}. Around the time of the development of the EHD process by Monsanto and Asahi, several other companies researched into this process, including BASF and Phillips. These two companies, along with Monsanto, researched an undivided cell for the process, with the former using the capillary gap cell.

8.1.8.2 The Undivided Cell Process. For an undivided cell to be effective the anode material had to be stable in the electrolyte and not significantly degrade the organic reagents and product. Lead was not suitable because of the latter reason, whilst DSA and precious metal anodes were subject to significant corrosion. The corrosion, as well as incurring high replacement cost, would also lead to contamination of the cathode and a subsequent loss in process current efficiency. A mild steel anode was found to be effective as a low cost stable material when used in conjunction with corrosion inhibitors (*e.g.* borax and EDTA).

The Monsanto design used cadmium cathodes instead of lead as they offered comparable efficiency whilst being more suitable as a material for the bipolar electrodes. These electrodes were steel sheet (anode side) with cadmium smoothly electroplated on one side (Figure 4).

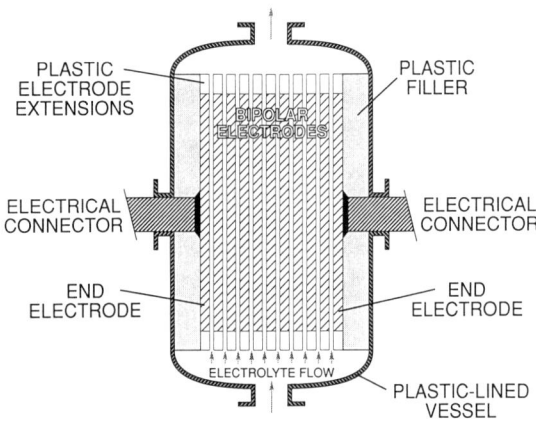

Figure 4   *Undivided bipolar reactor for ADN synthesis*

The EDTA present in the electrolyte, as well as sequestering metal ions, has the additional feature that it encourages slow corrosion of the cathode, which maintains an extensively clean surface, free of metal ion contaminants. This therefore served to maintain high yields throughout the operation of the cell.

The Monsanto cell design consists of an electrode bank of 50–100 bipolar plates, with a small interelectode gap (1–2 mm), maintained by plastic spacers. Plastic extensions are placed at the ends of each electrode to reduce the amount of current bypass. The electrode bank is placed in an insulated cylindrical steel reactor vessel which is fitted with electrode feeders and suitable electrolyte flow distributors. This design is in principle amenable to high pressure and temperature operation. This general design concept has subsequently been copied by several companies adopting undivided cell operation.

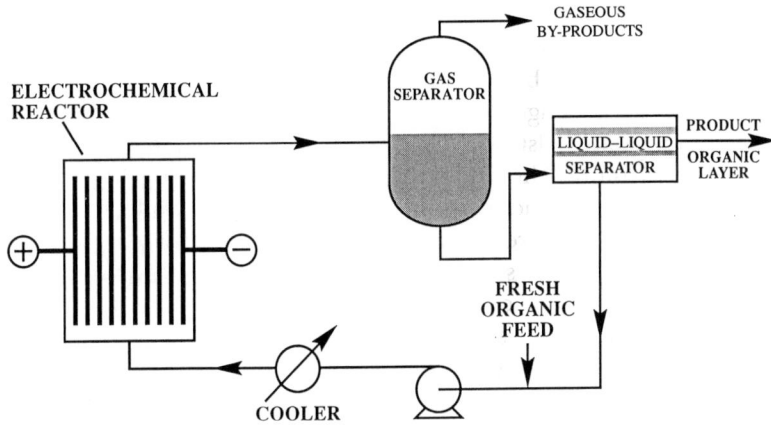

Figure 5 *Recycle reactor loop for the ADN process*

At the heart of the ADN process is the reactor loop necessary for its synthesis (see Figure 5). Around this loop circulates the electrolyte [aqueous Na_2HPO_4 (15%), hexamethylene bis(ethyldilbutylammonium) ions (0.4%), Na_4EDTA (0.5%), borax (2%) and other additives], which contains a dispersion of AN and ADN. The aqueous phase is thus saturated with AN, which is fed continuously into the recycle loop to maintain this concentration. AN is a good solvent for ADN, and the byproducts, and thus serves as a useful way of recovering the product directly from the aqueous phase. Separation of the organic phase and the aqueous phase is achieved in a decantor. The organic phase then proceeds to product recovery and reagent recycling.

The undivided cell produces large quantities of oxygen gas containing hydrogen in relatively small amounts. This is separated from the electrolyte, otherwise it would increase the resistivity of the electrolyte. The off-gas from the gas/liquid separator, which also contains the volatile AN, is an explosive hazard and is toxic. The AN is removed from the gas by scrubbing, with ADN, which is non-volatile. The gas is diluted with air to a composition outside the explosion limits and purged to the atmosphere.

A third feature of this reactor loop is that a purge stream of aqueous electrolyte is taken from the recycle loop to prevent the build-up of electrode corrosion products and organic byproducts. These latter species are removed from the stream, whilst the salts and AN and ADN are returned to the process.

Overall the electrochemical route to ADN operates at a current density of 2000 A m^{-2} with a cell voltage of less than 4 V and an electrical energy consumption of around 2500 kWh per tonne. The selectivity of ADN is of the order of 88–90%.

8.1.8.3 Ethylene Glycol. The electrohydrodimerisation of formaldehyde to ethylene glycol:

$$2 \ CH_2O + 2 \ H_2O + 2e^- \rightarrow HOCH_2 \ CH_2 \ OH + 2 \ OH^- \qquad (50)$$

is analogous to that of AN. It is a bulk chemical with an estimated market world-wide of *ca.* 9 million tonne per annum. At low concentrations of formaldehyde, hydrogen and methanol are favoured products. Thus the preferred electrolyte is an aqueous solution of 30– 50% by weight formaldehyde containing a supporting electrolyte (Na_2HPO_4 or sodium formate) and quaternary ammonium salt. As with the early ADN process a cation-exchange membrane is also used. Careful control of the pH to values between 5–8 is required otherwise polymerisation of formaldehyde occurs (at pH < 5) or the Canizzaro reaction to methanol and formic acid (pH > 8). In addition, water and alcohols (ethylene glycol and methanol) form electrochemical non-reducing hemiacetal with formaldehyde

$$CH_2O + ROH \Leftrightarrow ROCH_2OH \qquad (51)$$

This ties up the formaldehyde and limits the maximum concentration of glycol, which can be formed in the electrolyte. Formaldehyde solutions must also have a low methanol content. Operating temperatures of 80–100 °C favour the formation of free formaldehyde.

The only effective cathode material for ethylene glycol formation is graphite. This is preferably in the pre-oxidised form and current efficiencies of almost 99% can be achieved. Most metal cathodes with the exception of Hg, are ineffective.

The ethylene glycol process is as yet only at the pilot plant stage (in Canada, UK and South Africa). The current density of operation is *ca.* 4000 A m^{-2}, realising an energy consumption of 3.4 kWh kg^{-1}. The relatively low selling price of ethylene glycol (£3–10 per kg) makes the prospective process particularly sensitive to the price of oil. Thus alternative anode reactions to that of oxygen evolution have been proposed for operation in the membrane cell. These include the formation of the formaldehyde feedstock from methanol and the formation of terephthalic acid form *p*-xylene.

8.1.8.4 Electrochemical Synthesis in Aprotic Solvent. Many electroorganic reactions require the use of aprotic solvents to extend the electrochemical potential window beyond that for aqueous based electrolytes. However, the use of solvents such as acetonitrile and dimethylformamide (DMF) can cause technical problems with the use of membranes and the choice of a suitable anode or cathode. A method which overcomes these limitations, in the case of oxidations, is the use of consumeable anodes recently introduced at a commercial level by the SNPE company in the electrochemical synthesis of fenoprofen.[10] Fenoprofen is an anti-inflammatory

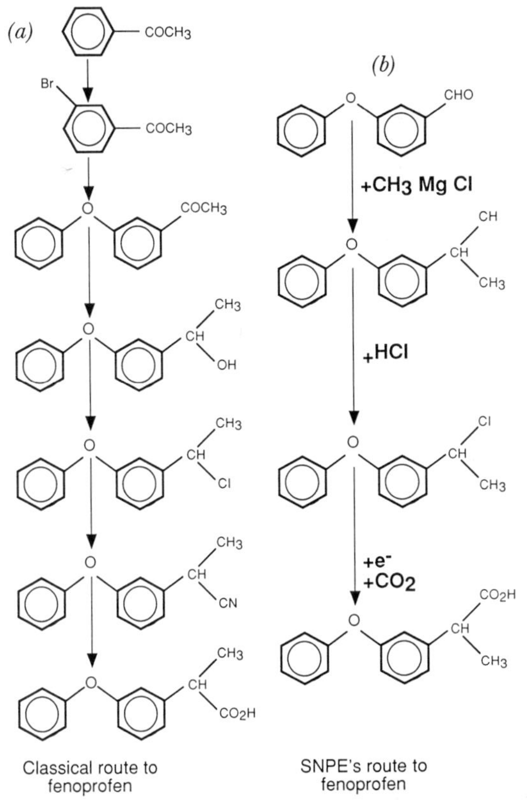

Figure 6 (a) *Electrochemical and* (b) *non-electrochemical routes to fenopropen*

drug of the arylpropionic acid family and its manufacture requires many steps. The usual synthetic route shown in Figure 6, starts from acetophenone, and is a seven-step procedure involving bromination, the Ulman reaction with sodium phenoxide to give the acetyldiphenyl ether, reaction with sodium cyanide to give the nitrile which is then hydrolysed. The electrochemical route starts with 3-phenoxybenzaldehyde to give the acetyldiphenyl ether. This then undergoes electrocarboxylation, as shown in Figure 6.

The first reported attempt at this synthesis used an expensive nickel salt to catalyse the reduction of the halide. Additionally a Hg cathode was used and also an expensive organic solvent mixture with at least a molar equivalent of tetrabutylammonuim bromide. For the process to be technically feasible an alternative cathode and electrolyte composition was required. The resultant solution was found in the use of consumable metal anodes, DMF as solvent and a stainless steel cathode. The procedure offered several advantages:

1   no cell separator is required;
2   the consumable anode which could be Mg, Zn or Al, caused no
     problem. Mg was preferred because aluminium salts are much less
     soluble in DMF and zinc salts are somewhat reducible to metallic
     zinc;
3   DMF is a cheap and safe solvent;
4   toxic Hg cathodes are avoided, stainless steel or zinc are suitable. In
     practice the base metal passivates after use and thus a coating of Pb
     (Zn or Cd) is used to hinder the passivation;
5   only small amounts of expensive tetraalkylammonium halide is
     required. Product recovery is *via* the magnesium salt and not the
     tetraalkylammonium salt, which is only required to impart initial
     conductivity.

   An effect obtained from the use of the Mg anode is that it is believed to
prevent the reduction of CO_2, as the reduction potential of CO_2 is shifted
cathodically in the presence of magnesium ions.
   The required reactor technology for this synthesis was developed to
offer several characteristics:

(a)   the consumeable anode must be fed continuously, during the reac-
      tion cycle;
(b)   the interelectrode gap must be small because of the low electrolyte
      conductivity;
(c)   be capable of operation under moderate pressure (5 bar) to main-
      tain a reasonable concentration of CO_2 in solution.

   The use of commercial electrolysers is unattractive with this synthesis
involving consumeable metal anodes because of the increase in cell gap,
which would occur during operation. A packed-bed cell using particulate
material was unsuccessful because of poor electrical contact between the
particles caused by the formation of insulating oxide layers. A cell with an
adjustable anode is necessary, the 'simplest' uses cheap Mg cylindrical
bars, 40 cm in diameter (Figure 7) in a cylindrical pressure vessel using a
conical cathode. The tip of the Mg bar is machined like a pencil and is
separated from the cathode (60 cm^2 area) by a thin plastic mesh (a few
mm's in thickness). The Mg is thus fed by gravity, at the rate of 10 cm per
week, and the interelectrode gap is self regulating — any tendency for
narrowing of the interelectrode gap is counteracted by a reduced ohmic
potential drop and higher localised current density.
   The plant is operated on a 600 A, 6 tonne per year scale using two reactors, in a batch mode. Each batch takes 28 h and gives 50 kg of product at
an 80% chemical yield and a 60% Faradaic yield — 9 kg of Mg are consumed in each batch. The SNPE process has been applied to several other
synthesis:

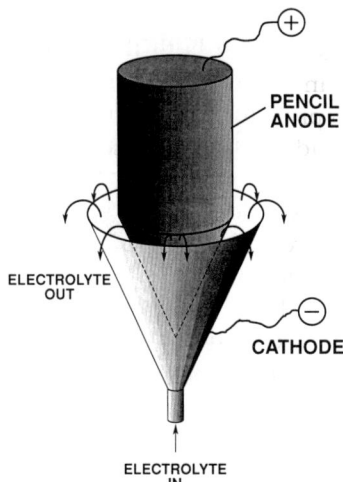

Figure 7   *The pencil sharpener cell*

- Biaryls, such as difluoro — or dimethoxybiphenyl, from electrodimerization of aromatic halides in the presence of a special catalyst.
- Alcohols from Grignard-like electrolytic coupling of organic halides with carbonyl compounds. Dimethylbenzylmethanol, a perfume component with a rose fragrance was prepared in high yield using an aluminium anode without the formation of troublesome bibenzyl (bibenzyl is usually formed as a byproduct in the Grignard synthesis).
- Formylation of aromatic halides by electroreduction, in DMF, used as a reactant as well the solvent. Various aromatic aldehydes *e.g.* trifluoromethylbenzaldehyde or fluorobenzaldehyde are thus obtained in reasonable yields.
- Diphenylacetic 2- and 4-trifluoromethylbenzoic acids prepared by electroreduction of the corresponding chloride.

Electroacetylation by electroreduction of benzylic halides in the presence of acetic anhydride as a route to various arylacetones. These are valuable starting materials for the manufacture of amphetamine-like drugs. For example, trifluoromethylphenylacetone, an intermediate in the synthesis of an anorectic drug, is prepared from trifluoromethylbenzyl chloride.

The use of a sacrificial anode is a promising route for the synthesis of numerous compounds in aprotic solvent. Both zinc and aluminium have been used in selected syntheses.

8.1.8.5 Paired Electroorganic Synthesis. An area which, in principle, has great potential in improving the efficiency of electrosynthesis is the pairing of anodic and cathodic 'synthesis' in one cell. Paired electrochemical syntheses are processes in which both the anode and cathode reactions simultaneously contribute to the formation of the final products. The classic example is the simultaneous production of chlorine and sodium hydroxide in the chlor-alkali industry. The chemistry associated with this industry, *i.e.* generation of halogen and caustic, has also been applied to paired electrosynthesis of alkene oxide from alkene and epichlorohydrin from allyl chloride (see chapter 4). Paired electrosynthesis can be generally classified in terms of the following :-

1   The generation of two distinct products from two reagents.
2   The generation of one product by coupling of anode and cathode reactions.
3   The generation of one common product from the anodic and cathodic reactions of two reagents (*e.g.* glyoxylic acid from glyoxal and oxalic acid).
4   The generation of two products from one reagent, *e.g.* sorbitol and gluconate from glucose.
5   The generation of one product in which the intermediate species is formed by a reaction at the counter electrode, *e.g.* the formation of butan-1,2-one from butane-2,3-diol.[14]

In the electrosynthesis of butane-2,3-one, shown in Figure 8, the first step is the oxidation of butane-2,3-diol to acetoin by bromine, which is electrogenerated at the anode. This takes place in bulk solution in a stirred tank reactor. The acetoin is then cathodically reduced to butan-2 - one. Thus the electrolyte must pass from the anode to the cathode in the external flow circuit *via* the reactor. In this synthesis the electrodes are both in the form of packed beds, graphite for the anode and amalgamated zinc for the cathode.

Cathode

$$CH_3COCHOHCH_3 + 2H + 2e \rightarrow CH_3COC_2H_5 + H_2O$$
$$\quad\quad Acetoin \quad\quad\quad\quad\quad\quad Butan\text{-}2\text{-}one$$

$$2H_2O + 2e^- \rightarrow H_2 + 2\,OH$$
$$Br_2 + 2e^- \rightarrow Br^-$$
$$BrO_3^- + 6H^+ + 6e^- \rightarrow Br^- + 3H_2O$$

Anode

$$2Br^- \rightarrow Br_2 + 2e^-$$
$$Br^- + 3H_2O \rightarrow BrO_3^- + 6H^+ + 6e^-$$

Bulk solution

$$CH_3\,CH(OH)\,CH(OH)CH_2 + Br_2 \rightarrow CH_3COCHOHCH_3 + 2HBr$$
$$Butane\text{-}2\text{-}3\text{-}diol \quad\quad\quad\quad\quad\quad\quad\quad Acetoin$$

Figure 8   *The synthesis reactions of butanone*

There are a number of potential loss reactions for this system, which include the formation of bromate. The bromate and bromine generated at the anode can also be reduced back to bromide, which results in a loss of current efficiency.

8.2   Inorganic Electrochemical Processes

The flagships of industrial inorganic synthesis are the production of chlorine and the production of aluminium metal. Technologically these are very different processes and typify the versatility of electrochemistry, on one hand functioning effectively at very high temperatures for molten salt electrolysis and on the other producing highly reactive gases at ambient temperature. Table 9 gives an overall view of inorganic electrochemical processes which are now discussed.

Table 9   *Inorganic electrochemical processes*

Al, Na, Mg, Li	Molten salt electrowinning
Cu, Zn, Co, Ni, Cr, Pb	Hydrometallurgy
Cd, Mn, Tl, Ga, In, Ag, Au	Electrowinning or refining
Chlorine; caustic	Noble metal oxide anode, brine electrolyte
Chlorate	Noble metal oxide anode, brine electrolyte
Perchlorate	Pt/Ti, PbO_2 anodes, chlorate electrolyte
Persulfate	Pt/Ti anode, conc. H_2SO_4
Hypochlorite	DSA^R, aqueous NaCl
Permanganate	Ni, monel anode, $KMnO_4$ electrolyte
Fluorine	Carbon anode, KF–2HF eutectic
Manganese dioxide	C, Pb, Ti anodes, MnSO4
Water electrolysis (H_2,O_2)	Ni on steel, KOH
Hydrogen peroxide	Carbon cathodes, NaOH
Ozone	Vitreous carbon anode, conc. aqu. HBF_4
Bromate	C, Pt/Ti, PbO_2, aqu. NaBr
Chromic acid	Lead anode, Cr^{III} in H_2SO_4
Copper(I) oxide	Copper, aqu. NaCl
Potassium stannate	Anodic dissolution
Chlorine dioxide	DSA^R, carbon cathode, sodium chlorate and HCl

8.2.1   Chemicals from the Electrolysis of Halides

There are several chemical species which can be produced by the electrolysis of halides in aqueous solutions. The reactions involved in the chemistry and electrochemistry of the halides (Br, Cl and I) are quite similar but the values of their kinetic and thermodynamic parameters are quite different. The major area of activity is in chlorine electrochemistry. In the electrolysis of a solution of sodium chloride the two desirable electrode reactions are the generation of chlorine

$$2 \; Cl^- \; -> \; Cl_2 \; + \; 2 \; e^- \tag{52}$$

and the formation of hydroxide ions and hydrogen gas

$$2 \; H_2O \; + \; 2 \; e^- \; -> \; 2 \; OH^- \; + \; H_2 \tag{53}$$

If the products of these reactions are kept separate then they will not react and this forms the basis of the chlor-alkali industry. However, if the products of both these cell reactions are allowed to mix then the dissolved chlorine can undergo further reactions. The main products of these reactions are hypochlorite, used in water treatment, and chlorate. The formation of these species depends on the electrolyte concentration, pH and temperature.

8.2.1.1 Chlorine and Sodium Hydroxide (Chlor-alkali). There are three electrochemical routes to chlorine and caustic soda in this industry, based on either mercury, diaphragm or membrane cells. The important technological and performance features of the three processes is summarised in Table 10. The technology used for the production of sodium hydroxide from NaCl is also applied industrially to the production of KOH from KCl, but on a much smaller scale.

8.2.1.2 Mercury Cells. Mercury cells are different to the other cell designs in several ways, notably in the absence of a separator between the anode and cathode and in the method of formation of the caustic soda product. In mercury cells the cathode reaction is the formation of Na–Hg amalgam,

$$Na^+ \; + \; e^- \; + \; x \, Hg \; -> \; NaHg_x \tag{54}$$

which has a reversible potential of -1.868 V. In comparison, the reversible potential for the hydrogen evolution reaction in diapragm or membrane cells is approximately 1.0 V more positive, thus putting Hg cells at a disadvangtage in terms of equilibrium potentials. However the absence of a separator in Hg cells tends to redress this imbalance through the reduction in the cell internal voltage loss. In addition the final product from the Hg cell does not require further evaporation — making mercury cells generally competitive.

The formation of the sodium hydroxide in Hg cell technology takes place in a separate reactor to the cell, called a denuder, a packed bed of graphite spheres impregnated with transition metal catalysts for the decomposition of the amalgam. Pure water is fed to this vessel at a controlled rate and in the presence of the catalyst serves to decompose the

Table 10 *Comparison of cell processes for chlorine manufacture*

Process parameter	Mercury cell	Diaphragm cell	Membrane cell
Anode material	Mixed precious metal oxide coating on Ti	Mixed precious metal oxide coating on Ti	Mixed precious metal oxide coating on Ti
Cathode material	Mercury	Ni or catalyst coated steel	Ni or catalyst coated steel
Separator	None	Asbestos (polymer)	Ion-exchange membrane
Cathode products	NaHg	NaOH in brine +H_2	NaOH solution +H_2
Final products	Conc. NaOH + H_2	NaOH + solid NaCl + H_2	Pure NaOH + H_2
Current density/ kA m^{-2}	2–12	2.8–3.0	3–4
Electrical power/ kWh tonne^{-1}	2 722–3 100	2 500	2 200–2 450
Steam required/ kWh tonne^{-1}	0	1179	318
Electrolysis + evaporation energy/ kWh tonne^{-1}	2 722–3100	3 679	2 520–2 770
Cell voltage/V at j/A m^{-2}	4.4 (10 000)	3.5 (2 000)	2.95 (4 000)
Current efficiency for Cl_2/%	97	96	98.5
purity Cl_2/%	99.2	98	99.3
purity H_2/%	99.9	99.9	99.9
O_2 in Cl_2/%	0.1	1–2	0.3
Cl⁻ in 50% NaOH(%)	0.003	1–1.2	0.005
NaOH(%) in cell product	50	11	35

amalgam to give the 50% sodium hydroxide product, hydrogen gas and the Hg for re-use in the cells.

The cells consist of large shallow troughs (15 × 2 m × 0.3 m) with a downward sloping steel baseplate over which the Hg cathode flows. A series of some 250 anodes, 30 cm × 30 cm, are held horizontally above the Hg surface, at a cell gap of 1 cm, from which the chlorine gas is liberated. All Hg cells use expanded titanium dimensionally stable anodes (DSA), which give the required stability, coupled with appropriate electrocatalytic activity, towards chlorine evolution. The thermodynamics of the anode reactions, in brine solution does not favour chlorine evoution, the equi-

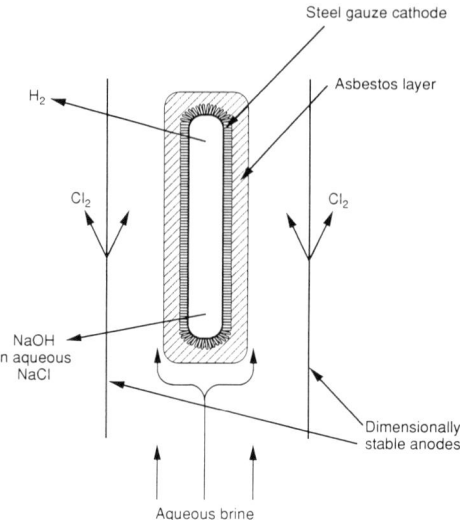

Figure 9   *Design of diaphragm cell separator*

librium potential of Cl_2 evolution is some 0.3 V greater than that of oxygen evolution.

8.2.1.3 Diaphragm Cells. The important component of the diaphragm cell is the separator between the anode and cathode used to 'isolate' the respective electrode reactions. This separator is generally based on asbestos which is deposited onto the steel gauze cathode, which serves to minimise the interelectrode gap (see Figure 9). Cell design is a compact unit of anodes and cathode/diaphragm composites. Designs by Oxytech Systems and PPG Industries are based on a 'toaster' design and an interlocking finger design respectively (see Figure 10). The former is a monopolar cell design in which the diaphragm /cathode acts as the grill surrounding the expanded metal anodes (toast). The interlocking finger design is a bipolar connected cell unit.

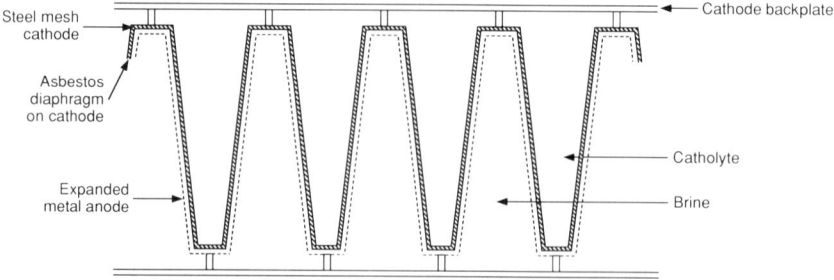

Figure 10   *Design features of "toaster" type diaphragm cells*

In diaphragms cells (as shown in Figure 3, chapter 3) a slow flow of anolyte into the catholyte is achieved under the influence of a hydrostatic pressure gradient. Thus the sodium ions required to form the sodium hydroxide solution pass into the catholyte under the influence of diffusion, migration and convection. However, the transport of chloride ions into the catholyte is, as a consequence, also large leading to a product contaminated with chloride. Furthermore the non-selectivity of the asbestos diaphragm limits the concentration of caustic formed in the catholyte to around 12% otherwise OH⁻ ions will migrate to the anolyte leading to loss of chlorine, by hydrolysis to hypochlorite, and possible oxygen evolution. The low concentration of caustic formed in diaphragm cells requires extensive and expensive evaporation to produce the final 50% aqueous solution required by industry.

8.2.1.4 Membrane Cells. Membrane cells are similar to diaphragm cells in some respects, although no bulk flow of electrolyte occurs through the separator, which in this case is a cation-exchange membrane. The cell design has gone as far as is physically possible to minimise the internal resistance by operating in the 'zero-gap' mode. This entails the anode and the cathode both being in physical contact with the membrane (see Figure 11).

A cation-exchange membrane, in principle, represents the ultimate separator for the chlor-alkali industry, that is one which will allow the transport of Na⁺ ions into the catholyte with the exclusion of Cl⁻ ions. To be effective this membrane must be stable to both the desired 50% caustic

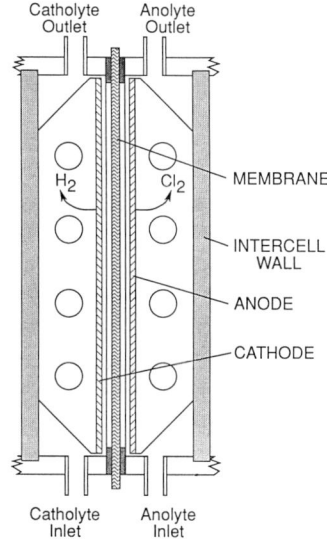

Figure 11 *Zero-gap configuration of membrane cells*

soda concentration produced in the catholyte and to the wet chlorine produced at the anode and have a low electrical resistance. No single membrane offers all the required characteristsics for operation in chlor-alkali cells and thus cell designs are based on modern bilayer membranes (see Figure 12), which consist of four components.

1   A thin weak acid cation-exchange polymer, capable of operating at caustic concentrations of 30–40% without a significant loss in current efficiency. Being thin counteracts the relatively poor electrical conductivity.
2   A thicker, strong acid polymer which in principle offers superior properties regarding resistance and efficiency in comparison to the weak acid type. However, the performance is limited to relatively low caustic concentrations (12%), and thus it is placed away from the cathode side.
3   A reinforcing fluorocarbon net in the strong polymer membrane side to give mechanical stability.
4   A surface coating on the anolyte side of the membrane to encourage gas release from an otherwise hydrophobic surface and the contacting anode surface.

The electrode design in the zero-gap configuration must ensure easy gas release from the back, otherwise the bubbles would constitute a large resistance to the flow of current. Typically this is achieved by using expanded metal or by louvered, or otherwise contoured anodes. Both the anode and the cathode will have appropriate electocatalyst coatings.

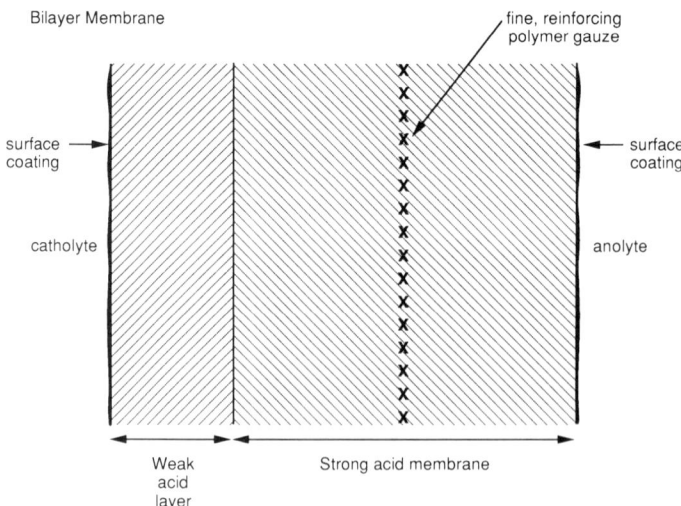

Figure 12 *A bilayer membrane for the chlor-alkali industry*

There are over twenty companies throughout the world who offer membrane cell technology, typically based on the filter press concept and there is a gradual move towards the use of membrane cells in new plant.

8.2.1.5 Oxygen Cathodes. The use of air cathodes as a replacement for hydrogen evolving cathodes in the chlor-alkali industry has been proposed as a method of reducing the cell voltage, a theoretical saving of 1.23 V. Performance estimates of pilot scale chlorine electrolysers achieve reductions in applied cell voltages of over 1.0 V at current densities of up to 3000 A m^{-2}. The cathodes are largely based on the use of high surface area platinum dispersed in a carbon–Teflon matrix. The use of air cathode cells will largely depend on the status of the hydrogen produced in standard cells — whether it is a valuable product, as fuel, or a potential hazard. Other proposed uses of gas diffusion electrodes include:

(i) The use of a hydrogen/chlorine fuel cell to consume a fraction of the byproduct hydrogen and some chlorine to produce hydrochloric acid while co-generating electrical energy. The hydrochloric acid is used in brine acidification in the chlorine cells to reduce the possible formation of hypochlorite due to hydroxide ion migration from the catholyte.

(ii) The use of an electrochemical concentrator, in conjunction with the membrane chlorine cell. The 30% caustic soda solution, formed in the membrane cell, is passed to the anode and the cathode compartments of an alkaline fuel cell fed with hydrogen from the membrane cell and air. In the fuel cell the consumption of water and generation of hydroxide ions at the cathode combined with the selective transport of sodium ions through the cation-exchange membrane into the catholyte results in a depletion of the OH^- ion concentration in the anolyte and an increase in the concentration of Na^+ in the catholyte to the give the required 50% solution. The depleted anolyte is returned to the membrane electrolyser as catholyte. This procedure can save in the thermal energy required to concentrate the 30% caustic soda solution produced in standard membrane cells.

(iii) The use of an alkaline fuel cell to generate electricity from the hydrogen produced from the membrane cell.

8.2.1.6 Hypochlorite. Hypochlorite solutions (of sodium) are produced by the reaction of chlorine with hydroxide ions in aqueous solutions. Hypochlorite solutions are used for a variety of water treatment, sterilisation and bleaching applications, as discussed in chapter 7.

8.2.1.7 Chlorate. Sodium chlorate is produced by the reaction of hypochlorous acid with hypochlorite in aqueous solution and is used predominantly in the papermaking industry, in North America, Scandinavia and Japan, and as a weed killer. The formation of chlorate is achieved by the undivided cell electrolysis of a high concentration of aqueous NaCl (*ca.* 3.3 kmol m^{-3}). The conditions of operation of the reactor are thus optimised to the chemical transformation of the generated chlorine into chlorate.

1   The anodic oxidation of hypochlorite can also lead to the formation of chlorate but as this route consumes nine Faraday's of charge per mol of chlorate it is less efficient than the chemical reaction of hypochlorous acid and hypochlorite, which consumes six Faraday's per mol. Thus conditions which minimise this anodic reaction (which also forms oxygen) are required. This requires a cell design with turbulent flow to transport the hypochlorite rapidly away from the electrode surface.

2   The reaction is quite slow and thus requires a fairly high temperature of the order of 90 °C.

3   Control of the pH of operation is essential as the hydrolysis of chlorine requires a pH above 6.0 whilst the disproportionation of hypochlorite requires slightly acidic conditions.

Modern industrial cells use parallel plate electrode configurations with anodes of coated titanium and cathodes, usually, of steel. To minimise the cathodic reduction of hypochlorite and chlorate, sodium dichromate is added to the electrolyte, which leads to the formation of a protective cathode film. Two general types of chlorate reactor are used in practice, one with an external vessel for the chemical formation of chlorate and the other with an internal circulation of the electrolyte between the cell region and the chemical reaction region. In the latter case the hydrogen gas generated at the cathode provides the gas lift for recirculation. The electrolyte returned to the cells contains a significant proportion of NaClO$_3$, which is stable under electrolysis conditions. The performance of chlorate cells, which operate in plants of capacities of 5000–40 000 tonnes per year, gives energy consumptions in the region of 4500–6500 kWh kg^{-1} at current efficiencies of *ca.* 95%. Current densities are between 2000–3000 A m^{-2}.

8.2.1.8 Bromate. An important method for the production of sodium and potassium bromates is by the electrolysis of the associated aqueous bromide solution which follows similar chemistry to that of chlorate although the disproportionation reaction between bromite ion and hypobromous acid is much faster. This reduces the required size of the reaction zone for the formation of bromate. In operation, temperatures in excess of 60 °C are used to promote the disproportionation reaction of the hyprobromite, which is formed in an undivided cell. The typical electrolysis conditions are current densities between 1000–2500 A m^{-2} and cell voltages of 2.4 –4.3 V. The reaction process is operated in a batch mode starting with an electrolyte containing *ca.* 200–260 g dm^{-3} of NaBr and 80–95 g dm^{-3}

$NaBrO_3$ and produces an electrolyte containing *ca.* 85 g dm^{-3} NaBr and 250 g dm^{-3} $NaBrO_3$. The sodium bromate is recovered from the final electrolyte by crystallisation, and the mother liquor from the crystalliser is recycled to electrolysis. Cell technology is similar to that of chlorate production.

8.2.1.9 Perchlorate and Perchloric Acid. The manufacture of perchlorates by electrosynthesis can proceed by one of two methods:

1   electrolysis of hydrochloric (or hypochlorous) acid to form perchloric acid, which can then be chemically converted to perchlorate;
2   anodic oxidation of chlorate in an aqueous solution.

The usual method for the preparation of any perchlorate is *via* anodic oxidation of sodium chlorate.

$$ClO_3^- + H_2O \rightarrow ClO_4^- + 2\,H^+ + 2\,e^- \qquad E^o = 1.19 \text{ V} \qquad (55)$$

The standard potential of this process is close to that of oxygen evolution, which is the major competing reaction. In the presence of perchlorate, in the chlorate solution, the current efficiency of perchlorate formation is found to increase with increasing current density. Increasing the amount of perchlorate in the electrolyte is also found to increase the current efficiency of the reaction. Perchlorate is fortunately an effective inhibitor for the oxygen evolution reaction especially on platinised titanium, the usual material used as the anode. Commercial electrolysis is carried out in undivided cells using cheap cathode materials such as steel on which the cathode reaction is hydrogen evolution, perchlorate reduction is kinetically hindered.

8.2.1.10 Iodate and Periodate. The chemistry of the anodic oxidation of iodide solutions is analogous to that of chloride and bromide although the competing discharge of hydroxide ions to oxygen is not significant. The chemical formation of iodate from the disproportionation reaction is faster than the analogous reaction for bromate and no hypoiodite builds up in solution. The anodes reported to be used in the production of iodate are either lead dioxide (coated on graphite) or graphite. The scale of production is not clear but is not comparable to chlorate or bromate.

Periodate can be formed by the oxidation of iodate in an analogous way to perchlorate.

8.1.2.11 Fluorine. Fluorine production is carried out electrochemically in tank electrolysers made from mild steel fitted with carbon anodes and steel cathodes. The fluorine gas is produced in conjunction with hydrogen gas at the cathode by the electrolysis of a fused HF–KF molten salt mixture at a temperature between 80 to 100 °C. In operation the electrolyte is continuously fed with HF at a concentration of *ca.* 40%, to enable continuous production. Two established cells are the BNFL/ICI and the Rhone-Poulenc/ISC designs which have comparable operating performance. Thus typically fluorine is produced at a rate of *ca.* 3.4 kg h^{-1} at a current efficiency of 90–95% from cells with nominal current capacities of 5000–6000 A, at anode current densities of 1000–1800 A m^{-2} and with cell voltages of 10–12 V. Typical energy requirements of the process are 15 kWh kg^{-1}. The BNFL design contains 24 mild steel cooling coil cathodes and 12 porous carbon anodes slotted between the cathodes and separated from them by monel skirts to prevent the mixing of the two product gases. The Rhone-Poulenc design is similar to the BNFL design but uses non-porous carbon anodes and a monel diaphragm to separate the anodes and cathodes. The use of a bipolar plate and frame design using an electrolyte mixture of HF, NH$_4$F and KF at a temperature of 37 °C is claimed to reduce the energy consumption by 40%, achieving an operating voltage of 6 V at a current density of 1500 A m^{-2}.

The markets for fluorine include the nuclear industry for the manufacture of uranium hexafluoride and the electronics industry as inert fluids.

8.2.2   Peroxydisulfate Electrosynthesis

The peroxydisulfates of ammonium, potassium and sodium are principally manufactured by electrochemical means and find a variety of uses as oxidants in the chemical and pharmaceutical industries; as initiators, etchants and pickling agents. Peroxydisulfate is formed by the anodic oxidation of a sulfate electrolyte in which the following reactions are principally involved:

$$2SO_4^{2-} \rightarrow S_2O_8^{2-} + 2e^- \qquad\qquad E^o = 2.01 \text{ V} \qquad\qquad (56)$$

$$2HSO_4^- \rightarrow S_2O_8^{2-} + 2H^+ + 2e^- \qquad E^o = 2.12 \text{ V} \qquad\qquad (57)$$

In the aqueous electrolyte the anodic formation of oxygen will be a competitive reaction and thus the electrosynthesis requires high overpotentials. Furthermore the efficiency of the process is affected by the H$^+$ ion catalysed hydrolysis of persulfate ions

$$S_2O_8^{2-} + H_2O \rightarrow HSO_4^- + HSO_5^- \qquad\qquad (58)$$

following which the peroxymonosulfate ions can be oxidised according to:

$$HSO_5^- + H_2O \rightarrow HSO_4^- + 2H^+ + O_2 + 2e^- \qquad E^o = 0.64 \text{ V} \qquad (59)$$

The general conditions for efficient peroxydisulfate manufacture are:

1   High over-potential and thus high current densities of the order of 5–10 kA m⁻², to inhibit oxygen evolution.

2   A high oxygen over-potential electrode which is chemically stable. Smooth platinum is the preferred material.

3   An optimum sulfuric acid concentration of *ca.* 8 mol dm⁻³ at which current efficiency exhibits a maximum value at the high current densities used.

4   Low electrolyte temperatures (15–25 °C). This primarily serves to minimise the formation of peroxymonosulfate.

8.2.2.1 Industrial Electrolysers. The industrial cells for this process must meet several requirements for efficient and selective synthesis:

- A high current concentration (space time yield) to allow low residence times whilst achieving the target final concentrations.
- A high current efficiencies requires a high anodic current density, while to keep the cell voltage low, low current densities at the cathode, and elsewhere in the cell, are required.
- A low temperature serves to increase current efficiency whilst increasing electrolyte resistivity and cell voltage.
- The formation of the gas phase fraction in the cell should be minimised to maintain a low cell voltage and prevent backmixing which reduces the current efficiency.

The typical industrial cells for this synthesis are divided, with specially designed anodes with areas much less than those of the cathode to achieve high anodic current densities relative to that of the cathode. The anode is in the form of thin platinum strips (foil) welded onto a tantalum frame (see Figure 13) and central rod, with a copper core for improved conductivity. The cathodes are of carbon(impregnated)-cooling is introduced internally in the cell structure. The internal design of the cell uses

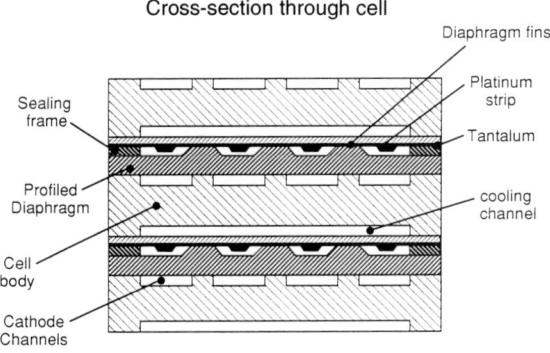

Figure 13 *Electrode design for peroxydisulfuric acid manufacture*

profiled diaphragms, which are shaped to form individual cell channels around each vertical platinum strip. This achieves the required gas-lift effect used for electrolyte circulation, while achieving the required low energy consumption. Operating characteristics of recent industrial cells are given in Table 11.[15]

8.2.2.2 Application in Etching of Printed Circuit Boards. Peroxydisulfates, because of their high oxidation potentials, are used as Cu etchants in the manufacture of printed circuit boards and in surface finishing in which the peroxydisulfate ions are reduced to sulfate ions:

$$Cu + H_2S_2O_8 \rightarrow CuSO_4 + H_2SO_4 \tag{60}$$

Regeneration of the spent solutions, which recovers Cu as a cathodic deposit and converts sulfate back to peroxydisulfate has been considered.

Table 11 *Electrolysis cell performance for peroxydisulfuric acid synthesis*

	Eilenburg cell *EZ II* *DD-P* 99548 (1972)	*USSR cell* *SU-P* 311502 (1973)	*Prototype of* *Eilenburg cell* *EZ III like EZ II*
Anode	Pt-strips on Ta-foil	Pt strips on Ti-cooling body	Pt-strips on Ta-foil
Cathode	Impregnated graphite cooled	Impregnated graphite cooled	Impregnated graphite cooled
Diaphragm	PVC–SiO$_2$ gel planar/profiled	PVC–SiO$_2$ gel planar	PVC–SiO$_2$ gel profiled
Electrical connection	Bipolar	Monopolar	Bipolar
Current density per anode / kA m^{-2}	5	5	5
Current density per cathode / kA m^{-2}	1.0	1.2	0.7
Cell voltage / V	3.7–3.8	4.6	3.7
Current efficiency (%)	78	75	80
Specific DC-consumption / kWH kg^{-1}	1.4	1.7	1.3
Current capacity per diaphragm in kA per electrolyser in kA per unit of basic area in kA m^{-2}	0.55 14 8	2 12–25	1.2 62 17

In practice this requires a two-stage process in which most of the dissolved Cu is removed following which the solutions passing through the regeneration cell. The flow in the cell, which is divided, is first through the cathode compartment to take out more of the Cu (deposit) and then through the anode compartment to oxidise the sulfate.

8.2.3   Permanganate

The electrochemical production of potassium permanganate is established as the only production route and is based on the anodic oxidation of potassium manganate.

$$MnO_4^{2-} \rightarrow MnO_4^- + e^- \tag{61}$$

with hydrogen being evolved at the cathode.

The electrolyte used in this process is a 100–250 kg m^{-3} solution of potassium manganate in potassium hydroxide (1.4 kmol m^{-3}) at a temperature of *ca.* 60 °C.

Other cell reactions which can occur in the manufacture of permangate are the cathodic reduction of permanganate

$$MnO_4^- + e^- \rightarrow MnO_4^{2-} \tag{62}$$

and the anodic oxidation of hydroxide ions to oxygen at low manganate concentrations. Chemical reduction of permanganate at high hydroxide concentration is also possible.

The cathodic reduction of permanganate is the major determining factor in the cell design and would generally require the use of a diaphragm or membrane. However, the anodic oxidation requires low current densities of the order of 50–150 A m^{-2} which enables an alternative strategy, using a cathode current density a factor of 100 greater than that of the anode. The predominant cathode reaction is then hydrogen evolution and the concept results in an ingenious cell design.

The continuous process for permanganate manufacture, operated by Carus Chem. Co. (USA), is based on a bipolar plate and frame design using Ni (or monel) anodes and iron or steel cathodes. The bipolar electrode design consists of a steel baseplate onto which a monel (anode) gauze or mesh is welded to one side. On the other side of the plate is the steel cathode, consisting of small perpendicular protrusions, between which are plastic non-conducting insulation. One complete cell unit consists of three groups of twenty cells fed separately from a manifold arrangement (see Figure 14). Within each cell grouping the electrolyte is pumped through the individual cell compartments in a serpentine flow pattern at relatively high velocities to ensure a high degree of turbulence and mixing and thus prevent crystallisation of the product in the cell. The

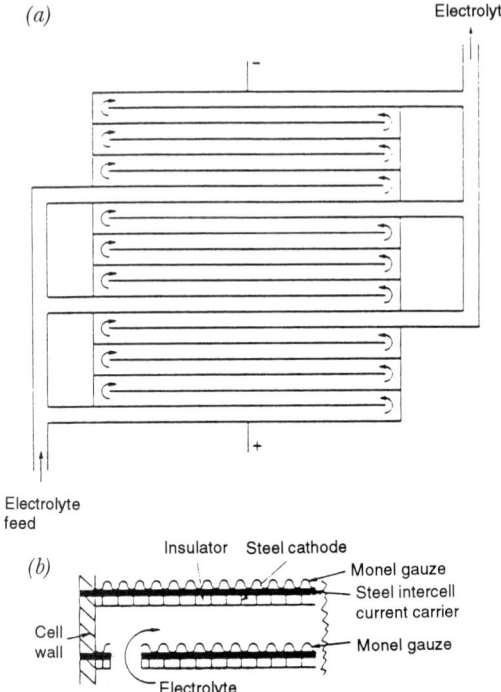

Figure 14 *Cell design concept for permanganate manufacture* (a) *arrangement of electrodes and flow,* (b) *individual cell arrangement*

performance of the cell gives a current efficiency of around 90% at an energy consumption of 500 kWh per tonne. The cell current is quoted at 1200–1400 A and the total cell votage at 140–170 V.

The smaller scale production of permanganate is typically carried out in cylindrical batch reactors using concentric cylindrical anodes between which are strategically located steel rod cathodes covered by porous PVC diaphragms. The latter serve to reduce the transport of permanganate to the cathode surface.

8.2.4   Water Electrolysis

The electrolysis of water represents a convenient and simple route to the production of hydrogen and oxygen gases.

$$4\ H_2O\ +\ 4\ e^-\ ->\ 4\ OH^-\ +\ 2\ H_2 \tag{63}$$

$$2\ OH^-\ ->\ O_2\ +\ 2\ H^+\ +\ 4\ e^- \tag{64}$$

The electrolysers are simple to operate and require little maintenance producing very pure products. It is for the latter reason that electrolytic hydrogen plants are operated, to produce H_2 for use in semiconductor manufacture, hydrogenation of food products and the production and refining of high purity metals. Otherwise the electrolytic generation of hydrogen is more expensive than its separation from synthesis gas, unless a cheap source of hydroelectric power is available. The demand for pure (electrolytic) oxygen has largely been taken over by improvements in cryogenic separation.

Conventional commercial electrolysers operate at current densities in the region 1000–3000 A m^{-2}, with electrolytes of *ca.* 30% KOH (approximately the maximum conductivity), at temperatures of between 70–90 °C. Most of the commercial cells are designed on the bipolar electrode filter press principle with the exception of one built on a monopolar tank electrolyser. Other cell technologies are based on elevated pressure operation or the use of solid polymer electrolytes.

8.2.4.1. Solid Polymer Electrolytes (SPE). Solid polymer electrolyte cells use typically cation-exchange fluoropolymer membranes (notably Nafion) as the ionic conducting medium between the anode and cathode. In operation SPE cells are fed with pure water at the anode which decomposes to O_2 and protons. These pass through the membrane in a hydrated form and are subsequently converted to hydrogen gas. The anode and cathode are deposited onto the membrane as thin coherent layers — a PTFE/carbon/ electrocatalyst porous structure with a Pt catalyst for the cathode and a mixture of Pt, RuO_2 and transition metals for the anode. Other components of the cell are a porous platinised titanium anode support, a carbon paper cathode support and a graphite current collector.

The reported performance of SPE water electrolysers is promising with cell voltages of less than 2 V at 10 000 A m^{-2} (80 °C) for extended tests of many thousands of hours. Such high operating current densities are likely to be required in practice to justify the higher cell cost, due mainly to the electrocatalyst and membrane material of this type of electrolyser.

8.2.5 Hydrogen Peroxide

Hydrogen peroxide can be produced by the direct cathodic reduction of oxygen, by the indirect reduction of oxygen using a cathodically regenerated organic redox couple and by the hydrolysis of peroxydisulfate formed by the anodic oxidation of sulfate or sulfuric acid. The latter was practiced on a large scale in the 1940s and 1950s. The resurgence in interest in the electrochemical production of H_2O_2 is a result of the demand for on-site production and has lead to processes based on the cathodic reduction of oxygen.

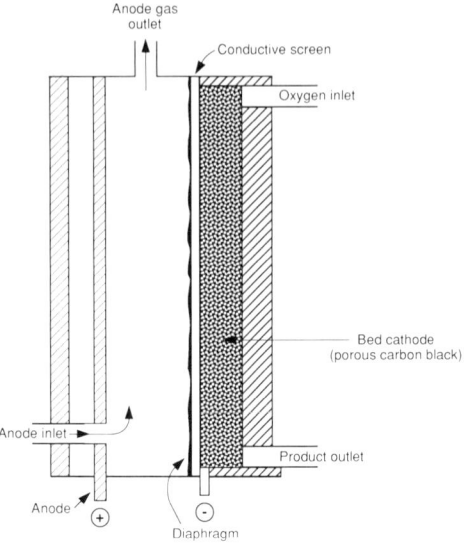

Figure 15 *Trickle flow cell for hydrogen peroxide synthesis*

$$O_2 + H_2O + 2 e^- \rightarrow HO_2^- + OH^- \quad (E^0 = 0.065 \text{ V}) \qquad (65)$$

This reaction can be carried out using a trickle-bed cell (see Figure 15), in which oxygen and a sodium hydroxide electrolyte flow down through a porous carbon cathode bed separated from a planar anode by an ion-exchange membrane. The membrane prevents the migration of the peroxide to the anode where it would be oxidised. The anodic reaction is the generation of oxygen gas. A key to the process is the effectiveness of carbon as an electrocatalyst for oxygen reduction, coupled with its poor peroxide decomposition properties. The carbon must be free from metallic components which would otherwise catalytically decompose the peroxide. The current efficiencies of the reduction of oxygen, in 2 mol dm^{-3} solutions of NaOH, on both graphite and reticulated vitreous carbon (RVC) approach values of 100% at potentials more positive than -0.7 V (*vs.* SCE) and current densities of < 250 A m^{-2}. At more negative potentials (higher current densities) the current efficiency falls as a result of the cathodic reduction of H_2O_2 to hydroxide ions.

$$HO_2^- + H_2O + 2 e^- \rightarrow 3 OH^- \qquad (66)$$

Another source of inefficiency in the process is the cathodic evolution of hydrogen. RVC was found to a more effective cathode material than graphite giving a solution of 10 g dm^{-3} H_2O_2 with a current efficiency of 75% at a current density of 500 A m^{-2} (*cf.* for graphite 7.5 g dm^{-3}, 50% current efficiency).

A commercial reactor based on the trickle flow concept shown in Figure 15 is part of a joint venture operation between The Dow Chemical Co. and Huron Technologies. The electrode material used in this case is a high specific surface area PTFE-bonded carbon black on a graphite particle substrate. The trickle flow of the catholyte is achieved by a controlled seepage of anolyte through the porous diaphragm into the catholyte. The reactor with an electrode area of 7.74 m^2, can produce a 30–50 g dm^{-3} solution of peroxide at *ca.* 765 kg per day, with a cell voltage of 2 V, a current density of 670 A m^{-2} and a current efficiency of 95%.

Gas diffusion electrodes are also being tested in the ICI FM21 reactor as a means of cathodically reducing oxygen. Additionally co-generation of peroxide with sodium chlorate and also with ozone is also of interest.

8.2.6 Ozone

Ozone is an effective and powerful oxidant in the treatment of potable water and waste waters and is seen as an alternative to chlorine, which can result in the formation of haloforms after reaction. However, because of its high reactivity it cannot be stored for any length of time and must be produced at the site of use. The established method for the production of ozone is by electrical (corona) discharge of pure oxygen and of dry air. There is, however, a high capital cost associated with this method which produces a relatively low concentration in solution. These two factors initiated interest in the electrochemical based production for small scale operation.

The electrolytic formation of ozone is typically achieved by the decomposition of water in acid electrolyte ($E^0 = -1.51$ V *vs.* NHE at 25 °C). The electrode potentials for the reaction are in the same range as for oxygen evolution, and the preferential evolution of oxygen is likely due to the lower standard potential. Thus a suitable electrode material must have a high oxygen over-potential and be stable at high anode potentials in acid media.

Several anode materials have been tested for the production of ozone, including gold, palladium, platinum, lead dioxide and tin oxide. In sulfuric acid electrolyte (5 mol dm^{-3}) at a temperature of 0 °C the β form of PbO$_2$ gave significantly higher efficiencies than the other anode materials. The variation in the current efficiencies for the electrode materials is accredited to the different reactive intermediates which are formed during ozone generation. It is believed that the accumulation of hydroxide radicals at the electrode surface is responsible for the formation of ozone by the sequence of reactions

$$OH^{\bullet}(ads) + O_2(ads) \rightarrow HO_3^{\bullet}(ads) \tag{67}$$

$$HO_3^{\bullet}(ads) \rightarrow HO_3^+ + e^- \tag{68}$$

$$HO_3^+ \rightarrow O_3 + H^+ \qquad\qquad (69)$$

At Pt, for example, only small traces of hydroxide radicals are found and thus ozone formation is low, while at lead dioxide OH radicals are found.

The search for electrode materials for ozone production has encountered problems of anode stability. Lead dioxide is more stable in H_3PO_4 than in H_2SO_4, $HClO_4$ and HBF_4. However, the use of solid polymer electrolyte (Nafion 120) was found to give good stability of the PbO_2 for over 2500 h continuous operation at a current density of $10\,000$ A m^{-2} and a temperature of 30 °C. Another suitable material for ozone generation is glassy carbon, with a suitable electrolyte, being HBF_4. These latter two anode–electrolyte combinations have formed the basis for the commercial development of electrolysers for ozone production.

1   The ABB-Membrel cell (Aseo Brown Boveri Ltd) uses a solid polymer electrolyte [see Figure 16(a)] and produces O_3 from a stream of relatively pure water at a porous anode adjacent to the membrane. High current densities (10 kA m^{-2}) can be used but the current efficiencies are low (<14%). An alternative anode material is carbon. This is formed into a porous structure using a fluorocarbon binder and gives current efficiencies of up to 35%.

2   Oxytech (UK) Ltd have developed a cell [Figure 16(b)] using hollow, cylinderical fluorocarbon impregnated anodes, surrounded by the electrolyte, and then by a gas diffusion air cathode. In operation the anode requires cooling and the product gas must be diluted with air to keep the ozone concentration below the explosive limit.

8.2.7   Manganese Dioxide

The electrolytic manufacture of manganese dioxide is based on the anodic oxidation of Mn^{II} ions in aqueous solution.

$$Mn^{2+} + 2\,H_2O \rightarrow MnO_2 + 4\,H^+ + 2\,e^- \qquad\qquad (70)$$

The evolution of hydrogen is the cathode reaction. The typical electrolysis conditions are a manganese sulfate electrolyte (0.5–1.2 kmol m^{-3}) in sulfuric acid, a temperature of 90–100 °C, a current density of 50–150 A m^{-2} and a cell voltage of 2.2–3.0 V. Electrolysis is carried out in tank electrolysers with alternating rows of cathodes and anodes of inert materials such as titanium, lead dioxide or graphite. The MnO_2 is formed as a deposit on the anode and thus a cell separator is not required. Removal of the MnO_2 is achieved mechanically outside the cell. Improvements in the operation of cells have focused on the formation of MnO_2 as a slurry to enable higher space time yields and reduced cell voltages and energy consumptions.

(a)

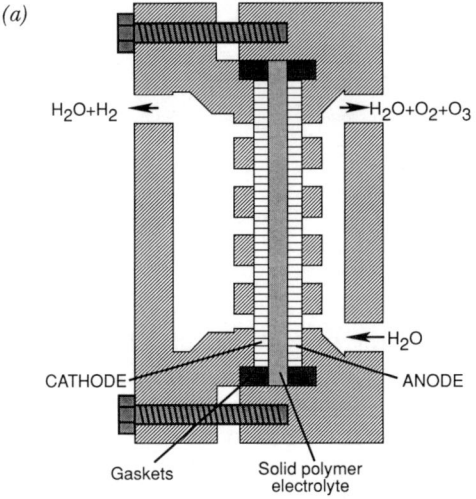

(b)

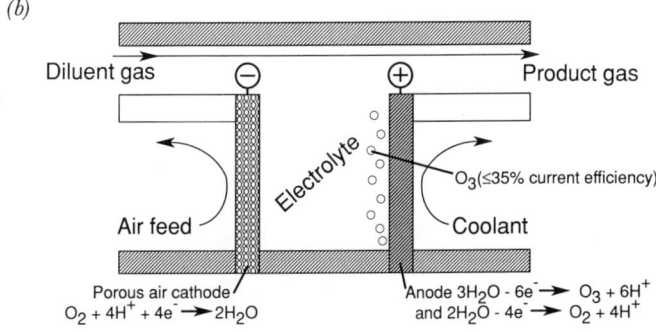

Figure 16 *Electrochemical ozone cells* (a) *the ABB-membrel cell;* (b) *the Oxytech cell*

The major market for electrolytic manganese dioxide is in the manufacture of batteries owing to its relatively high activity (electrochemical) in comparison to the product from the mineral processing route. Manganese dioxide is one of several electrochemically formed oxidants used in industrial organic synthesis.

8.2.8 Copper(II) Oxide

Copper(II) oxide is used as precursor for copper chemicals, as a pigment in marine antifouling paints and ceramics, and as a reducing agent in agricultural chemicals. The scale of production is small with plant capacities < 1000 tonnes per year.

Copper(II) oxide is formed as a powder by the anodic oxidation of solid copper anodes in a sodium chloride electrolyte at a pH of 8–10. The principle reactions in the electrochemical process are:

anode: $Cu + n\ Cl^- - e^- = CuCl_n^{(1-n)+}$ $(n=2,3)$ (71)

cathode: $2\ CuCl_n^{(1-n)+} + 2\ OH^- = Cu_2O + 2\ n\ Cl^- + H_2O$ (72)

$2\ Cu + 2\ OH^- \rightarrow Cu_2O + H_2O + 2\ e^-$ (73)

The cell technology is of the traditional parallel plate tank unit with monopolar connected copper anodes and cathodes. The cathode reaction is hydrogen evolution. Current densities are in the range 100–300 A m^{-2} and cell voltages from 2–2.6 V. The copper oxide product falls (or is scraped) from the anode to the base of the cell, enabling the cell potential to be reversed at frequent intervals whereby the Cu cathode then becomes the anode. This procedure prevents possible passivation of the electrodes and gives a more uniform product.

There are certain difficulties associated with the melting and casting of bulk sheet anodes and it is possible to use small pieces of high grade copper scrap contained within a titanium basket anode in this process. This approach is used in the electrorefining of scrap copper and has potential in the electrorefining of tin.[16]

8.2.9   Dichromate and Chromic Acid

There are several industrial oxidation processes that involve the formation of Cr^{III} solutions, which have to be subsequently processed for economic and environmental reasons. This arises from the operation of etching and polishing baths, electroplating baths and in the oxidation of organic compounds. The oxidation of chromic ions can only effectively take place on lead dioxide anodes.

$2\ Cr^{3+} + 7\ H_2O \rightarrow Cr_2O_7^{2-} + 14\ H^+ + 6\ e^-$ (74)

The lead dioxide is formed *in-situ* on lead or lead/antimony alloy. The mechanism of the reaction is believed to be *via* anodic oxidation of an adsorbed Cr–O complex. The reaction requires high over-potentials and thus oxygen gas is a common byproduct. Cell designs employed are based on tank electrolysers or filter press designs both fitted with either diaphragms or membranes to prevent the back reduction of dichromate. Further details of cell designs are discussed in chapters 6 and 7.

8.2.10   The Electrowinning and Refining of Metals

The production of metals can generally be divided into two areas; electrowinning and electrorefining. The former is involved with the electrochemical reduction step in the extraction of metal from its ore. The latter is an electrochemical method for the purification of otherwise impure metals. The metal to be purified is the anode of the cell which undergoes dissolution to its ionic form, whilst the anode impurities are produced as solid anode sludge, which is kept separate from the cathode. The dissolved metal ions are then electrodeposited at the cathode as the pure (refined) metal.

Electrochemical cells used for the refining of metal in aqueous solution are similar in design to the tank house cells used in electrowinning. The choice of operating conditions and electrolyte is determined by requirements of high efficiencies, for both the cathodic and anodic processes, prevention of impurities transferring to the cathode region and a good quality, highly crystalline deposit. The important metals, which are electrorefined from aqueous solution, are Cu, Sn, Pb, Ni, Co, Au and Ag although many others can, and are now refined to a certain extent. Refining or dissolution is, however, now featuring more and more in processes developed for the recycling of metal. Electrorefining from molten salt electrolytes is largely confined to aluminium, although other processes for the refining of beryllium, magnesium, molybdenum and uranium are known.

8.2.10.1 Electrowinning. Electrowinning of metals can be divided into two areas, molten salt electrolytes and hydrometallurgical (aqueous) processes. The large scale use of electrolysis is generally confined to the very electropositive metals such as Al, Na, Mg and Li (in molten salts) and to metals where a high purity is required or where other reduction processes realise environmental problems.

8.2.10.2 Aqueous Electrolytes. The major processes for electrowinning from aqueous solutions are for Cu, Zn and Ni. Often the process plants are located at a source of cheap hydroelectric power. Other metals are electrowon on a much smaller scale (see Table 12). The typical concentration of the dissolved metal ion in the electrolytes are between $0.5–1.0$ kmol m^{-3}. The current densities of operation are relatively modest, usually less than 1000 A m^{-2}.

In a typical acid/sulfate electrolyte the anode reaction is the evolution of oxygen. Typical anode materials are alloys of lead with Ag, Sb (5%) or Ca (0.05%). The alloyed metals in the lead, essentially serve as electrocatalysts which lower the oxygen over-potential and reduce the extent of anode corrosion, and subsequent contamination of the metal deposit with Pb. Another procedure which is used to lower the over-potential in Cu electrowinning is to add Co^{II} ions (10 ppm) to the electrolyte.

A competing reaction for the electrical energy is that of hydrogen evolution. For some metals (Ni, Co) cells operating with chloride electrolyte have been introduced in which the anode reaction is the evolution of chlorine. Chloride electrolytes offer several advantages over sulfate electrolytes, including higher conductivity, lower anode over-potentials (on

Table 12 *Operating characteristics of hydrometallurgical metal recovery*

Metal	Electrolyte	Additives	Temp /°C	Current density / A m⁻²	Anode	Current efficiency (%)	Voltage /V	Energy consumption / kWh kg⁻¹	Separator
Cu	Acid sulfate	Additives	40–60	150–1500	Pb–Sb Pb–Ca	80–96	1.9–2.5	1.9–2.5	No
Zn	Acid sulfate	Additives	35	300–750	Pb–Ag	90	3.3	3–3.5	No
Ni	Sulfate pH 3.5 chloride		65	200	Pb–DSA DSA	94	3.7	3.7	
Co	Sulfate pH 7	Air agitated	50–65	400–500			5.0	6.5	
Cr	Sulfate pH 2.4		55	700	Pb–Ag	45	4.2	18	Yes
Mn	$(NH_{42}SO_4)$ pH 7.2	Additives		400–600	Pb–Ag	60	5.1	8-9	Bagged anode
Ga	Alkaline		90	6000		Low	7–9	High	
Ta	Thallous sulfate		30		Pt				
Cd	Acid sulfate		20–30	80	Pb–Ag	90	2.5–2.7	1.5	
Te	Sodium tellunite pH >14		45	70	Stainless steel				

DSA) and higher solubilities.

The cathodes used in electrowinning cells are of two types, re-usable cathode blanks, used mainly in Zn and Co cells and starter sheets, *i.e.* prepared sacrificial metal sheets of the metal to be electrowon. Electrowinning cell technology has been designed for its simplicity of operation and there is ample scope for improvements in performance when new plant or cells are required and when lower concentrations of electrolyte are used. These improvements can be summarised as:

1   Attention to the mass-transfer characteristics of the cell can lead to improved and more uniform mass-transfer rates. Methods include, tapered anodes to direct oxygen gas more towards the cathode surface, electrolyte circulation, ultrasonics and gas sparging of the electrode surface.

2   Reducing the applied cell voltage by reducing the interelectrode gap, which would require a more frequent removal of the cathode blanks and greater control of the distribution in the thickness of the metal deposit.

3   A reduction in over-potential by changing the electrocatalyst and/or by increasing the surface area of the electrode. Significant savings in over-potential, of the order of 0.5 V at 250 A m⁻², have been achieved with the use of DSA (RuO_2/TiO_2) in place of Pb/Sb alloy

anodes in sulfuric acid electrolyte. Similar savings are possible with RuO_2/valve metal catalysed lead electrodes. More recently the use of IrO_2/Ti and a IrO_2 (70%)-Ta_2O_4 (30%) coated titanium anode have given impressive results.

4   Use an alternative anode reaction with a lower equilibrium potential, *e.g.* chlorine evolution from chloride electrolytes, hydrogen oxidation and the oxidation of sulfite.

8.2.10.3 Molten Salt Electrolytes. The principle metals electrowon from moten salt electrolytes are Al, Na, Mg and Li. Aluminium is widely used in construction and engineering applications. Sodium is used in the maufacture of several organic and inorganic substances including lead alkyls and the isolation of Ti. Magnesium is used as an alloy and in organic syntheses and lithium is used in batteries and in organic syntheses. Table 13 summarises the operating conditions and performance of cells for the production of these four metals. It will be noted that all these metals are electrowon from chloride electrolytes, although by far the greatest tonnage production is for aluminium production from alumina.

8.2.10.4 Sodium from Chloride Salts. The electrochemical cell for production of sodium metal is known as the Down's cell after its inventor who patented the process in 1921. The electrolyte is a mixture of NaCl and $CaCl_2$ at an operating temperature of 600 °C. The purpose of the $CaCl_2$ in the electrolyte is to reduce the solubility of Na, which would otherwise recombine with the Cl_2 and reduce the current efficiency. However, a drawback of the process with $CaCl_2$ in the electrolyte is that some calcium

Table 13 *Operating conditions for fused salt electrolysis*

	Electrolyte	Temp /°C	Current density / A m^{-2}	Voltage / V	Current efficiency (%)	Energy consumption / kWh kg^{-1}
Al	Al_2O_3	970	10 000	4–4.5	85–90	14–16
	AlCl in NaCl–LiCl undivided	700	10 000	4		12.5–14.5
Na	NaCl (40%) $CaCl_2$(60%) diaphragm cell	600	10 000	7	80	9–10
Mg	$MgCl_2$–KCl diaphragm cell	700–800	10 000	7.5	90	18.5
Li	LiCl–KCl diaphragm cell	450				

metal is formed as there is only a 30 mV difference between the decomposition voltages of Na and Ca. In addition to the loss of current efficiency due to Ca formation, other side reactions can occur at the cathode. These include the preferential reduction of any water in the melt and other species such as sulfate, carbonate and oxide. To avoid such reactions the feed salt needs to be of a purity greater than 99.7% on a dry basis. The lifetime of the cell operation is controlled by the oxidation of the anode (to CO and CO_2) caused by the presence of these impurities.

A Down's cell typically operates at a current density of 10 000 A m^{-2} and with a cell voltage of 7.0 V. The reversible cell voltage is 3.6 V and the voltage drop in the electrolyte is *ca.* 2.1 V.

8.2.10.5 Aluminium Poroduction. The majority of aluminium is produced from its bauxite ore in Hall–Heroult cells by the reaction:

$$Al_2O_3 + 1.5\ C \rightarrow 2\ Al + 1.5\ CO_2 \tag{75}$$

The cell shown in Figure 17, consists of a horizontal carbon cathode at which the molten Al is formed. Above this is the cryolite (Na_3AlF_6) melt containing the alumina. Carbon anodes are suspended above the cathode and are gradually consumed as the reaction takes place forming CO_2 (and some CO), and thus are lowered throughout their lifetime to maintain the interelectrode gap. The interelectrode gap is large (5 cm) because of instabilities of the melt/aluminium interface and the risk of shorting. The *IR* drop in the electrolyte is therefore relatively high at 1.5 V, which is significant in comparison to the overall voltage of 4.3 V for the operating current density of 10 000 A m^{-2}.

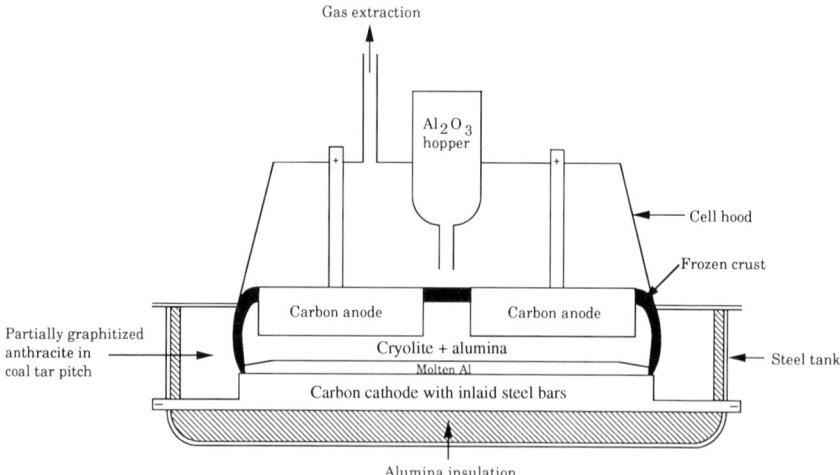

Figure 17 *The Hall–Heuralt aluminium cell*

An alternative aluminium cell has been developed by Alcoa in the USA based on a chloride electrolyte. The cell uses bipolar carbon electrodes for the production of Al and Cl_2 gas. The major difference between this cell and the Hall–Heroult cell is the lower operating temperature of 700 °C. It is claimed that this cell gives a 10% lower energy consumption than the Hall–Heroult cell.

8.2.10.6 Other Metal Extraction. The extraction and refining of many other metals by fused salt electrolysis has either operated at a small scale or undergone research and development, *e.g.* calcium from a molten chloride melt. With a suitable source of carbon, calcium carbide can also be prepared electrolytically. Beryllium, columbium, gallium, molybdenum, niobium, zirconium, tantalum, titanium from fused salt electrolysis in alkali metal chloride. Germanium, manganese, molybdenum and scandium from fused salt electrolysis of their oxide. Plutonium from fused salt electrolysis of plutonium trichlorides.

Boron and silicon from the oxidation of their carbides as anodes in fused salts with cathodic deposition of B or Si. The electrorefining of crude boron metal and the preparation of a boron silicon alloy, from boron carbide and silicon carbide, can be achieved by a similar process.

8.2.11 Metal Recovery from Base Metal Sulfide

The direct recovery of base metals such as Cu, Pb and Zn from their sulfide minerals was introduced by the Dextec Metallurgical Ltd[17] using an electrolytic diaphragm cell. In one example of the Dextec process for copper, the following reactions occur between air bubbles, mineral particles (*e.g.* chalcopyrite), electrolyte and anodic species in a strong chloride solution.

$$CuFeS_2 + 0.75\ O_2 + 0.5\ H_2O \rightarrow CuS + FeOOH + S \quad (76)$$

$$Cu^{2+} + CuS \rightarrow 2\ Cu^+ + S \quad (77)$$

The anode reaction is:

$$2\ Cu^+ \rightarrow 2\ Cu^{2+} + 2\ e^- \quad (78)$$

Thus in this process the mineral is leached in the anode compartment of a cell in a one-electron transfer to form the complex chloride $(CuCl_3)^{2-}$ and some $CuCl^+$. This leachate is then passed to the cathode compartment where the copper is recovered by electrodeposition. The process thus produces base metal, iron oxide and elemental sulfur from the mineral in one operation.

The industrial cells operate on a continuous basis, producing the metal as a powder, which either freely falls from the cathode or is relatively easily dislodged. Cylindrical copper pipe is used as the cathode which is covered in sections by heat shrink plastic, producing a cathode surface of spaced 2 mm diameter circles. The growth of the Cu from the surface as 'trees' results in a high stress concentration at the surface, requiring only minimal vibration to dislodge the particles which fall to the bottom of the cell and are collected as a slurry. The copper cathodes are contained within a diaphragm bag which starts from the centre of the reactor and has 36 branches radiating out into the anode/leachate side.

Other applications of the Dextec cell are in the selective leaching of lead from mixed lead–zinc–copper and iron sulfides and the recovery of zinc, lead and copper from complex mixtures of Cu–Pb–Zn–Fe–Ag–Au–As–Sb–Bi *etc.*

8.2.12 Other Processes

There are several electrochemical processes that are currently practiced on a small scale, or in the past have been considered to be feasible. Some of these are now discussed.

8.2.12.1 Metal Salt Preparation. There are several small scale processes in operation for the manufacture of metal salts *e.g.* gold, silver and tin by anodisation. The electrochemical method offers the feature of controlled purity and is based on the overall 'simple' formation of a soluble metal ion.

$$M \rightarrow M^{n+} + n\,e^- \tag{79}$$

The possible reaction of the dissolved metal ion at the cathode is prevented by the use of suitable membranes in the cell. The use of anodic dissolution is practiced in the following examples:

1   The production of potassium gold cyanide solutions from gold. The process uses a divided parallel plate flow-through cell with a potassium cyanide electrolyte operating at a temperature of 60 °C. The process is practiced by Englehard Sales in the UK at the modest rate of *ca.* 0.35 tonne per annum.

2   Potassium stannate $[K_2Sn(OH)_6]$. The production of potassium stannate from solid tin is by the overall cell reaction:

$$Sn + 2\,OH^- + 3\,H_2O + \tfrac{1}{2}\,O_2 \rightarrow Sn(OH)_6^{2-} + H_2 \tag{80}$$

The cell consists of a basket, which holds the tin anode bars, a cation-exchange membrane and a steel cathode for the evolution of oxygen in potassium hydroxide electrolyte. Under ambient conditions the tin dis-

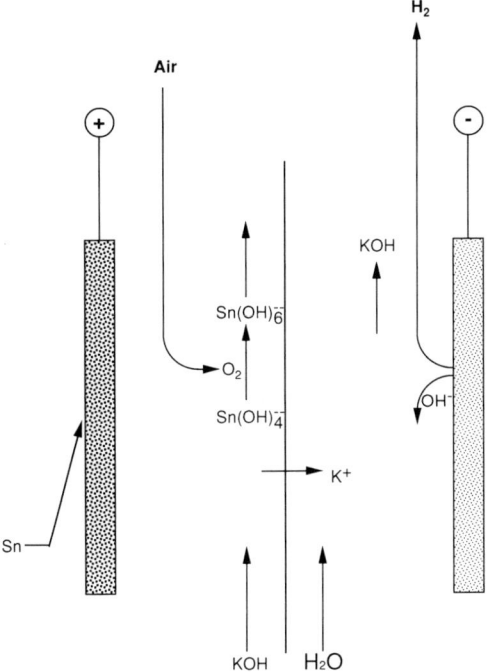

Figure 18 *Schematic cell concept for potassium stannate production*

solves in the potassium hydroxide electrolyte to stannite, which reacts rapidly with the oxygen in air, sparged to the cell, to form the stannate (see Figure 18).

$$Sn(OH)_4^{2-} + H_2O + \tfrac{1}{2} O_2 \rightarrow Sn(OH)_6^{2-} \qquad (81)$$

The potassium stannate solution is continuously withdrawn from the anolyte while the potassium hydroxide catholyte is passed into the anolyte. The migration of K^+ ions from the anolyte, into the catholyte, through the membrane provides part of the KOH required for the overall cell reaction.

Cathodic reduction is also used to produce metal salts. For example vanadium (II) formate is produced for the nuclear industry by the cathodic reduction of vanadium (V) in undivided parallel plate cells. Production is believed to be in the range 1–10 tonne per annum, based on the intermittent use of the cells, of total area 2 m^2, operating at a current density of 1000 A m^{-2} and cell voltage of 10 V.

A range of metal salts can be produced by electrochemical means as shown in Table 14.

Table 14 *Preparation of metal salts*

---

1 Potassium gold cyanide solutions from gold
2 Silver nitrate liquors by the anodic dissolution of Ag in nitric acid
3 Titanium(III) chloride
4 Nickel acetate, carbonate, chloride *etc.*
5 Potassium and sodium stannate (from tin/lead solder)
6 Cuprammonium nitrate from copper scrap dissolution in ammonium hydroxide
7 Copper acetate and pyrophosphate

8.2.13   Electrochemical Generation of Arsine

The use of arsine gas (AsH_3) is essential in several applications in the electronics industry. These include mixtures of 1–100 ppm arsine in hydrogen in the defining of silicon wafers *via* thermal diffusion or ion implantation methods, vapour phase epitaxy (VPE) and metal organic chemical vapour deposition (MOCVD) using concentrations in the range 2–50% (v/v), and gas source molecular beam expiatory (MBE) using high concentrations (80–100%) in the fabrication of GaAs and InGaAs materials.

Arsine is an extremely toxic material and stringent safety requirements are needed in its storage and use. These concerns have promoted the use of on-site, on-demand supply of arsine to eliminate the need for storage. One method[18] developed recently is based on the electrochemical reduction of arsenic to arsine.

$$As + 3\ H^+ + 3\ e^- = AsH_3 \tag{82}$$

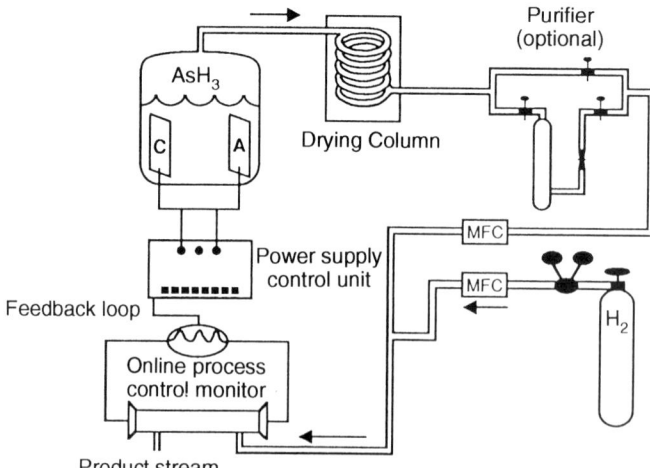

Figure 19 *Schematic flowsheet for arsine manufacture*

In acid and neutral electrolyte solutions, arsine generation occurs with copious co-evolution of hydrogen gas. Current efficiencies are typically *ca.* 10% in neutral (sodium sulfate) solutions and 1–2% in sulfuric acid solutions.

However, the generation of arsine in alkaline solution (1 mol dm^{-3} NaOH) from high purity (99.999%) arsenic cathodes can be carried out with current efficiencies in the range of 95–97%, over two orders of magnitude of current density. A prototype device, shown in Figure 19, has been used to supply high purity arsine in the manufacture of InGaAs materials. The device includes, the electrochemical reactor, a desiccant stage for water vapour removal and an online arsine concentration monitor for process control. This feedback controller ensures constant flow and concentration in feeds by regulating the current to the cell.

8.2.14 Sodium Dithionite Production

The production of dithionite, $S_2O_4^-$, by the electroreduction of HSO_3^- ions is practised on a small commercial scale (*ca.* 420 tonne per annum) in the USA by Olin Corp. A divided cell using a flow through metal felt electrode operating at 3000 A m^{-2} achieves a current efficiency of 90%. The bisulfite is supplied continuously as sulfur dioxide gas in an external vessel to the cell.

8.2.15 Manufacture of Dinitrogen Pentoxide

A commercially viable process for the production of dinitrogen pentoxide (N_2O_5) from nitric acid has been developed[19] by the Ministry of Defence using the ICI range of continuous flow electrolysers. A schematic diagram of the process based on membrane divided cell electrolysis is shown in Figure 20. Both electrode reactions are effectively utilised, in the following reactions:

Anode: $\quad N_2O_4 + 2\ HNO_3 \rightarrow 2\ N_2O_5 + 2\ H^+ + 2\ e^-$ $\qquad$ (83)

Cathode: $\quad 2\ HNO_3 + 2\ H^+ + 2\ e^- \rightarrow N_2O_4 + 2\ H_2O$ $\qquad$ (84)

Overall reaction: $\quad 4\ HNO_3 \rightarrow 2\ N_2O_5 + 2\ H_2O$ $\qquad$ (85)

The N_2O_4 generated at the cathode assists in the splitting of nitric acid into N_2O_5 and water. The water formed is separated from the anolyte by a membrane. Any water which is present in the anolyte will be converted to nitric acid and thus in the process the cathode product N_2O_4 must be purified of water before it is fed to the anode reaction.

The N_2O_5 generated by this system (in HNO_3) is seen as a replacement

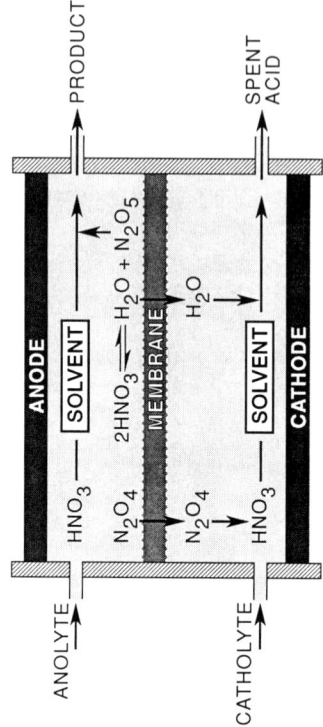

Figure 20 *Process for dinitrogen pentoxide production*

to sulfuric acid–nitric acid and oleum–nitric acid mixtures for the syntheses of nitro compounds, such as intermediates in the manufacture of pharmaceuticals, dyestuffs, pesticides and explosives. N_2O_5 offers a reduced reaction time, increased product yield and a simplified process time although these must be offset against higher reagent costs. It is also possible to use N_2O_5 in organic solvents, such as dichloromethane, to offer less aggressive media (compared to HNO_3) and more moderate nitrating properties.

8.2.16 Chlorine Dioxide

Chlorine dioxide is a major chemical used in the pulp and paper industry. It is used in conjunction with sodium hydroxide for purifying pulp and is typically produced on-site from the reaction of sodium chlorate with hydrochloric acid.

$$NaClO_3 + 2\ HCl \rightarrow ClO_2 + \tfrac{1}{2}\ Cl_2 + NaCl + H_2 \tag{86}$$

This method produces chlorine as a byproduct. The current and future imbalance in caustic soda and chlorine requirements, due to environmental limitations on the latter, has seen a new electrochemical route for ClO_2 generation developed.[10]

Chlorine dioxide can be formed at both high acidity (10 kmol m^{-3} H_2SO_4) and low acidity (0.1 kmol m^{-3} HCl) but by different mechanisms. The low acidity process, presently used in small scale ClO_2 generation, is based on chlorine participation in the chemical oxidation of chlorite as the ClO_2 production step.

$$2\ ClO_3^- + 2\ Cl^- \rightarrow 2\ ClO_2^- + 2\ ClO^- \qquad (87)$$

$$2\ ClO^- + 2\ Cl^- + 4\ H^+ \rightarrow 2\ Cl_2 + 2\ H_2O \qquad (88)$$

$$2\ ClO_2^- + 2\ Cl_2 \rightarrow 2\ ClO_2 + 2\ Cl^- \qquad (89)$$

The electrochemical behaviour of chlorine dioxide/chlorine redox couple in acidic media depends upon the cathode material. Low overpotential materials (*e.g.* platinum) will reduce chlorine close to the thermodynamic potential of 1.36 V *vs.* NHE.

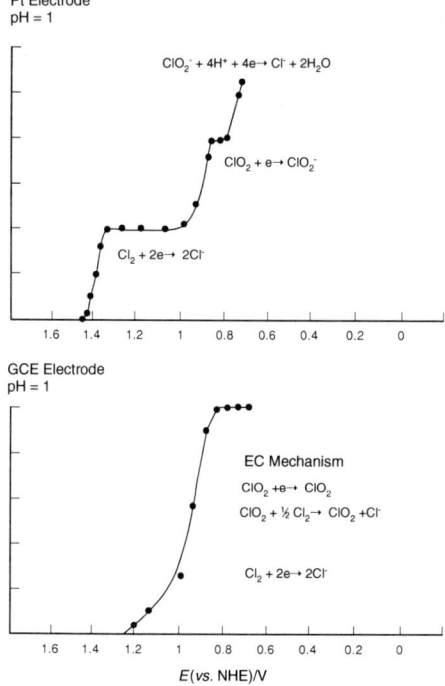

Figure 21 *Over-potential characteristics for chlorine dioxide oxidation*

$$Cl_2 + 2\ e^- \rightarrow 2\ Cl^- \tag{90}$$

This is then followed by ClO_2 reduction at over-potentials of *ca.* 0.95 V, as shown schematically in Figure 21, and then by ClO_2^- ion reduction.

$$ClO_2 + e^- \rightarrow ClO_2^- \tag{91}$$

$$ClO_2^- + 4\ H^+ + 4e^- \rightarrow Cl^- + H_2O \tag{92}$$

At high over-potential electrodes, *e.g.* vitreous carbon, chlorine reduction is at more negative potentials (0.85 V *vs.* NHE). In the presence of ClO_2, the reduction wave for chlorine is not observed (see Figure 21). This is explained by an EC catalytic mechanism in which the ClO_2^- formed electrochemically is re-oxidised near the electrode by chlorine in solution, until one is completely depleted,

$$\text{Electrochemical:}\quad ClO_2 + e^- \rightarrow ClO_2^- \tag{93}$$

$$\text{Chemical:}\quad\quad\ \ ClO_2^- + \tfrac{1}{2}\ Cl_2 \rightarrow ClO_2 + Cl^- \tag{94}$$

This reaction scheme therefore offers a method for electrochemically purifying ClO_2.

The electrochemical cell for this synthesis, shown in Figure 22, comprises of five compartments, with a central anode and two different cathodes providing two separate cell processes. Overall the cell has three functions:

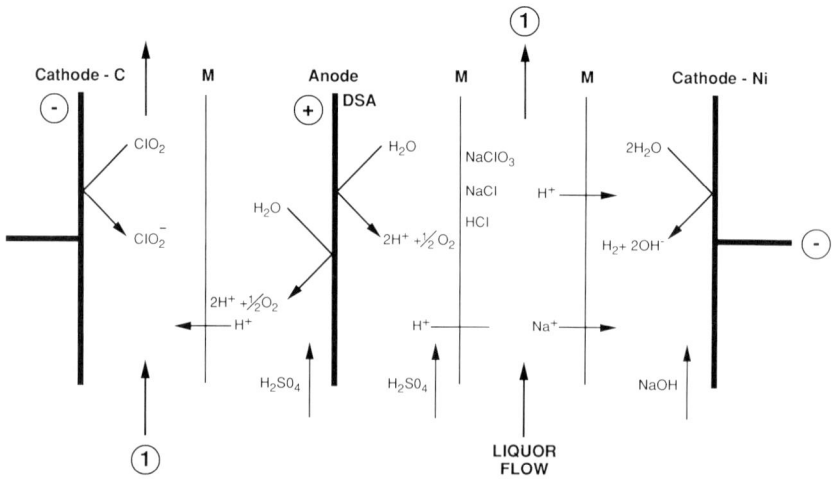

Figure 22 *Electrochemical cell for chlorine dioxide synthesis*

Table 15 *A selection of inorganic electrosyntheses*

1	Aluminium chlorohydroxides, $Al_2(OH)_nCl_{6-n}$, can be produced by the electrolysis of aqueous aluminium chloride solutions.
2	Cerium monosulfide from fused salt electrolysis of cerium sulfide.
3	Copper(II) hydroxide by the electrolytic dissolution of copper anodes.
4	Copper(II) carbonate from copper anodes.
5	Cyanogen chloride gas from anodic oxidation of ammonium chloride and HCN in a membrane cell.
6	Germane, GeH_4, by fused salt electrolysis using the reaction between germanium tetrachloride and lithium hydride produced *in-situ* in the cell.
7	Hydrazine, the electrolysis of solutions of liquid ammonia.
8	Hydroxylamine hydrochloride by the electrolytic reduction of nitric acid containing hydrochloric acid.
9	Lead borate from lead and boric acid.
10	Lead chromate from electrolysis of lead anodes in chromate solution.
11	Lead fluoroborate from lead anode oxidation in fluoroboric acid.
12	Lead sulfochromate. The electrolysis of sodium sulfate and sodium chromate solutions with lead anodes.
13	Magnesium dialkyls by electrolysis of magnesium anodes in molten sodium–aluminium alkyl salts.
14	Mercury(I) sulfate from mercury anodes.
15	Nitrogen trichloride from the electrolysis of ammonium chloride solutions.
16	Nitrogen trifluoride from molten salt electrolysis of ammonium bifluoride .
17	Phosphine from the electrolysis of phosphorus (molten) cathodes.
18	Potasssium manganate from manganese anodes.
19	Selenic acid, H_2SeO_4, by oxidation of selenious acid, H_2SeO_3.
20	Silane, SiH_4, *via* the electrolytic production of Li , converted to lithium hydride and reacted with silicon tetrachloride.
21	Silicon tetrafluoride by fused salt electrolysis of fluoride silicate melt.
22	Silver (argentous) oxide powder from silver anodes.
23	Alkali metal arsenate by anodic oxidation of arsenite.
24	Tetrafluorohydrazine from fused salt electrolysis.

1	Acid generation at a DSA–O_2 evolution anode.
2	Chlorine reduction at a fixed flow-by carbon bed cathode, for electrochemical chlorine scrubbing.
3	Sodium hydroxide generation by virtue of the hydroxide ion generation at a Ni cathode and Na^+ ion transport through an adjacent cation-exchange membrane.

In practice this electrochemical cell is linked to a commercial scale low acidity ClO_2 generator; which has a typical reaction mixture composition of 5 mol dm^{-3} $NaClO_3$, 0.1 mol dm^{-3} H^+ ions and 1.5 mol dm^{-3} NaCl. The fact that the molar ratio of Na$^+$: H$^+$ is *ca.* 65, enables the cation-exchange membrane, used in function (3) above, to favourably descriminate for Na$^+$ ion transport even though H$^+$ ion is approximately eight times more mobile than the Na$^+$ ion. The liquor flow from the reactor passes through the central compartment of the DSA/Ni cell, where the Na$^+$ ion is replaced by H$^+$ ions. It then passes to the chlorine electrochemical scrubber where one more acid equivalent is transferred through the cation-exchange membrane into the liquor. The reduction of Cl$_2$ and generation of ClO$_2$ take place at this stage. The liquor then recirculates back to the chemical reactor. Thus overall three solution flow loops exist; for NaOH generation, ClO$_2$ generator liquor and sulfuric acid. The process offers a method for generating ClO$_2$ of enhanced purity, simultaneously producing NaOH and avoids the need for a crystalliser to recover the NaCl byproduct formed by the commercial chemical method. At what are realistic operating current densities of 1000–2000 A m^{-2} the cell gave Cl$_2$ removal efficiencies of 83%–75%, NaOH membrane efficiencies of 81%–70% and resulted in ClO$_2$ chemical efficiencies and purity of around 90 and 87%, respectively. The latter compares with values of 88 and 51% for the commercial chemical generator.

8.2.17   The Scope of Inorganic Electrosynthesis

In conclusion and to illustrate the potential of inorganic electrosynthesis it is interesting to note that there are many examples of proposed applications which have appeared in the technical and patent literature (see for example ref. 20). Although generally not applied on a 'commercial' scale, the processes (listed in Table 15) may stand up to re-investigation at some time in the future.

REFERENCES

1   'Electrochemical Technology in Industry, A UK Status Report', ed. S.J.D. Tait, SCI, London, 1991.
2   'Comprehensive Treatise of Electrochemistry', ed. J. O'M Bockris, B.E. Conway, E. Yeager and R.E. White, Plenum Press, New York, 1981, vol. 2.
3   D. Pletcher and F.C. Walsh, 'Industrial Electrochemistry', Chapman and Hall, London, 2nd edn., 1990.
4   'Technique of Electroorganic Synthesis. Part III Scale-up and Engineering Aspects', ed. N.L. Weinberg and B.V. Tilak, J. Wiley, New York, 1982.
5   D. Degner, 'Organic Electrosynthesis in Industry', *Top. Curr. Chem.*, 1988, **148**.

6   B.V. Tilak, S. Sarangapani and N.E. Weinberg, 'Performance of Two-phase-Electrolyte Electrolysis', ch. IV, p.95, ref. 4.

7   H. Wendt, *Electrochim. Acta*, 1984, **29**, 1513.

8   E. Steckhan, *Angew. Chem., Int. Ed. Engl.*, 1986, **25**, 683.

9   J.J. Jow, A.C. Lee and T.C. Chou, *J. Appl. Electrochem.*, 1987, **17**, 753.

10   'Electrosynthesis, From Pilot, To Plant, To Production', ed. J.D. Genders and D. Pletcher, The Electrosynthesis Co Inc. New York, 1990, ch. 10, p.187.

11   F. Beck, B. Wermeckes and E. Zimmer, Dechema Monograph, 1988, vol. 112, p.257.

12   A.M. Couper, D. Pletcher and F.C. Walsh, *Chem. Rev.*, 1990, **90**, 837.

13   W.V. Childs, 'The Philip's Electrochemical Fluorination Process', ch. VII, ref. 4.

14   J.C. Yu, M.M. Baizer and K. Nobe, *J. Electrochem. Soc.*, 1988, **135**, 1400.

15   W. Thiele and H. Matschiner, 'Electrochemical Cell Design and Optimisation Procedures', Dechema Monograph, 1990, vol. 123, p.133.

16   M.G. Figueroa, R.E. Gano and W.C. Cooper, *J. Appl. Electrochem.*, 1993, **23**, 308.

17   P.K. Everett, 'The Dextec Copper Process, Extraction Metallurgy', The Inst. Min. Met. London, 1981, p.149.

18   J.L. Valdes, G. Cadet and J.W. Mitchell, *J. Electrochem. Soc.*, 1991, **138**, 1654.

19   'Processing and Manufacturing Dinitrogen Pentoxide', Defence Technology Enterprises Ltd, Norfolk House London.

20   M. Sittig, 'Inorganic Chemical and Metallurgical Process Encyclopedia', Noyes Dev. Corp., New Jersey, USA, 1968.

Appendix

NOTATION

* indicates dimensions are dependent on order of reaction

A	area of electrode	m^2
a	electrode area per unit volume	m^{-1}
a_j	activity	
b	Tafel slope	V or mV
CE	current efficiency or current yield	—
C_1	concentration of vacant sites	mol m^{-2}
C_j	concentration of species, j	mol m^{-3}
C	dimensionless concentration C_j/C_{jo}	—
C_{js}	concentration of species, j, at the surface	mol m^{-3}
C_T	sum of the concentrations	mol m^{-3}
D	diffusivity	$m^2\ s^{-1}$
D_{eff}	effective diffusivity	$m^2\ s^{-1}$
d	interelectrode gap or electrode thickness	m
d_s	diaphragm thickness	m
E_e	equilibrium potential of electrode reaction	V
E	electrode potential	V
E_m	metal potential	V
E_s	solution potential	V
E^o	standard electrode potential	V
E_o	equilibrium cell potential or open circuit voltage	V
EOD	electrochemical oxygen demand	
EOI	electrochemical oxidation index	
e	voidage or porosity	
e_g	bubble void fraction	
e	permittivity	
F	Faraday	C(g equiv)$^{-1}$
f	$= nF/RT$	V^{-1}
I	current	Amp
I_-	cathodic current	Amp
j	current density	A m^{-2}
j_-	cathodic current density	A m^{-2}
j_+	anodic current density	A m^{-2}
j_o	exchange current density	A m^{-2}

j_l	limiting current density	A m^{-2}
j_k	kinetic current density	A m^{-2}
j_d	diffusion limiting current density at a rde	A m^{-2}
K_j	equilibrium constant	*
K'	hydraulic permeability coefficient	*
k	rate constant or velocity constant	*
K_j	adsorption equilibrium constant	*
K_f	equilibrium formation constant	*
k_{bj}	electrochemical rate constant of step i	L
k_{fj}	electrochemical rate constant	*L
k_L	mass transfer coefficient	m s^{-1}
k_j	rate or velocity constant	m *L
k_+	forward electrochemical rate constant	m s^{-1}
k_-	reverse electrochemical rate constant	m s^{-1}
k_o	standard rate constant	*
M	relative molar mass	kg kmol^{-1}
m	amount of deposition or product	mol or kg
N	flux	mol m^{-2} s^{-1}
N_M	MacMullin number	—
n	number of electrons	—
R	Universal gas constant	J mol^{-1} K^{-1}
R_e	electrolyte resistance	ohm
r	reaction rate	mol m^{-2} s^{-1}
r_A	conversion rate of A	mol m^{-2} s^{-1}
R_{cell}	cell resistance	ohm
P	pressure	N m^{-2}
p	partial pressure	N m^{-2}
P	order of reaction	
q	charge	coulombs
Q	heat or heat flow	J s^{-1}
r_p	particle radius	mol m^{-3} s^{-1}
S	cross-sectional area	m^2
STY	space time yield	kg m^{-3} s^{-1}
T	temperature	°C or K
t	time	s
t_j	transport number	—
U_p	electrophoretic mobility	
U_j	mobility	
u	velocity	m s^{-1}
u_F	velocity of water in pore	m s^{-1}
u_p	velocity of particle	m s^{-1}
V	volume	m^3
v	volumetric flowrate	m^3 s^{-1}
V_m	volume of mixing tank	m^3
x	coordinate dimension	m
z_j	charge number	

Dimensionless Groups

Da	Damkohler number
Ha	Hatta number
Pr	Prandtl number
Pe	Peclet number
Re	Reynolds number
Sc	Schmidt number
Sh	Sherwood number
St	Stanton number
Wa	Wagner number

Greek Symbols

α	transfer coefficient	—
β	$= \alpha n \text{F} / RT$	—
ζ	zeta potential	
δ	thickness of diffusion layer	m
δ_{Pr}	thickness of Prandtl layer	
η	over-potential	V
κ	electrolyte conductivity	$\Omega^{-1} \text{ m}^{-1}$
κ_{e}	effective electrolyte conductivity	$\Omega^{-1} \text{ m}^{-1}$
ϕ	potential	V
ρ	density	kgm^{-3}
ρ	specific resistance	$\Omega \text{ m}$
σ	electrode area per unit length	m
υ	kinematic viscosity	$\text{m}^2 \text{ s}^{-1}$
υ_{i}	stoichiometric coefficient	—
μ	dynamic viscosity	$\text{kg m}^{-1} \text{ s}^{-1}$
τ	resistance time	s
τ_{m}	resistance time for mixer	s
τ_{r}	resistance time for a reactor	s
w	rotation speed	s^{-1}

Subscripts

a	anode
ads	adsorbed state
b	bulk
c	cathode
g	gas
j	species or step
m	metal or membrane
R	reactor

s   solution
max   denotes maximum value
x   local position
-          cathodic
+   anodic

Others

Δ denotes change in quantity

GENERAL REFERENCES

'Electrochemical Technology in Industry — A UK Status Report', ed. S.J.D. Tait, SCI Electrochem. Tech. Group, London, 1991.

D. Degner, 'Organic Electrosynthesis in Industry', in 'Electrochemistry III', ed. E. Steckhan, Springer, Berlin, Heidelberg NY, 1988, p.1 (*Top. Curr. Chem.*, 1988, **148**).

'Fuel Cell Systems', ed. L.J.M.J. Bolmen and M.N. Mugerwa, Plenum Press, NY, 1993.

M.S. Antelman and F.J. Harris, 'The Encyclopedia of Electrode Potentials', Plenum Press, NY, 1982.

'Organic Electrochemistry', ed. M.M. Baizer and H. Lund, Marcel Dekker, New York, 1991.

A.J. Bard, 'Encyclopedia of Electrochemistry of the Elements', Marcel Dekker, New York, 1973–1985, vols. 1–15.

A.J. Bard and L.R. Faulkner, 'Electrochemical Methods, Fundamentals and Applications', Wiley, New York, 1980.

'Modern Aspects of Electrochemistry', ed. J.O'M. Bockris, Plenum Press, NY, 1980, vol. 1.

'Comprehensive Treatise of Electrochemistry', ed. J.O'M. Bockris, B.E. Conway and E. Yeager, Plenum Press, NY, in ten volumes:
  'The Double Layer', 1980, vol. 1;
  'Electrochemical Processing', with R.E. White, 1981, vol. 2;
  'Electrochemical Energy Conversion and Storage', with R.E. White, 1981, vol. 3;
  'Electrochemical Materials Science', with R.E. White, 1981, vol. 4;
  'Thermodynamic and Transport Properties of Aqueous and Molten Electrolytes', 1982, vol. 5;
  'Electrodics: Transport', with S. Sarangapani, 1982, vol. 6;
  'Kinetics and Mechanisms of Electrode Processes', with R.E. White and S.V.M. Khan, 1983, vol. 7;

'Experimental Methods in Electrochemistry', with R.E. White, 1984, vol. 8;

'Electrodics', with S. Sarangapani, 1984, vol. 9

and 'Bioelectrochemistry', with S. Srinivasan and Y.A. Chizmadzhev, 1984, vol. 10.

J.O'M. Bockris and A.K.N. Reddy, 'Modern Electrochemistry', Plennum Press, NY, 1973, vols. 1 and 2.

P.H. Rieger, 'Electrochemistry', Prentice-Hall, Hemel Hempstead, 1988.

T. Shono, 'Electroorganic Synthesis', Academic Press, New York, 1990.

'Electrochemistry', ed. E. Steckham, Springer-Verlag, Berlin, 1987–1989, vols. I–IV.

'Electrochemical Hydrogen Technologies', ed. H. Wendt, Elsevier, Amsterdam, 1990.

G. Wrangler, 'Introduction to Corrosion and Protection of Metals', Chapman and Hall, London, 1985.

T.R. Crompton, 'Small Batteries', Macmillan, London, 1982.

D.R. Crow, 'Principles and Applications of Electrochemistry', Chapman and Hall, London, 3rd edn., 1988.

U.R. Evans, 'The Corrosion and Oxidation of Metals', Edward Arnold, London, 1960.

A.J. Fry, 'Synthetic Organic Electrochemistry', Wiley, New York, 1989.

'Topics in Organic Electrochemistry', ed. A.J. Fry and W.E. Britton, Plenum Press, NY, 1986.

K.S. Goto, 'Solid State Electrochemistry and Its Applications to Sensors and Electronic Devices', Elsevier, Amsterdam, 1987.

F. Gutmann and H. Keyzer, 'Modern Bioelectrochemistry', Plenum Press, NY, 1985.

J. Koryta, 'Ions, Electrodes and Membranes', Wiley, New York, 2nd edn., 1992.

J. Koryta, J. Dvorak and L. Kavan, 'Principles of Electrochemistry', Wiley, New York, 2nd edn., 1993.

'Superionic Solids and Solid Electrolytes: Recent Trends', ed. A. Laskar and S. Chandra, Academic Press, New York, 1989.

D.G. Lovering and R.J. Gale, 'Molten Salt Techniques', Plenum Press, NY, 1983, vol. 1.

'Organic Electrochemistry: An Introduction and Guide', ed. H. Lund and M.M. Baizer, Marcel Dekker, New York, 3rd edn., 1990.

'Polymer Electrolyte Reviews' ed. J.R. MacCallum and C.A. Vincent, Elsevier Applied Science, London, 1987, vol. 1; 1990, vol. 2.

'CRC Handbook Series in Inorganic Electrochemistry', ed. L. Meites, CRC Press, 1980.

J. Newman, 'Electrochemical Systems', Prentice-Hall, New Jersey, 2nd edn., 1991.

Environmental Electrochemistry

P.M. Bersier, L. Carlsson and J. Bersier, 'Electrochemistry for a Better Environment', Top. Curr. Chem., 1994, **170**, 114

J.D. Genders and N.L. Weinberg, 'Electrochemistry for a Cleaner Environment 1992', The Electrosynthesis Co. Inc., Baffalo, USA, 1992.

C.A.S. Sequeira, 'Environmentally Oriented Electrochemistry', Elsevier, Amsterdam, 1993.

Electrochemical Engineering

F.C. Walsh, 'A First Course in Electrochemical Engineering', The Electrochemical Consultancy Ltd, Romsey, UK, 1993.

F. Goodridge and K. Scott, 'Electrochemical Process Engineering. A Guide to the Design of Electrolytic Plant', Plenum Press, NY, 1995.

I. Rousar, K. Micka and A. Kimla, 'Electrochemical Engineering', Elsevier, Amsterdam, 1986, vols. 1 and 2.-

W.S. Ho and K.K. Sirkar, 'Membrane Handbook', Van Nostrand Publishing Co., 1992.

'Advances in Electrochemistry and Electrochemical Engineering', ed. P. Delahay and C.W. Tobias, Wiley, New York, 1960.

T.Z. Fahidy, 'Principles of Electrochemical Reactor Analysis', Elsevier, Amsterdam, 1985.

'Advances in Electrochemical Science and Engineering', ed. H. Gerischer and C.W. Tobias, VCH Publishers, Berlin, 1990, vol. 1.

'Electrochemical Engineering 1989', ed. H. Gerischer and C.W. Tobias, Hemisphere Publishing, New York, 1989.

D.B. Hibbert and A.M. James, 'Dictionary of Electrochemistry', Macmillan, London, 2nd edn., 1984.

Subject Index

Also published by
The Royal Society of Chemistry . . .

Ref No 1260
Industrial Inorganic Chemicals: Production and Uses
Edited by R. Thompson, CBE, *Consultant*
Softcover   xviii + 408 pages   ISBN 0 85404 514 7   1995   Price £39.50

Ref No 1152
Insights into Speciality Inorganic Chemicals
Edited by David Thompson, *Consulting Chemist, Reading, UK*
Softcover   xxiv + 506 pages   ISBN 0 85404 504 X   1995   Price £39.50

Ref No 1263
Trace Element Medicine and Chelation Therapy
By David M. Taylor, *University of Wales, Cardiff*
David R. Williams, OBE, *University of Wales, Cardiff*
RSC Paperbacks Series   Softcover   x + 124 pages   ISBN 0 85404 503 1   1995   Price £15.50

Ref No 1051
Medicinal Chemistry: Principles and Practice - 1st Reprint 1995
Edited by Frank D. King, *SmithKline Beecham Pharmaceuticals, Harlow, UK*
Softcover   xxiv + 314 pages   ISBN 0 85186 494 5   1994   Price £39.50

Ref No 1144
Food Microbiology
By M. R. Adams, *University of Surrey*
M. O. Moss, *University of Surrey*
Softcover   xiv + 398 pages   ISBN 0 85404 509 0   1995   Price £22.50

Ref No 987
Chemistry and Light
By Paul Suppan, *University of Fribourg, Switzerland*
Softcover   xiv + 296 pages   ISBN 0 85186 814 2   1994   Price £19.50

Ref No 981
The Chemistry and Physics of Coatings
Edited by A.R. Marrion, *Courtaulds Coatings (Holdings) Ltd*
RSC Paperbacks   Softcover   x + 206 pages   ISBN 0 85186 994 7   1994   Price £15.00

Ref No 1012
Ion Exchange: Theory and Practice - 2nd Edition
By C.E. Harland, *The Permutit Company Limited, UK*
RSC Paperbacks   Softcover   xvi + 286 pages   ISBN 0 85186 484 8   1994   Price £16.95

Ref No 1258
Computing Applications in Molecular Spectroscopy
Edited by W. O. George, *University of Glamorgan*
D. Steele, *University of London*
Softcover   Approx 260 pages   ISBN 0 85404 519 8   1995   Price £39.50

Prices subject to change without notice.

To order please contact:
Turpin Distribution Services Ltd., Blackhorse Road, Letchworth, Herts SG6 1HN, UK.
Telephone +44 (0) 1462 672555   *Fax* +44 (0) 1462 480947
Telex 825372 TURPIN G

RSC Members should order from Membership Administration at our Cambridge address.

For further information please contact:
Sales and Promotion Department, The Royal Society of Chemistry,
Thomas Graham House, Science Park, Milton Road, Cambridge CB4 4WF, UK.
Telephone +44 (0) 1223 420066   *Fax* +44 (0) 1223 423623
E-mail (Internet) RSC1@RSC.ORG   *World Wide Web* http://chemistry.rsc.org/rsc/

THE ROYAL
SOCIETY OF
CHEMISTRY

Information
Services